BOOKS BY A. ALBERT KLAF

Calculus Refresher for Technical Men
Trigonometry Refresher for Technical Men

ARITHMETIC REFRESHER

By A. ALBERT KLAF, B.S., M.E.

New York
DOVER PUBLICATIONS, INC.

Published in Canada by General Publishing Company, Ltd., 30 Lesmill Road, Don Mills, Toronto, Ontario.
Published in the United Kingdom by Constable and Company, Ltd.

Arithmetic Refresher was first published by Dover Publications, Inc., in 1964 under the title *Arithmetic Refresher for Practical Men.*

International Standard Book Number: 0-486-21241-6
Library of Congress Catalog Card Number: 64-18856

Manufactured in the United States of America
Dover Publications, Inc.
31 East 2nd Street
Mineola, N.Y. 11501

FOREWORD

My father wrote this *Arithmetic Refresher for Practical Men* for the mass audience of professionals and laymen who are frequently faced with numerical problems. The book includes the knowledge and practical experience gathered during a lifetime of searching curiosity. He completed the manuscript a year before his passing. It is the testament of a career dedicated to public service and mathematical enlightenment.

I wish to express my deep appreciation to my father's colleague, Mr. Victor Feigelman, B.C.E., M.C.E., for solving the sample problems and checking the manuscript. Thanks are also due to Mr. Hayward Cirker, President of Dover Publications, Inc., who was my father's valued friend as well as his publisher.

This book was to have been one of a series that began with his *Calculus Refresher for Technical Men*, and progressed to his *Trigonometry Refresher for Technical Men.* The succeeding volumes will, of course, remain unwritten. But the best has been said. Now it must be used by those who seek to experience the joy of mathematics my father so deeply felt.

FRANKLIN S. KLAF, M.D.

CONTENTS

CHAPTER		PAGE
	Introduction	1
I	Addition	13
II	Subtraction	24
III	Multiplication	37
IV	Division	57
V	Factors—Multiples—Cancellation	75
VI	Common Fractions	82
VII	Decimal Fractions	107
VIII	Percentage	125
IX	Interest	140
X	Ratio—Proportion—Variation	159
XI	Averages	186
XII	Denominate Numbers	200
XIII	Power—Roots—Radicals	217
XIV	Logarithms	241
XV	Positive and Negative Numbers	262
XVI	Progressions—Series	266
XVII	Graphs—Charts	278
XVIII	Business—Finance	297
XIX	Various Topics	336
XX	Introduction to Algebra	356
	Appendix A—Answers to Problems	403
	Appendix B—Tables	417
	Index	429

INTRODUCTION

1. What is arithmetic?

The science of number and the art of computation.

2. What is our numerical system called and why is it so called?

It is called the Arabic system because it was given to us by the Arabs, who developed it from the Hindu system.

3. What is a digit?

Any whole number from 1 through 9 is called a digit. Thus 1, 2, 3, 4, 5, 6, 7, 8, 9 are called digits.

4. What is a cipher and what is its symbol?

The word "cipher" comes from an Arabic word meaning "empty" and means "no digit." The symbol for a cipher is 0.

5. What other commonly used words may be substituted for the word "cipher"?

"Zero" and "nought" may be used for "cipher."

6. What is the foundation of the Arabic numerical system?

The foundation consists of the nine symbols called digits—1, 2, 3, 4, 5, 6, 7, 8, 9—and one symbol called a cipher, zero, or nought.

7. What is a decimal point and what is its symbol?

A decimal point is a point that is used to separate the fractional part of a number from a whole number, and its symbol is a dot [.].

8. What is meant by computation or calculation?

Computation or calculation is the process of subjecting numbers to certain operations. The word "calculation" comes from a Latin word meaning "pebble," as reckoning was done with counters or pebbles.

9. How many fundamental operations are there in arithmetic?

There are six operations, all growing out of the first.

The six operations are divided into two groups: (*a*) three direct operations and (*b*) three inverse operations, each of which has the effect of undoing one of group (*a*).

Group (*a*) Direct Operation	Group (*b*) Inverse Operation
1. Addition	4. Subtraction
2. Multiplication	5. Division
3. Involution	6. Evolution

10. What are the symbols for (*a*) "equals to" or "equals," (*b*) addition, (*c*) subtraction, (*d*) multiplication, (*e*) division, (*f*) involution, (*g*) evolution, and (*h*) "therefore"?

(*a*) The equals sign [=] means "equals to" or "equals."

$1 + 1 = 2$; one plus one *equals* two.

(*b*) The plus sign [+] means "plus," "and," or "added to."

$2 + 2 = 4$; two *plus* two equals four, or two *and* two equals four, or two *added to* two equals four.

(*c*) The minus sign [−] means "minus," "subtracted from," or "from."

$5 - 3 = 2$; five *minus* three equals two, or three *subtracted from* five equals two, or three *from* five equals two.

(*d*) The multiplication sign [×] means "multiplied by," or "times."

$5 \times 3 = 15$; five *multiplied by* three equals fifteen, or five *times* three equals fifteen.

"Times" may also be indicated by a dot in the center of the line between the two numbers.

$5 \cdot 3 = 15$; five times three equals fifteen.

(*e*) The division sign [÷] means "divided by."

$10 \div 2 = 5$; ten *divided by* two equals five.

The signs $\underline{\big/\quad}$ or $\overline{)\quad}$ mean "divided into."

$2\underline{/10}$; two *divided into* ten equals five.
$\quad 5$

$$2\overset{5}{\overline{)10}}$$; two *divided into* ten equals five. This form is used in long division.

$\dfrac{10}{2} = 5$ expressed as a fraction means "ten *divided by* two equals five."

(*f*) A small number placed in the upper right-hand corner of a number is used to indicate the number of times the number is to be multiplied by itself.

$$2^5 = 2 \times 2 \times 2 \times 2 \times 2 = 32$$

Read: Two to the fifth power equals thirty-two. The process of finding a power of a number is involution.

(*g*) The radical sign [$\sqrt{}$] means "root of." A figure is placed above the $\sqrt{}$ to indicate the root taken. It is omitted in the case of the square root.

It is the inverse operation of involution and is called evolution.

$\sqrt[5]{32} = 2$. The fifth root of thirty-two is two, which is the number that when multiplied by itself five times will give thirty-two.

(*h*) The sign [$\therefore$] means "therefore."

11. What is the significance of parentheses enclosing numbers?

The presence of parentheses means that the operations within the parentheses are to be performed before any operations outside. A number preceding parentheses means that the final figure within parentheses is to be multiplied by that number.

EXAMPLE: $3(5 + 2) = 21$. The operation $(5 + 2)$ is performed first $= 7$. Then $3 \times 7 = 21$. The operation of 3 times is then performed.

12. What is meant by a unit?

Any one thing is called a unit.

13. What is meant by a number?

A unit or collection of units is called a number.

14. What is meant by an integer, whole number, or an integral number?

Numbers representing whole units are called integers, whole numbers, or integral numbers.

EXAMPLES: 12, 8, 75, 134, 2,659 are integers or whole numbers.

15. What symbols are used to express numbers?

Digits or figures are used to express numbers.

The symbol $0 =$ zero is used to express "no digit."

16. How are digits used to express numbers in our Arabic system?

The *value* of the digit is fixed by its position, starting from the right and going towards the left.

The first position is that of "units." The next position is that of "tens." The third position is that of "hundreds." These are called the three "*orders*." A group of three orders is called a *period.*

17. How are the orders and periods arranged in the Arabic system?

(Rarely is there use for any number larger than "trillions.")

Periods →	5 Trillions			4 Billions			3 Millions			2 Thousands			1 Units		
Orders →	Hundreds	Tens	Units	Hundreds	Tens	Units	Hundreds	Tens	Units	Hundreds	Tens	Units	Hundreds	Tens	Units
			7	6	5	3	4	6	0	5	3	4	6	4	6

18. How do we read a number written in the Arabic system?

Separate the numbers by commas into "periods" or groups of three figures beginning at the *right.*

Now begin at the *left* and read each period as if it stood alone, adding the name of the "period."

EXAMPLE: 7, 653, 460, 534, 646 (above).

Read: Seven *trillion,* six hundred and fifty-three *billion,* four hundred and sixty *million,* five hundred and thirty-four *thousand,* six hundred and forty-six.

Note that the word "and" may in all cases be omitted.

19. What is the relation of a unit of any period to that of the next lower period?

The unit of any period = 1,000 units of the next lower period.

EXAMPLE: One thousand = 1,000 = 1,000 units.
One million = 1,000,000 = 1,000 thousands.
One billion = 1,000,000,000 = 1,000 millions.
One trillion = 1,000,000,000,000 = 1,000 billions.

20. How would you write a number in figures?

Begin at the *left* and write the hundreds, tens, and units of each "period," placing zeros in all vacant places and a comma between each two periods.

EXAMPLE: 400, 536, 080, 209.

Four hundred billion, five hundred thirty-six million, eighty thousand, two hundred nine.

21. How do zeros before or after a number affect the number?

A zero in *front* of a number does not affect it.

EXAMPLE: 000 8, 060 = eight thousand sixty.

A zero *after* a number moves the number one place to the *left* or multiplies it by 10.

EXAMPLE: 8,060. Now add a zero after the number, or 80,600. The eight thousand sixty becomes eighty thousand six hundred.

Two zeros added at the right moves the number two places to the left or multiplies it by 100.

EXAMPLE: 80,600. Add two zeros getting 8,060,000 = eight million sixty thousand. And so on with added zeros.

For another method of writing very large numbers, see Question 534.

22. What are the names of the periods beyond trillions up to and including the twelfth period?

5. Trillions	9. Septillions
6. Quadrillions	10. Octillions
7. Quintillions	11. Nonillions
8. Sextillions	12. Decillions

23. How may we think of the orders of the successive periods as being built up of bundles of lower units?

Take the "units" period. The largest digit that can appear in the units *order* is 9. Now add 1 to 9 and it becomes *a bundle of ten* = 10. This means digit 1 in the "tens" order and zero in the units order. Note that the "tens" position is 10 times the units position.

The largest number that can appear in the "tens" and "units" orders is 99. Now add 1 to 99 and it becomes *a bundle* of *one hundred* = 100. This means digit 1 in the "hundreds" order and zero in both the tens and units orders. 100 may also be thought of as made up of 10 bundles of "tens." Note that the "hundreds" position is ten times the "tens" position.

Now take the "thousands" period. The largest number that can appear in the "units" period is 999. Now add 1 to 999 and it becomes a bundle of one thousand = 1,000. This means digit 1 in the "units" order of this period and zeros in the orders of the units period. The "thousands" position is ten times the "hundreds" position. 1,000 may also be thought of as made up of 10 bundles of one hundreds or 100 bundles of tens.

The largest number that can appear in the "units" order of *this* period together with the *units* period is 9,999. Now add 1 to 9,999 and it becomes a bundle of ten thousand = 10,000. This means digit 1 in the "tens" order of this period and zeros in all the other places. 10,000 may also be thought of as made up of 10 bundles of one thousands, 100 bundles of one hundreds or 1,000 bundles of tens. The "ten thousands" position is ten times the "thousands" position.

The largest number that can appear in the tens and units orders of this period, together with entire units period is 99,999. Now add 1 to 99,999 and it becomes a bundle of one hundred thousand = 100,000. This means digit 1 in the "hundreds" order of this period and zero in all the other places. 100,000 may also be thought of as made up of 10 bundles of ten thousands, 100 bundles of one thousands, 1,000 bundles of one hundreds, or 10,000 bundles of tens. The "hundred thousand" position is ten times the "ten thousands" position.

Follow a similar procedure in the "millions" period. Add 1 to 999,999, getting a bundle of one million = 1,000,000. Digit 1 is in the units order of this period. 1,000,000 may be gotten by 10 bundles of one hundred thousands, 100 bundles of ten thousands, 1,000 bundles of thousands, 10,000 bundles of hundreds or 100,000 bundles of tens. The "millions" position is ten times the "hundred thousand" position.

Add 1 to 9,999,999, getting a bundle of ten million = 10,000,000. 10,000,000 may also be gotten by 10 bundles of one millions, 100 bundles of one hundred thousands, 1,000 bundles of ten thousands, 10,000 bundles of thousands, 100,000 bundles of hundreds, or 1,000,000 bundles of tens. The "ten millions" position is ten times the "millions" position.

Add 1 to 99,999,999, getting a bundle of one hundred million = 100,000,000 which may also be gotten by 10 bundles of ten millions, 100 bundles of millions, 1,000 bundles of one hundred thousands, 10,000 bundles of ten thousands, 100,000 bundles of one thousands, 1,000,000 bundles of hundreds, 10,000,000 bundles of tens.

$$100,000,000 = 10 \times 10,000,000.$$

This procedure can be continued to the other periods which follow this one.

Note the relation of the bundles. Any bundle is ten times the size of the bundle on its right and one tenth that of a bundle at its immediate left.

24. When is a decimal point used?

It is used to express values less than one.

EXAMPLES:

$$0.2 = \text{two tenths of one unit} = \frac{2}{10} \text{ in fraction form.}$$

$$0.02 = \text{two hundredths of one unit} = \frac{2}{100} \text{ in fraction form.}$$

$$0.002 = \text{two thousandths of one unit} = \frac{2}{1{,}000} \text{ in fraction form.}$$

$$0.0002 = \text{two ten thousandths of one unit} = \frac{2}{10{,}000} \text{ in fraction form.}$$

For another method of writing decimals, see Question 536.

25. What are the names of the decimal or fractional places?

Units	Decimal point	Tenths	Hundredths	Thousandths	Ten thousandths	Hundred thousandths	Millionths	Ten millionths	Hundred millionths	Billionths
0	.	6	8	0,	0	5	7,	9	2	3

Note the value of the decimal becomes smaller and smaller as you advance to the right. Also there is no units place after the decimal point. This reduces the number of places by 1 as compared with a whole number.

26. How is a decimal read?

Read exactly as if it were a whole number but with the addition of the fractional name of the *lowest* place. The above number is read as "six hundred eighty million, fifty seven thousand, nine hundred twenty-three *billionths.*" The *lowest,* or *smallest,* place here is *billionths.*

27. What is the relation of each place in a decimal to the place that precedes it?

Each place is one-tenth ($\frac{1}{10}$) of the preceding place. It is thus a ten (10) times smaller fraction.

EXAMPLE:

$$0.247{,}897 = \frac{2}{10} + \frac{4}{100} + \frac{7}{1{,}000} + \frac{8}{10{,}000} + \frac{9}{100{,}000} + \frac{7}{1{,}000{,}000}$$

Read: Two hundred forty-seven thousand, eight hundred ninety-seven *millionths.*

28. Can you show that zeros, added after the last digit, do not affect the value of the decimal?

EXAMPLE: $0.4 = \frac{4}{10}$ $0.40 = \frac{40}{100} = \frac{4}{10}$

$0.400 = \frac{400}{1,000} = \frac{4}{10}$ again

29. How does a zero placed before a digit affect the value of the decimal?

The value of a digit is *divided* by ten as you move from left to *right.* So, adding a zero before the digit moves the digit one place to the right and makes its value one tenth of what it was.

EXAMPLE: $0.3 = \frac{3}{10}$ $0.03 = \frac{3}{100} = \frac{1}{10} \times \frac{3}{10}$

Adding two zeros moves the digit two places to the right and makes its value one hundredth of what it was.

EXAMPLE: $0.5 = \frac{5}{10}$ $0.005 = \frac{5}{1,000} = \frac{1}{100} \times \frac{5}{10}$

Each additional zero reduces its former value by one tenth again.

30. How is a number read that consists of a whole number and a decimal?

The point separates the whole number from the decimal. The decimal point is read "and."

EXAMPLE: 24.51. Read: Twenty-four and fifty-one hundredths. It may also be read: Twenty-four point fifty-one.

To avoid any possibility that the decimal point will be overlooked, write 0.6 instead of .6 (= six tenths).

31. How do we write dollars and cents?

Place a decimal point between the dollars and cents: \$16.43 = sixteen dollars forty-three cents.

Numbers to the left of the decimal point are dollars, to the right of it are cents in the first two places with a number in the third place as *mills*: \$16.437 = sixteen dollars forty-three cents seven mills.

Note: 10 mills = 1 cent = \$0.01. Therefore forty-three cents seven mills = four hundred thirty seven mills.

When the number of cents is less than 10 write a zero in the tenths place at the right of the decimal point:

\$3.08 = three dollars eight cents
\$3.10 = three dollars ten cents

32. What are the essential symbols in the Roman system of numeration?

Inherited from the Etruscans, the Roman system of notation uses seven capital letters of the alphabet and combinations of these letters to express numbers.

I	V	X	L	C	D	M
1	5	10	50	100	500	1,000

A bar over a letter multiplies its value by 1,000:

$$\overline{\text{V}} = 5{,}000 \qquad \overline{\text{X}} = 10{,}000$$

33. What are the rules for the values of the symbols when used in combinations?

(*a*) Each repetition of a letter repeats its value.

EXAMPLES: II = 2 III = 3 XX = 20 XXX = 30
CCC = 300 MM = 2,000

(*b*) A letter *after* one of greater value is added to it.

EXAMPLES:

$$\begin{aligned} \text{VIII} &= 8 \\ = (5 + 3) &= 8 \\ \text{XII} &= 12 \\ = (10 + 2) &= 12 \\ \text{XVII} &= 17 \\ = (10 + 5 + 2) &= 17 \\ \text{CXXXII} &= 132 \\ = (100 + 10 + 10 + 10 + 2) &= 132 \\ \text{DCCV} &= 705 \\ = (500 + 100 + 100 + 5) &= 705 \end{aligned}$$

(*c*) A letter *before* one of greater value is subtracted from it.

EXAMPLES:

$$\begin{aligned} \text{IX} &= 9 \\ = (10 - 1) &= 9 \\ \text{XL} &= 40 \\ = (50 - 10) &= 40 \\ \text{XC} &= 90 \\ = (100 - 10) &= 90 \end{aligned}$$

(*d*) A letter between two letters of greater value is subtracted from the letter which follows it.

EXAMPLES:

$$\begin{aligned} \text{XIX} &= 19 \\ [= 10 + (10 - 1)] &= 19 \\ \text{CXC} &= 190 \\ [= 100 + (100 - 10)] &= 190 \\ \text{MCMLXII} &= 1962 \\ = [1{,}000 + (1{,}000 - 100) + 50 + 10 + 2] &= 1962 \end{aligned}$$

PROBLEMS*

1. How many units in 3, 7, 9?

2. How many tens in 30, 40, 60?

3. How many tens and units in 19, 37, 46, 72, 96?

4. How many bundles of hundreds in 300, 500, 700, 900?

5. How many bundles of hundreds, tens, and units in 765, 234, 489, 536, 977, 658, 8, 54, 567, 98, 548, 958, 842, 891, 346, 738?

6. What is 1,000 called and how many bundles of hundreds are in it?

7. How many bundles of thousands, hundreds, tens, and units are there in 7,486; 8,090; 9,935; 5,803; 2,500; 2,925; 7,623; 9,260; 4,087; 6,079; 7,850; 3,374; 7,839; 5,974; 9,294?

8. What is 10,000 called and how many bundles of thousands are in it?

9. How many bundles of ten thousands, thousands, hundreds, tens, and units are in: 60,308; 46,951; 37,568; 45,382; 89,465; 63,895; 34,956; 92,857; 98,975; 20,306; 45,951; 99,358; 34,925; 98,872; 29,573?

10. How many bundles of thousands are in 100,000, and what is this number called?

11. How many bundles of hundred thousands, ten thousands, thousands, hundreds, tens, and units are in: 369,243; 780,979; 703,148; 282,297; 503,005; 386,470; 460,007; 386,364; 117,008; 204,951; 596,382; 245,520; 498,287; 995,193; 579,697?

12. What is 1,000,000 called and how many bundles of thousands, ten thousands, and tens are in it?

13. How many bundles of millions, hundred thousands, ten thousands, thousands, hundreds, tens, and units are in: 1,753,002; 75,206,008; 285,239; 42; 895; 947; 23,795,000; 9,460,280; 1,737,311; 142,755,000; 58,303,100; 473,285,000; 155,903,892; 14,237,295; 296,086,000; 82,930,711; 839,286; 2,863,401?

14. What is 1,000,000,000 called and how many bundles of hundred millions and thousands are in it?

15. How many bundles of billions, hundred millions, ten millions, millions, hundred thousands, ten thousands, thousands, hundreds, tens, and units are there in: 27,392,496,000; 1,406,762,001; 7,002,406,010; 407,841,075; 107,396,432,570; 1,900,800,005?

16. How would you express the following in figures, using a comma to separate the periods?

(*a*) Five hundred eighty-four
(*b*) Three hundred seventeen
(*c*) Six hundred ninety-nine
(*d*) Three hundred seven
(*e*) One thousand four hundred eighty-three

* Answers to odd-numbered problems begin on p. 403.

(*f*) Eight thousand sixty
(*g*) Nine thousand four hundred
(*h*) Fourteen thousand six hundred forty
(*i*) Eighty-eight thousand six
(*j*) Sixty-six thousand eighteen
(*k*) Three hundred seven thousand two hundred forty
(*l*) Eight thousand eight
(*m*) Four thousand ninety-nine
(*n*) Seventy thousand twenty-three
(*o*) Seven hundred ninety-four thousand three
(*p*) Sixty-two thousand two hundred three
(*q*) Two million two hundred eighty-five thousand
(*r*) Thirty-eight million one hundred forty-eight thousand
(*s*) Seven million two
(*t*) Sixty-one million fifty-eight thousand six
(*u*) One hundred twenty-two billion seventy thousand seven
(*v*) Five billion seven million eight thousand nine hundred nine
(*w*) Eighteen billion one million two hundred three thousand sixteen
(*x*) Ten trillion two billion one million seven hundred six
(*y*) One hundred million twenty
(*z*) Sixty million six hundred thousand six hundred

17. How are the following expressed as decimals?

(*a*) Seventy-three thousand five hundred eighty-six hundred-thousandths
(*b*) Eight thousand and eight thousandths
(*c*) Five tenths, three tenths, two and one tenth
(*d*) Seven and nine thousandths, twelve millionths
(*e*) Two hundred thirty-five thousandths, four hundred ninety-one thousandths, six ten-thousandths, three hundred and three hundredths
(*f*) Four and seven tenths, nine and two tenths, eighty-six hundredths, five hundred and five thousandths
(*g*) $\frac{4}{10}, \frac{9}{10}, \frac{3}{10}, \frac{6}{10}, \frac{1}{100}, \frac{7}{1{,}000}, \frac{8}{10{,}000}$
(*h*) Three hundred sixty-four thousand five hundred seventy-five millionths
(*i*) Nine hundred eight million six thousand thirty-four billionths

18. What is the name of the place at the right of tenths, at the right of hundredths, at the right of thousandths, the fourth place, the fifth, the sixth, the seventh?

19. How are the following read?

(*a*) 16.005	(*b*) 50.607	(*c*) 0.0002	(*d*) 87.9375
(*e*) 35.201	(*f*) 86.5392	(*g*) 2.3441	(*h*) 200.3487
(*i*) 20.2074	(*j*) 206.10057	(*k*) 30.564	(*l*) 97.4356

20. How are the following read in dollars, tenths, and hundredths of a dollar?

(*a*) \$4.57 (*b*) \$5.55 (*c*) \$6.66 (*d*) \$9.99

21. How is \$356.356 read?

22. How are the following read as dollars, dimes, and cents, and as dollars and cents?

(*a*) \$6.52 (*b*) \$3.44 (*c*) \$5.55 (*d*) \$9.75 (*e*) \$4.44 (*f*) \$8.88

23. How are the following written as cents, using the dollar sign:

(*a*) Sixty-six hundredths of a dollar?
(*b*) Eighty hundredths of a dollar?
(*c*) Forty-seven hundredths of a dollar?
(*d*) Ten hundredths of a dollar?
(*e*) One dollar and twenty hundredths?
(*f*) Seven dollars and twelve hundredths?

24. How are the following written in decimal form:

(*a*) $\frac{6}{10}$? (*b*) $\frac{8}{100}$? (*c*) $\frac{66}{1{,}000{,}000}$? (*d*) $\frac{1{,}285}{100{,}000}$? (*e*) $\frac{4{,}346}{100}$?
(*f*) Five hundredths? (*g*) Fifty-six ten-thousandths?
(*h*) Eleven thousand and thirty-six tenths?
(*i*) Five hundred hundredths?
(*j*) Six hundred forty-three ten-thousandths?

25. How many mills are there in:

(*a*) \$0.475? (*b*) \$5.621? (*c*) \$0.022? (*d*) \$10.54?
(*e*) \$1.0765? (*f*) \$0.2555? (*g*) \$0.10? (*h*) \$0.4444?

26. How are the following expressed in Arabic notation:

(*a*) XI? (*b*) VIII? (*c*) XX? (*d*) XIV? (*e*) XXX?
(*f*) XXXV? (*g*) XL? (*h*) LXXV? (*i*) XVI? (*j*) XCIV?
(*k*) LV? (*l*) DCCC? (*m*) MCMXX? (*n*) LXXXIII?
(*o*) $\overline{\text{VI}}$? (*p*) XLIX? (*q*) MDCCCXCVI? (*r*) XCV?
(*s*) MDLXXXIX? (*t*) MCXLV? (*u*) MCXL? (*v*) CDIX?
(*w*) DCIX? (*x*) MDLIV? (*y*) MDLX? (*z*) MDXLVII?
(*a′*) MMDCCXCII? (*b′*) $\overline{\text{V}}$CCCLXXVI? (*c′*) $\overline{\text{D}}$XLIX?
(*d′*) $\overline{\text{X}}$DCCXCIX? (*e′*) MMMDCCXIX? (*f′*) $\overline{\text{MD}}$CCLXXXI?
(*g′*) $\overline{\text{MMCCC}}$CLXIX?

27. How would you express the following in Roman notation:

(*a*) 12? (*b*) 18? (*c*) 19? (*d*) 43? (*e*) 33? (*f*) 28? (*g*) 56?
(*h*) 82? (*i*) 76? (*j*) 97? (*k*) 117? (*l*) 385? (*m*) 240?
(*n*) 512? (*o*) 470? (*p*) 742? (*q*) 422? (*r*) 942? (*s*) 1,426?
(*t*) 1,874? (*u*) 5,872? (*v*) 24,764? (*w*) 257,846?
(*x*) 1,450,729? (*y*) 4,840,005? (*z*) 10,562,942?

CHAPTER I

ADDITION

34. Why is addition merely a short way of counting?

If we have four apples in one group and five in another, we may count from the first object in one group to the last object in the other and obtain the result, nine. But seeing that $4 + 5 = 9$ under all conditions, we make use of this fact without stopping to count each time we meet this problem.

$$\begin{array}{r} 4 \\ + \\ 5 \\ \hline 9 \end{array}$$

The addition of two numbers is thus seen to be a process of regrouping. We do not increase anything, we merely regroup the numbers.

35. What is our standard group or bundle?

Our number system is based on groups or bundles of *ten*.

EXAMPLE: $9 + 8 = 17$. Two groups of 9 and 8 are regrouped into our standard arrangement of 17, or one bundle of 10 and 7 units. While we say "seventeen" we must think "ten and seven" or "1 ten and 7 units."

36. What is thus meant by addition?

It is the process of finding the number that is equal to two or more numbers grouped together.

37. What is meant by sum?

It is the *result* obtained by adding numbers.

38. Of the total number of 45 additions of two digits at a time for all the nine digits, which give single numbers as a sum and which give double numbers?

(*a*) The following 20 pairs result in one-number sums:

	1	2	3	4	5	6	7	8		2	3	4	5	6	7		3	4	5	6		4	5
	1	1	1	1	1	1	1	1		2	2	2	2	2	2		3	3	3	3		4	4
Sums	2	3	4	5	6	7	8	9		4	5	6	7	8	9		6	7	8	9		8	9

(*b*) The following 25 pairs give double numbers:

	9		8	9		7	8	9		6	7	8	9
	1		2	2		3	3	3		4	4	4	4
Sums	10		10	11		10	11	12		10	11	12	13

	5	6	7	8	9		6	7	8	9
	5	5	5	5	5		6	6	6	6
Sums	10	11	12	13	14		12	13	14	15

	7	8	9		8	9		9
	7	7	7		8	8		9
Sums	14	15	16		16	17		18

39. What is the rule for addition?

Write the numbers so that units stand under units, tens under tens, hundreds under hundreds, etc. Begin at the right and add the units column. Put down the units digit of the sum and carry the "tens" bundles to the next column representing the "tens" bundles. Do the same with this column. Put down the digit representing the number of tens and carry any "hundreds" bundles to the hundreds column. Continue in the same manner with other columns.

40. What is the proper way of adding?

Add without naming numbers, merely sums.

EXAMPLE:

8	49
9	41
8	32
7	24
9	17
8	
49	

If you start from the bottom, say to yourself "17, 24, 32, 41, 49." *Do not* name the individual numbers as "8 and 9 are 17 and 7 makes 24" and so on. It is even better merely to *think* 17, 24, 32, 41, 49.

41. What is the simplest but slowest way of adding?

Column by column and one digit at a time. Add from the top down or from the bottom up; each way is a check on the other.

EXAMPLE:

7, 9 3 7	30	15	29	28
8, 8 4 9	23	12	20	21
5, 2 2 8	14	8*	12*	13*
$6_2\,9_1\,3_3\,6$	↑	↑	↑	↑
2 8, 9 5 0				
Ten thousands, Thousands, Hundreds, Tens, Units	Add up Units column	Add up Tens column	Add up Hundreds column	Add up Thousands column

* These numbers include the bundles carried over from the previous column sum.

42. What is a variation of the above?

Add each column separately. Write one sum under the other, but set each successive sum one space to the left. A subsequent addition gives the total or sum.

EXAMPLE: (as above)

7,9 3 7	Sum of units column	=	3 0	
8,8 4 9	Sum of tens column	=	1 2	place to left
5,2 2 8	Sum of hundreds column	=	2 8	place to left
6,9 3 6	Sum of thousands column	=	2 6	place to left
2 8,9 5 0	Total	=	2 8 9 5 0	

43. How can grouping of numbers help you in addition?

Add two or more numbers at a time to two or more others in the columns.

EXAMPLE:

		Add down
6, 4, 3	13 group	13
8, 2	10 group	23
5, 3	8 group	31
9, 4	13 group	44
Total 44		

Grouping applies to any other column in addition. When adjacent numbers are large only *one* number for that group may suffice and do much to help accuracy in adding. Groups of two or three numbers may often be recognized as equal to 10.

44. How is addition accomplished by multiplication of the average of a group?

When you have a group of numbers whose middle figure is the average of the group, then:

sum = average number times number of figures in the group

EXAMPLES:

(*a*) Of 4, 5, and 6 number 5 = average of the three
$\therefore$ Sum $= 5 \times 3 = 15 = (4 + 5 + 6)$

(*b*) Of 8, 9, and 10 9 = average
$\therefore$ Sum $= 9 \times 3 = 27 = (8 + 9 + 10)$

(*c*) Of 12, 13, and 14 13 = average
$\therefore$ Sum $= 13 \times 3 = 39 = (12 + 13 + 14)$

(*d*) Of 6, 7, 8, 9, and 10 8 = average
$\therefore$ Sum $= 8 \times 5 = 40 = (6 + 7 + 8 + 9 + 10)$

(*e*) Of 11, 12, 13, 14, and 15 13 = average
$\therefore$ Sum $= 13 \times 5 = 65$

Note that whenever an odd number of equally spaced figures appears, you can immediately spot the center one or average and promptly get the sum of all by multiplying the average by the number of figures in the group.

45. What is the procedure for adding two columns at a time?

37	Start at bottom. Add 96 to 80 of above, then the 2 getting 178. Add 178 to the 20 above, then the 4 getting 202. Add 202 to the 30 above, then the 7 getting 239 = sum.
24	
82	
96	
239	

A variation would be to add the units of the line above it first and then the tens, as

(units)	(tens)
96 + 2 =	98 + 80 = 178
178 + 4 =	182 + 20 = 202
202 + 7 =	209 + 30 = 239

46. How are three columns added at one time?

Start at bottom. Add hundreds, then tens, then units as you continue up.

EXAMPLES:

(*a*) 968 + 886. Place in columns as

$$\begin{array}{r} 968 \\ \underline{886} \end{array}$$

$$886 + 900 = 1{,}786 + 60 = 1{,}846 + 8 = 1{,}854$$

	(hundreds) ↓		(tens) ↓		(units) ↓	
(*b*) 652	876 + 300	= 1,176 + 40		= 1,216 + 8		= 1,224
348	1,224 + 600	= 1,824 + 50		= 1,874 + 2		= 1,876
876						
1,876						

47. What is a convenient way of adding two small quantities by making a decimal of one of them?

Make a decimal of one by adding or subtracting and reverse the treatment for the other.

EXAMPLE: 96 + 78. Add 4 to 96 getting 100 = decimal number.
Subtract 4 from 78 getting 74.
∴ Sum = 174 at once.

48. How may decimalized addition be carried out to a fuller development?

Reduce each number to a decimal. Add the decimals. Add or subtract the increments.

EXAMPLE:

$$\begin{array}{rcrcl} 452 + 48 & = & 500 & = & \text{decimal number} \\ 987 + 13 & = & 1{,}000 & = & \text{decimal number} \\ 413 - 13 & = & 400 & = & \text{decimal number} \\ \hline 1{,}852 + 48 & = & 1{,}900 & - 48 = & 1{,}852 \end{array}$$

49. How may sight reading be used in addition?

By use of instinct you get an immediate result.

EXAMPLES:

(*a*) Add 26 to 53.

$$\begin{array}{r} 2\cdot6 \\ 5\cdot3 \\ \hline 7\ 9 \end{array}$$

(*b*) Add 67 to 86.

$$\begin{array}{r} 6\cdot7 \\ 8\cdot6 \\ \hline 13 \\ 14\ \ \\ \hline 153 \end{array}$$

Fix eyes between the two columns where the dots are and at once see a 7 and a 9 or a 13 and a 14 to make 153. Actually 70 is added to 9 and 140 to 13 but each is done instinctively.

50. What simple method is used to check the correctness of addition of a column of numbers?

First begin at the bottom and add up. Then begin at the top and add down. When the columns are long it is often better to write down the sums rather than to carry the "bundles" from column to column. Place sums in proper columns.

EXAMPLE:

2 2 1 carried over	Add *UP*.	Add *DOWN*.
1,8 9 6	1,896 Begin at *right*.	1,896 Begin at *left*.
4 3 2	432	432
8 4	84	84
9,8 7 6	9,876	9,876
1 2,2 8 8	18	10
	27	2 0
	2 0	27
	10	18
	12,288	12,288

51. What is meant by a check figure in addition?

One which, when *eliminated* from each number to be added and from the sum, will give a *key* number that may indicate the correctness of the addition. The check numbers 9 and 11 are generally used.

52. What are the interesting facts on the use of the check number 9?

(1) The fact that the *remainder* left after *dividing any number* by 9 is the same as the *remainder* of the *sum of the digits* of that number *divided* by 9.

Ex. (*a*) $\frac{654}{9} = 72 + \frac{\textcircled{6}}{9}$ = Remainder

$\frac{6 + 5 + 4}{9} = \frac{15}{9} = 1\frac{\textcircled{6}}{9}$ = Remainder

Ex. (*b*) $\frac{2{,}677}{9} = 297 + \frac{\textcircled{4}}{9}$ = Remainder

$\frac{2 + 6 + 7 + 7}{9} = \frac{22}{9} = 2 + \frac{\textcircled{4}}{9}$ = Remainder

(2) Also note that the sum of the digits alone will give the same number as a remainder as the division of the number by 9. Thus in (*a*) $6 + 5 + 4 = 15$ and $1 + 5 = \textcircled{6}$. In (*b*) $2 + 6 + 7 + 7 = 22$ and $2 + 2 = \textcircled{4}$.

(3) Also the fact that 9's can be discarded when adding the digits. Thus in (*a*) $6 + 5 + 4$, discard $4 + 5$ right away and the remainder is again $\textcircled{6}$. In (*b*) $2 + 6 + 7 + 7$, discard $2 + 7$ but add $6 + 7 = 13$ and $1 + 3 = \textcircled{4}$.

53. What is the procedure in checking addition by the use of the check figure 9, often called "casting out nines"?

(*a*) Add the digits in each number horizontally and get each remainder.

(*b*) Add the digits of these remainders and get the *key* figure.

(*c*) Add the digits horizontally of the answer and get the same key figure if the answer is correct.

EXAMPLE:

	Remainders	
5 5 5 4 ←(carried over)		
9·6,2 8 6 =6+2+8+6=22, 2+2=	4·	
8 5,9·7 8 =8+5+7+8=28, 2+8=10, 1+0=	1	1
7 7,4 9·3 =7+7+4+3=21, 2+1=	3	+3
4 6,9·6 7 =4+6+6+7=23, 2+3=	5·	
1·8;9·6 7 =6+7=13, 1+3=	4·	
8 5,8·9·1· =8+5=13, 1+3=	4	+4
9·4;8 5·8 =8+8=16, 1+6=	7	+7
2·6;1·3 2 =3+2=	5·	
5 3 2;5 7·2 =5+3+5+2=15, 1+5= ⑥ =	Key figure check	15, 1+5= ⑥

In practice it is sufficient to add the numbers mentally to get the remainders.

Note that all 9's and digits that add up to 9 are discarded right away. Each digit so discarded is shown with a dot at the upper right corner.

54. Why is "casting out nines" not a perfect test of accuracy in addition?

It is possible to omit or add nines or zeros without detection. Also figures may be transposed; 27 is quite different in value from 72 although the sum of the digits is the same.

This method is not generally recommended as a practical test in addition work but has its greatest value in multiplication and division work. However, it is sometimes useful as a quick check of addition.

55. What are the interesting facts on the use of the check number 11?

(1) The *remainder* left after *dividing any number* by 11 is the same as the *remainder* left after *subtracting* the *sum* of the digits in the *even places* from the *sum* of the digits in the *odd places.* If the subtraction cannot be made add 11 or a multiple of it to the odd-places sum.

EXAMPLES:

(*a*) $\frac{654}{11} = 59 + \frac{⑤}{11}$ = Remainder

Sum of odd places = 4 + 6 = 10

Sum of even places = 5   10 − 5 = ⑤ = Remainder

(*b*) $\frac{8{,}677}{11} = 788 + \frac{⑨}{11}$ = Remainder

Sum of odd places = 7 + 6 = 13 (add 11) = 24
Sum of even places = 7 + 8 = 15 = −15
⑨ = Remainder

(2) The *same remainder* is also obtained by starting with the *extreme left digit* in the number and *subtracting* it from the digit to its *right.* When necessary add 11 to make the subtraction possible. Subtract the *remainder* from the next digit. Again add 11 if necessary. Repeat the process of subtraction until all the digits of the number have been used.

In Ex. (*a*) 654 — 6 from (11 + 5) = 10
10 from (11 + 4) = ⑤ = Remainder

In Ex. (*b*) 8,677 — 8 from (11 + 6) = 9
9 from (11 + 7) = 9
9 from (11 + 7) = ⑨ = Remainder

56. Why is the checking of addition work by the use of the check figure 11 (often called "casting out elevens") superior to that of "casting out nines"?

"Casting out elevens" can indicate an error due to transposition of digits which is not possible with the "nines" method.

EXAMPLE: Suppose our number is 8,706

8 from (11 + 7) = 10 10 from (11 + 0) = 1
1 from 6 = ⑤ = Remainder = Check number

Now suppose the transposed number is 8,076

8 from (11 + 0) = 3 3 from 7 = 4
4 from 6 = ② = Remainder = Check number

The check numbers are seen to be different and we have uncovered a transposition of digits.

57. What is the procedure in checking addition by the use of the check figure 11?

(*a*) Cast out elevens from each row and get each remainder.

(*b*) Add the remainders and cast out elevens from this sum, getting the *key* figure.

(*c*) Cast out elevens from the answer and get key figure. Compare.

EXAMPLE:

8,354 → 6, 10,	5 = Remainder	
4,689 → 2, 6,	3 = Remainder	
2,987 → 7, 1,	6 = Remainder	
5,386 → 9, 10,	7 = Remainder	
21,416 → 10, 5, 7 → ⑩ = Key	Sum = 21 → 2 from 12 = ⑩ = Key	

PROBLEMS

1. Count from 3 to 99 by 3's.
2. Count from 4 to 100 by 4's.
3. Count from 6 to 96 by 6's.
4. Count from 9 to 99 by 9's.
5. Start with 3 and count by 2's, 4's, 6's, 8's to just below 100.
6. Start with 2 and count by 3's, 5's, 7's, 9's to just below 100.
7. Start with 9 and count by 4's, 7's, 9's, 2's to just below 100.
8. Start with 14 and count by 6's, 2's, 4's, 8's to just below 100.
9. Add 269, 745, and 983.
10. Add, using "carry overs."

(a)	(b)	(c)	(d)	(e)	(f)
48	89	25	36	132	707
26	54	19	6	285	459
67	32	49	88	911	420
86	8	99	75	84	871
29	62	7	34	346	385
5	86	89	51	438	42

11. Add \$5.25, \$17.60, \$0.85, \$175, \$4.565.
12. Find the sum of:

(a)	(b)	(c)
\$3,808.65	\$9,873.67	\$8,874.06
376.92	388.98	518.56
386.23	5,732.00	1,298.97
480.08	8,987.19	542.65
888.42	7,824.92	386.00
751.82	6,086.04	42.09

13. What is the sum of 10, 20, 30 by the average method?
14. What is the sum of 14, 15, 16 by the average method?
15. What is the sum of 17, 18, 19, 20, 21 by the average method?
16. What is the sum of 3, 4, 5, 6, 7, 8, 9 by the average method?
17. What is the sum of 5, 7, 9 by the average method?
18. What is the sum of 13, 15, 17 by the average method?
19. What is the sum of 14, 16, 18, 20, 22 by the average method?
20. What is the sum of 9, 12, 15 by the average method?
21. Add two columns at a time.

(a)	(b)	(c)	(d)	(e)	(f)
14	68	39	86	64	85
62	29	48	15	77	48
85	91	57	95	86	24
94	56	65	67	43	93

22. Add three columns at a time.

(*a*)	(*b*)	(*c*)	(*d*)	(*e*)
489	562	431	116	762
837	328	218	237	132
734	42	562	358	211
216	896	789	469	482

23. Add the following by the decimalizing method:
(*a*) 94 + 75 (*b*) 86 + 69 (*c*) 92 + 48 (*d*) 89 + 52
(*e*) 468 + 982 + 429 (*f*) 346 + 899 + 212
(*g*) 589 + 913 + 165 (*h*) 862 + 791 + 386

24. Add by sight reading:
(*a*) 27 + 56 (*b*) 21 + 43 (*c*) 32 + 65
(*d*) 49 + 57 (*e*) 68 + 87 (*f*) 76 + 82

25. A gasoline station owner had 275 gallons left after selling 632 gallons. How many gallons did he have originally?

26. One pipe from a tank discharges 76 gallons per second while another pipe from the same tank discharges 16 gallons per minute more than the first. How many gallons will both pipes discharge in a minute?

27. An automobile travels 386 miles on the first day and 416 miles the second day, at which time it is 237 miles from its point of destination. What is the distance from its starting point to its destination?

28. A suburban house was built with the following expenses: masonry, \$3,565; lumber, \$4,850; millwork, \$1,485; carpentry, \$3,800; plumbing, \$2,758; painting, \$679; hardware, \$1,508; heating, \$1,250; and electricity, \$687. What did the house cost when completed?

29. If a family of two persons spends \$135 for rent, \$205 for food, \$85 for clothing, \$35 for fuel, \$7 for light, \$22 for insurance, \$6 for carfare, \$12 for charity, and saves \$18, what is the income after taxes and other payroll deductions?

30. The twenty-second of February is how many days after New Year's? How many days from New Year's to the fourth of July?

31. Check the following by first adding up and then by adding down. Place check marks as proof.

(*a*)	(*b*)	(*c*)	(*d*)
28,751	28,537	45,122	61,413
86,757	93,106	10,793	74,892
57,682	52,877	97,515	54,637
93,864	62,534	83,462	20,145
13,724	58,209	55,798	31,806

32. Prove the following by use of the check figure 9.

(a)	(b)	(c)	(d)
79,432	32,676	43,678	78,973
4,365	5,789	10,925	1,329
679	735	4,576	402
19,807	19,253	4,474	35,029
5,974	20,671	1,661	8,911
473	8,341	893	906
42,978	62,147	11,025	45,414
2,674	4,128	8,901	7,657

33. Prove the following by use of the check figure 11.

(a)	(b)	(c)	(d)
33,766	28,593	70,513	30,189
73,412	13,747	34,526	4,153
4,762	92,061	1,517	259
6,513	7,284	2,028	74,105
805	768	204	7,123
132	974	715	856
8,913	51,631	89,374	16,734
6,842	3,587	4,675	423
60,419	94,575	5,123	8,573

34. Add horizontally and vertically.

(*a*)

$2 + 5 + 9 + 8 + 3 + 7 + 9 = ?$
$8 + 4 + 6 + 7 + 1 + 2 + 6 = ?$
$4 + 2 + 8 + 7 + 5 + 6 + 5 = ?$
$1 + 3 + 5 + 7 + 9 + 3 + 6 = ?$
$9 + 4 + 6 + 3 + 9 + 6 + 7 = ?$
$6 + 7 + 3 + 2 + 8 + 2 + 3 = ?$
$6 + 8 + 9 + 4 + 6 + 3 + 8 = ?$
? ? ? ? ? ? ? ?

(*b*)

$6 + 4 + 7 + 5 + 8 + 9 + 7 = ?$
$8 + 9 + 5 + 4 + 3 + 9 + 1 = ?$
$1 + 7 + 6 + 8 + 5 + 2 + 8 = ?$
$4 + 1 + 2 + 6 + 8 + 5 + 3 = ?$
$6 + 3 + 8 + 7 + 5 + 3 + 2 = ?$
$5 + 2 + 1 + 8 + 7 + 9 + 8 = ?$
$3 + 5 + 7 + 6 + 4 + 8 + 9 = ?$
? ? ? ? ? ? ? ?

CHAPTER II

SUBTRACTION

58. What is subtraction?

It is the reverse of addition. Since we know that five apples + three apples = eight apples, it follows reversely that taking five apples away from eight apples leaves three apples.

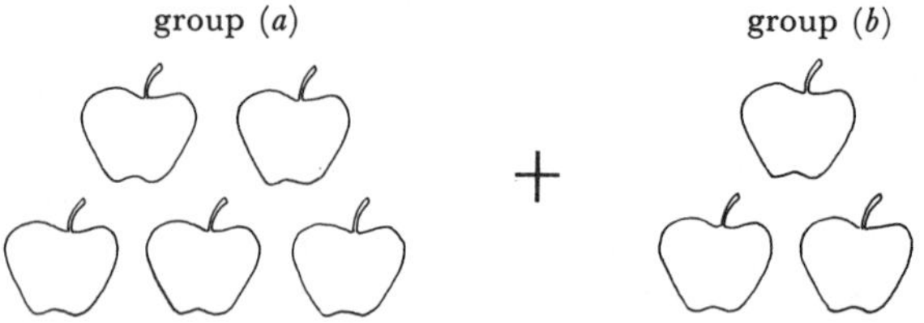

Or taking three apples away from eight apples leaves five apples.

$$8 - 5 = 3 \qquad 8 - 3 = 5$$

As with addition, subtraction is thus seen to be merely a regrouping:

$$\text{group } (a) + \text{group } (b) = \text{group } (c) = 8.$$

$$\text{group } (c) - \text{group } (a) = 3. \qquad \text{group } (c) - \text{group } (b) = 5.$$

59. Why may subtraction be said to be a form of addition?

Ex. (*a*) $9 - 4 = 5$. May be thought of as "4 and what make 9?"
4 and 5 make 9.

Ex. (*b*) $16 - 9 = 7$. 9 and what make 16?
9 and 7 make 16.

60. What three questions will lead to the process of subtraction?

(*a*) How much *remains*?
(*b*) How much more is required?
(*c*) By how much do they *differ*?

In (*a*) if Bert has \$10 and pays out \$6, how many dollars *remain*? Here the \$6 was originally a part of the \$10.

In (*b*) Bert has \$65 and would like to buy a 35-mm. camera that costs \$89. How much more does he require?

In (*c*) if Bert has \$10 and Charles has \$6, by how much do they *differ*? Here the \$10 and the \$6 are distinct numbers.

61. What are the terms of a subtraction?

$$\begin{array}{rl} 15 & = \text{"minuend"} = \text{the number to be reduced (usually on top)} \\ -\ 9 & = \text{"subtrahend"} = \text{the number to subtract (usually below)} \\ \hline 6 & = \text{"remainder" or "difference"} \end{array}$$

If the subtrahend was originally a part of the minuend then the answer is called the "remainder." If the minuend and subtrahend are distinct numbers the answer is called the "difference."

62. Why is it said that we can always add but we cannot always subtract?

Subtraction is not always possible. It is not, when the number of *things* which we wish to subtract is greater than the number of *things* we have.

Ex. (*a*) Addition: 5 apples + 3 apples = 8 apples
Subtraction: 8 apples − 3 apples = 5 apples
Addition: 5 apples + 7 apples = 12 apples
Subtraction: 5 apples − 7 apples = impossible.

There exist no negative apples. At best we can only express the relation as 2 apples missing.

Ex. (*b*)

7 foot-candles of illumination − 5 foot-candles = 2 foot-candles.

7 foot-candles − 9 foot-candles is impossible because there cannot be a negative illumination of 2 foot-candles. The limit is zero illumination, or darkness.

Ex. (*c*) From an electric cord of 8 feet we can cut off 3 feet leaving 5 feet, but we cannot cut off 10 feet leaving −2 feet of cord.

63. When is it possible to subtract with the number expressing the subtrahend greater than the number expressing the minuend?

By introduction of the concept of "direction" to the quantities expressed by the numbers and calling all numbers in one direction positive numbers and numbers in the reverse direction (from the starting point zero) negative numbers.

Ex. (*a*)

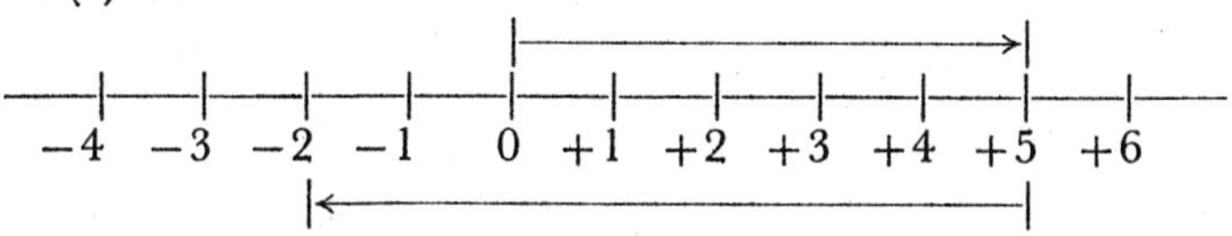

Now, if we step off 5 steps to the right and then step off 7 to the left we land at -2

$$\therefore 5 - 7 = -2$$

Ex. (*b*) If we let zero = freezing temperature, then $+5°$ is 5 degrees above freezing and if it falls 3 degrees it will be 2 degrees above freezing. If it falls 7 degrees it will be 2 degrees below freezing, or

$$\left.\begin{aligned} +5° - 3° &= +2° \\ +5° - 7° &= -2° \end{aligned}\right\}$$

Ex. (*c*) If zero is latitude, then $+5°$ lat. $-$ $7°$ lat. $= -2°$ lat. This would be in the Southern Hemisphere.

If we have \$5 in the bank and if we have credit, we may be able to draw out \$7 in which case \$5 − \$7 = − \$2 overdraft. Again, if we have \$10 in our pocket and buy something that costs \$25 we are in debt for \$15: \$10 − \$25 = − \$15 debt.

The negative number is not a physical but a mathematical conception which may or may not have a physical representation depending on how it is applied.

64. What is the subtraction table that should be studied until the answers can be given quickly and correctly?

Subtraction Table

1 − 1 =	2 − 2 =	3 − 3 =	4 − 4 =	5 − 5 =
2 − 1 =	3 − 2 =	4 − 3 =	5 − 4 =	6 − 5 =
3 − 1 =	4 − 2 =	5 − 3 =	6 − 4 =	7 − 6 =
4 − 1 =	5 − 2 =	6 − 3 =	7 − 4 =	8 − 5 =
5 − 1 =	6 − 2 =	7 − 3 =	8 − 4 =	9 − 5 =
6 − 1 =	7 − 2 =	8 − 3 =	9 − 4 =	10 − 5 =
7 − 1 =	8 − 2 =	9 − 3 =	10 − 4 =	11 − 5 =
8 − 1 =	9 − 2 =	10 − 3 =	11 − 4 =	12 − 5 =
9 − 1 =	10 − 2 =	11 − 3 =	12 − 4 =	13 − 5 =
10 − 1 =	11 − 2 =	12 − 3 =	13 − 4 =	14 − 5 =

6 − 6 =	7 − 7 =	8 − 8 =	9 − 9 =
7 − 6 =	8 − 7 =	9 − 8 =	10 − 9 =
8 − 6 =	9 − 7 =	10 − 8 =	11 − 9 =
9 − 6 =	10 − 7 =	11 − 8 =	12 − 9 =
10 − 6 =	11 − 7 =	12 − 8 =	13 − 9 =
11 − 6 =	12 − 7 =	13 − 8 =	14 − 9 =
12 − 6 =	13 − 7 =	14 − 8 =	15 − 9 =
13 − 6 =	14 − 7 =	15 − 8 =	16 − 9 =
14 − 6 =	15 − 7 =	16 − 8 =	17 − 9 =
15 − 6 =	16 − 7 =	17 − 8 =	18 − 9 =

65. What is the rule for subtraction?

(*a*) Write the subtrahend under the minuend, units under units, tens under tens, etc.

(*b*) Begin at the right and subtract each figure of subtrahend from the corresponding figure of the minuend, and write the remainder underneath.

(*c*) If any figure of the subtrahend is greater than the minuend, increase the minuend by 10 (which uses 1 unit of the next higher order) and subtract. Now reduce the minuend of the next higher order by 1 and continue to subtract until all the digits have been taken care of.

Ex. (*a*)	73	Add 10 to units 3, getting 13	8 from 13 = 5
	−48	Reduce minuend 7 to 6	4 from 6 = 2
Remainder =	25		

Ex. (*b*)	893	Add 10 to 3, getting 13	6 from 13 = 7
	−456	Reduce minuend 9 to 8	5 from 8 = 3
Remainder =	437		4 from 8 = 4

Ex. (*c*)	9,650	Add 10 to 0, getting 10	7 from 10 = 3
	−4,857	Reduce 5 to 4 and add 10, getting 14	5 from 14 = 9
Remainder =	4,793	Reduce 6 to 5 and add 10, getting 15	8 from 15 = 7
		Reduce 9 to 8	4 from 8 = 4

Ex. (*d*)	93,187	Add 10 to 7, getting 17	9 from 17 = 8
	−48,549	Reduce 8 to 7	4 from 7 = 3
Remainder =	44,638	Add 10 to 1, getting 11	5 from 11 = 6
		Reduce 3 to 2 and add 10, getting 12	8 from 12 = 4
		Reduce 9 to 8	4 from 8 = 4

Ex. (*e*)	3,000	Add 10 to 0, getting 10	2 from 10 = 8
	−2,292	You can't reduce 0 so you add 10 and reduce by 1 to get 9.	9 from 9 = 0
Remainder =	708	Again you can't reduce 0, so you add 10 and reduce by 1 getting 9.	2 from 9 = 7
		Finally reduce 3 to 2	2 from 2 = 0

Note that you do not actually add or take away anything from the number. You merely regroup a bundle by unscrambling it and placing it with the lower order to make the subtraction possible. In Ex. (*c*) above, we can see that we will need one thousands bundle to unscramble to 10 hundreds, one hundreds bundle to become 10 tens, and one tens bundle to become 10 units. The numbers then become

$$\begin{array}{r} 9{,}650 \\ -4{,}857 \\ \hline \end{array} \qquad \begin{array}{rrrr} & (10) & (10) & (10) \\ 8, & 5 & 4 & 0 \\ -4, & 8 & 5 & 7 \\ \hline 4, & 7 & 9 & 3 \end{array} = \text{Answer}$$

66. What is known as the method of "equal additions" in subtraction?

The method is based on the fact that the same number may be added to both minuend and subtrahend without changing the value of the difference.

Ex. (*a*)

$$\begin{array}{r} 9{,}650 \\ -4{,}857 \\ \hline 4{,}793 \end{array}$$

- Add 1 "tens" (= 10 units) to *minuend*; Add 1 "tens" to *subtrahend* (5 + 1) = 6 — 7 from 10 = 3
- Add 1 "hundreds" (= 10 tens) to *minuend*; Add 1 "hundreds" to subtrahend (8 + 1) = 9 — 6 from 15 = 9
- Add 1 "thousands" (= 10 hundreds) to minuend; Add 1 "thousands" to subtrahend (4 + 1) = 5 — 9 from 16 = 7; 5 from 9 = 4

Ex. (*b*)

$$\begin{array}{r} 93{,}187 \\ -48{,}549 \\ \hline 44{,}638 \end{array}$$

9 from 17 = 8
5 from 8 = 3
5 from 11 = 6
9 from 13 = 4
5 from 9 = 4

This method is quick and simple. All you need to remember is to add 1 to the next column in the *subtrahend* every time you add 10 to the minuend to make subtraction possible.

Ex. (*c*)

$$\begin{array}{r} 4{,}000 \\ -2{,}292 \\ \hline 1{,}708 \end{array}$$

2 from 10 = 8
10 from 10 = 0
3 from 10 = 7
3 from 4 = 1

67. What is the mode of thinking of subtraction that is called the Austrian method, or the method of making change?

A good deal of subtraction in the business world is concerned with making change. It consists in building to the subtrahend until the minuend is reached.

Ex. (*a*)	786	You think:	5 and 1 are 6	Write 1
	−435		3 and 5 are 8	Write 5
	351		4 and 3 are 7	Write 3

When subtraction is to be made possible in any column it becomes a modification of the above "equal addition" method.

Ex. (*b*)	76,053	9 and 4 are 13	Write 4
	−29,679	8 and 7 are 15	Write 7
	46,374	7 and 3 are 10	Write 3
		10 and 6 are 16	Write 6
		3 and 4 are 7	Write 4

68. How may subtraction be simplified?

Add or subtract a quantity to get a multiple of 10. It is easier to subtract a multiple of 10 from another quantity than to subtract any other double digit number.

EXAMPLE:

49	Add 3 to each, getting		52	Subtract 7 from each		42
−37			−40	(= multiple of 10) →		−30
12	Ans.	Ans.	12		Ans.	12

Note that the answer is the same when you add or subtract the same number from both the minuend and subtrahend and that it is easier to subtract when the subtrahend is made a multiple of 10.

69. How may the above be extended?

Divide the numbers into couples and make each couple a multiple of 10 (which is known as a decimal number).

			Add 3	Add 4				
Ex. (*a*)	8,974		89	74		92	78	
	−3,736	→	37	36	→	40	40	
	5,238		52	38		52	38	Ans.
			Subtract 7	Subtract 6				
			89	74		82	68	
			37	36	→	30	30	
						52	38	Ans.

If the subtrahend in one couple is larger than the minuend, there will be 1 to carry, which is subtracted from the differences of the couple next on the left.

```
            Add 2  Subtract 2
Ex. (b)      97       54              99  52
            -38       72      --->    40  70
            ---------------          -------
             58       82              58  82
```

In subtracting 70 from 52 borrow one (hundred), then subtract 1 from the difference of (99 − 40).

```
                  Add 3
Ex. (c)      38   63              38  66
           -  2   87      --->     2  90
           ---------              -------
                                  35  76
```

70. How can the subtraction of two-figure numbers be done by simple inspection using decimalization?

```
Ex. (a)    74
          -28
         ----
Ans.       46
```

Disregard units digits, making numbers decimal numbers. 20 from 70 = 50.

Now mentally add 4 and subtract 8, leaving 4 to subtract. Then 50 − 4 = 46.

Ex. (*b*) 89 − 47 = 40 + 9 − 7 = 42
98 − 36 = 60 + 8 − 6 = 62
95 − 22 = 70 + 5 − 2 = 73

71. How can inverted or left-hand subtraction be done?

Start from the left and subtract, noting whether there is one to carry from the column at the right.

```
Ex. (a) 8,927,862  Start at left
        7,184,249  7 from  8 = 1
        ---------
        1,743,613     (nothing to carry from column at right)
                   2 from  9 = 7   (carry 1 to make the 1 a 2)
                   8 from 12 = 4
                      (nothing to carry from column at right)
                   4 from  7 = 3   (none to carry)
                   2 from  8 = 6   (none to carry)
                   5 from  6 = 1   (carry 1 to 4, to make it 5)
                   9 from 12 = 3
```

```
Ex. (b)    9,993   (Carry 1)  9 from  9 = 0   (starting at left)
          -8,998   (Carry 1) 10 from 19 = 9
          ------
           0,995   (Carry 1) 10 from 19 = 9
                              8 from 13 = 5
```

72. What is meant by the arithmetical complement of a number?

Abbreviated a.c., arithmetical complement is the remainder found by subtracting the number from the next *highest* multiple of 10.

EXAMPLE: a.c. of 2 is $10 - 2 = 8$
a.c. of 57 is $100 - 57 = 43$
a.c. of 358 is $1{,}000 - 358 = 642$
a.c. of 0.358 is $1.000 - 0.358 = 0.642$

73. What is the simplest way of calculating the a.c. of a number?

Subtract its right-hand digit from 10 and each of the others from 9. This does away with carrying of 1's.

EXAMPLE: a.c. of 68,753 = 31,247

Start at left: 6 from 9 = 3
8 from 9 = 1
7 from 9 = 2
5 from 9 = 4
3 from 10 = 7

74. When and how is the a.c. used in subtraction?

When a quantity is to be subtracted from the sum of several others. To subtract by means of the a.c., add the a.c. of the subtrahend and subtract the multiple of 10 used in getting the a.c.

Ex. (*a*) Subtract 9,431 from 9,805 by a.c.

	9,805
a.c. of 9,431 =	0,569
Sum =	10,374
Subtract 10,000 (next highest multiple of 10)	10,000
Ans.	374

Nothing is gained by use of a.c. in so simple a case.

Ex. (*b*) Subtract 1,284 from the sum of 9,747, 1,283, and 1,292.

By a.c. method

	9,747
	1,283
	1,292
a.c. of 1,284 =	8,716
Sum =	21,038
Subtract (10,000)	10,000
Ans.	11,038

By ordinary method

	9,747
	1,283
	1,292
Sum =	12,322
	−1,284
Ans.	11,038

Ex. (*c*) From bank deposits of $226.80, $342.61, and $187.34 deduct a withdrawal of $560.79 to get the net increase.

	$ 226.80
	342.61
	187.34
a.c. of $560.79 =	439.21
Sum =	$1,195.96
Subtract 1,000	1,000.00
Ans.	$ 195.96

75. How do we proceed to give change to a customer by the use of the so-called "Austrian method" of subtraction?

Add from the amount of the purchase up to the next higher money unit, then to the next, and so on until you reach the amount of the bill tendered in payment.

EXAMPLE: If the bill given in payment is $5 and the purchase is $2.38, give customer the following as change: 2 cents to make $2.40, 10 cents to make $2.50, 50 cents to make $3.00, $2 to make $5.

Total change adds up to $2.62.

76. What is the best check in subtraction?

The sum of remainder and subtrahend must equal the minuend. This means: we have taken away a certain number; we now put it back and return to the original number. This check should always be made. It is done mentally.

EXAMPLES:

(*a*)		(*b*)	(*c*)
834,705	minuend	4,513	$9,207.68
−493,598	subtrahend	−2,057	−463.86
341,107	difference	2,456	$8,743.82
834,705	check	4,513	$9,207.68

77. Is "casting out nines" a practical check in subtraction?

It is not, and too much time must not be spent on this method.

Ex. (*a*)

			(add 9 to make subtraction possible)
3·5 6;3 7 5	Add digits	5 + 3 + 7 + 5 = 20, 2 + 0 =	2 + 9 = 11
2 0 9;9·8 8	Add digits	2 + 8 + 8 = 18,	1 + 8 = 9
1·4 6;3·8·7	Add digits	4 + 7 = 11, 1 + 1 = ②	9 from 11 = ②

It is seen that the difference between the remainders of the minuend and subtrahend = remainder of answer.

Ex. (*b*)

	(add 9 to make subtraction possible)	
\$9;8 6 4: 5·9·	14 = Remainder of minuend	6 from 14 = ⑧
\$8,4·5·7. 3·6·	6 = Remainder of subtrahend	
\$1,4 0 7·.2·3	⑧ = Remainder of answer	

78. May casting out of elevens be used as a check?

Yes, but here also too much time should not be devoted to this method.

Ex. (*a*)
356,375
209,988
146,387

Take the minuend. Start at left.
3 from 5 = 2
2 from 6 = 4
4 from 14 = 10
10 from 18 = 8
8 from 16 = ⑧
Add 11 = ⑲
to be able to subtract

Take the subtrahend. Start at left.
2 from 11 = 9
9 from 9 = 0
0 from 9 = 9
9 from 19 = 10
10 from 19 = ⑨

Remainder of minuend (19) − remainder of subtrahend (9) = ⑩

Take the difference. Start at left.
1 from 4 = 3
3 from 6 = 3
3 from 3 = 0
0 from 8 = 8
8 from 18 = ⑩

Ex. (*b*)

\$9,864.59	1	1 + 11 =	12
\$8,457.30	6		−6
\$1,407.29	⑥		⑥

PROBLEMS

Perform the following subtractions:

1.	67	94	93	85	79	98	96	89
	45	52	74	63	54	37	72	64

2.	798	946	893	765	986	897
	463	625	732	541	585	246
3.	889	795	986	957	965	894
	256	363	594	823	645	522
4.	265	247	258	287	254	237
	93	85	74	94	92	65

5.	2,489	1,847	2,369	1,698	2,637
	967	905	864	717	917
6.	2,658	2,577	3,786	4,386	1,597
	936	652	853	984	865

7. If we say a certain tree is in zero position and we take 8 steps to the right of the tree which we call the positive direction and then we step off 12 steps to the left, where will we land?

8. If zero is freezing temperature, what does +7 deg. mean? What does −8 deg. mean?

9. If your latitude is zero and you travel north to +11° lat. and then southward for 15°, what would be your last position?

10. If you had $85 in the bank and you issued a check for $97, what would be your overdraft?

11. If you had only $63 and you wanted to buy a 35-mm. camera that cost $87, how much would you be in debt?

12. Subtract:

(*a*)	(*b*)	(*c*)	(*d*)
594,795	847,806	965,943	798,432
469,836	695,973	891,647	636,858

(*e*)	(*f*)	(*g*)	(*h*)
$3,957.90	$9,207.69	$983,769.29	$792,529.36
898.83	468.86	595,276.75	186,737.57

13. Check the answers to problem 12 by addition. Check the answers by casting out nines. Check the answers by casting out elevens.

14. What is the subtrahend for each of the following sets of values?

(a)	(b)	(c)	(d)
89,684	86,458	78,325	$360,025.86
63,987	36,679	– 14,597	$139,873.36

(e)	(f)
$986,983.50	$567,899.20
– $ 13,367.40	– $345,678.98

15. Check the answers to problem 14 by addition and by casting out nines.

16. Use the simplified method of subtraction by making the subtrahend a multiple of ten.

(a)	(b)	(c)	(d)	(e)	(f)	(g)	(h)	(i)	(j)
48	81	56	93	89	76	63	36	96	74
36	44	28	57	37	48	38	19	47	36

17. Extend the simplified method of subtraction to two couples making each a multiple of ten or a decimal number.

(a)	(b)	(c)	(d)	(e)	(f)
87 76	78 43	67 54	93 42	62 53	52 25
45 48	36 29	24 35	36 83	28 96	30 60

18. Do the following subtractions of two-figure numbers by simple inspection, using decimalization:

(a)	(b)	(c)	(d)	(e)	(f)	(g)	(h)
75	87	99	96	67	57	78	84
38	46	37	28	29	39	49	67

19. Do the following by inverted, or left-hand, subtraction.

(a)	(b)	(c)	(d)
9,738,751	9,875	6,548	758,243
7,295,358	8,987	4,879	579,687

20. What is the arithmetical complement of:

(a) 7? (b) 69? (c) 472? (d) 1,282? (e) 0.472? (f) 79,864? (g) 864,348?

21. (a) Subtract 8,562 from 9,983 by a.c. method.

(b) Subtract 46,827 from 87,962 by a.c. method.

22. Subtract 4,976 from the sum of 8,432, 1,343, and 1,565 by a.c. method.

23. From bank deposits of $342.76, $562.59, and $134.59 deduct a withdrawal of $632.48 by a.c. method.

24. If a $20 bill is given in payment and the purchase is $12.89, what change will the customer get, using the so-called "Austrian" method of subtraction?

25. If a railroad carries 2,325,879 passengers one year and 3,874,455 passengers the following year, what is the increase?

26. If the Federal income tax collected one year is $67,892,762,945 and $71,432,652,982 the following year, what is the increase?

27. (*a*) Begin with 53 and subtract by 2's, 4's, 6's, 8's.
(*b*) Begin with 89 and subtract by 3's, 5's, 7's, 9's.
(*c*) Begin with 74 and subtract by 5's, 7's, 3's, 9's.

28. A man bought a farm for $17,500. He kept it two months, during which time he paid $439.50 in taxes and $782.75 for repair of fences. He then sold it for $21,500. What was his profit?

CHAPTER III

MULTIPLICATION

79. What is multiplication?

It is merely a simplified form of addition. Suppose we have eight apples in a row and there are four rows. We can add them as $8 + 8 + 8 + 8 = 32$ or we can say simply $4 \times 8 = 32$. Also if we have four apples in a row and there are eight rows, then

$$4 + 4 + 4 + 4 + 4 + 4 + 4 + 4 = 32 \quad \text{or} \quad 8 \times 4 = 32.$$

You see that $4 \times 8 = 8 \times 4 = 32$. In each case the sum is 32. When several equal numbers are to be added, it is much shorter to obtain the result by multiplication.

80. What are the terms of a multiplication?

(*a*) The number to be repeated is called the *multiplicand.*

(*b*) The number of times the multiplicand is to be repeated is called the *multiplier.*

(*c*) The *result* of the multiplication is called the *product.*

(*d*) The multiplicand and the multiplier are also known as the *factors* of the product.

EXAMPLE:

$$4 \times 8 = 32$$

$$\text{or} \quad \underbrace{\text{multiplicand} \times \text{multiplier}}_{\text{factors of the product}} = \text{product}$$

81. What is (*a*) a concrete number, (*b*) an abstract number, (*c*) the type of number of the multiplier in multiplication?

(*a*) A number that is applied to any particular object is called a *concrete* number. *Examples:* an apple, an auto, 2 hours, etc.

(*b*) A number that is not applied to a particular object is an *abstract* number. *Examples:* 1, 5, 62.

(*c*) In multiplication the multiplier is always an abstract number.

82. What are the most useful products that should be committed to memory?

Multiplication Table

$2 \times 2 = 4$	$3 \times 3 = 9$	$4 \times 4 = 16$	$5 \times 5 = 25$
$3 \times 2 = 6$	$4 \times 3 = 12$	$5 \times 4 = 20$	$6 \times 5 = 30$
$4 \times 2 = 8$	$5 \times 3 = 15$	$6 \times 4 = 24$	$7 \times 5 = 35$
$5 \times 2 = 10$	$6 \times 3 = 18$	$7 \times 4 = 28$	$8 \times 5 = 40$
$6 \times 2 = 12$	$7 \times 3 = 21$	$8 \times 4 = 32$	$9 \times 5 = 45$
$7 \times 2 = 14$	$8 \times 3 = 24$	$9 \times 4 = 36$	$10 \times 5 = 50$
$8 \times 2 = 16$	$9 \times 3 = 27$	$10 \times 4 = 40$	$11 \times 5 = 55$
$9 \times 2 = 18$	$10 \times 3 = 30$	$11 \times 4 = 44$	$12 \times 5 = 60$
$10 \times 2 = 20$	$11 \times 3 = 33$	$12 \times 4 = 48$	
$11 \times 2 = 22$	$12 \times 3 = 36$		
$12 \times 2 = 24$			
$6 \times 6 = 36$	$7 \times 7 = 49$	$8 \times 8 = 64$	$9 \times 9 = 81$
$7 \times 6 = 42$	$8 \times 7 = 56$	$9 \times 8 = 72$	$10 \times 9 = 90$
$8 \times 6 = 48$	$9 \times 7 = 63$	$10 \times 8 = 80$	$11 \times 9 = 99$
$9 \times 6 = 54$	$10 \times 7 = 70$	$11 \times 8 = 88$	$12 \times 9 = 108$
$10 \times 6 = 60$	$11 \times 7 = 77$	$12 \times 8 = 96$	
$11 \times 6 = 66$	$12 \times 7 = 84$		
$12 \times 6 = 72$			
	$10 \times 10 = 100$	$11 \times 11 = 121$	
	$11 \times 10 = 110$	$12 \times 11 = 132$	
	$12 \times 10 = 120$	$12 \times 12 = 144$	

83. When several numbers are multiplied does it matter in what order the multiplication is performed?

The order of multiplication does not matter.

EXAMPLE: $2 \times 6 \times 4 = 2 \times (6 \times 4) = (2 \times 4) \times 6 = 48$

The 2 may be multiplied by 6 and this result (= 12) may then be multiplied by 4 to get 48, or the 6 and 4 may first be multiplied and then the 2 used, etc.

84. What is the rule in multiplication when (*a*) the two signs of the numbers are both plus [+]; (*b*) both signs are minus [−]; (*c*) the two signs are unlike?

(*a*) Two pluses produce a plus product.
(*b*) Two minuses produce a plus product.
(*c*) Two unlike signs produce a minus product.

$(+4) \times (+6) = +24 \qquad (+4) \times (-6) = -24$

$(-4) \times (-6) = +24 \qquad (-4) \times (+6) = -24$

Note: It is not necessary to write the plus in front of the product.

85. What is the effect upon a number when you move it one, two, three places to the left in the period?

Moving a figure one place to the left has the same effect as multiplying it by 10. *Example:* $76 \times 10 = 760$. So, to multiply by 10, place a zero at the right of the multiplicand; thus moving each digit one place to the left and increasing its value 10 times.

To multiply by 100, place two zeros at the right of the multiplicand. *Example:* $76 \times 100 = 7{,}600$.

To multiply by 1,000 place three zeros at the right of the multiplicand, etc. *Example:* $76 \times 1{,}000 = 76{,}000$.

86. What is the rule for multiplying when either multiplier or multiplicand ends in zeros?

Multiply the multiplicand by the multiplier without regard to the zeros and *annex* as many zeros at the right of the product as are found at the right of the multiplier *and* multiplicand.

EXAMPLE:

$$\underbrace{6{,}37\underset{\substack{\uparrow\\ \text{one zero}}}{\mathit{0}} \times 2\underset{\substack{\uparrow\\ \text{two zeros}}}{\mathit{00}}}_{=\ \text{three zeros}} \qquad 637 \times 2 = 1{,}274 \quad \therefore\ \text{Product} = 1{,}274{,}\underset{\substack{\uparrow\\ \text{three zeros}}}{\mathit{000}}$$

87. How is ordinary, simple multiplication performed?

Write the multiplier under the multiplicand, placing the units of the multiplier under units of multiplicand, and begin at the right to multiply.

EXAMPLE:

```
            8,639
                7
           ------
               63
              21
             4 2
            56
           ------
Product =  60,473 = Sum
```

$7 \times 9 = 63$; you add up all the units $= 6$ tens $+ 3$ units.

$7 \times 3 = 21$; adding all the bundles of tens, result $= 2$ bundles of 100's $+$ 1 bundle of tens.

$7 \times 6 = 42$; adding all the bundles of 100's, result $= 4$ bundles of 1,000's $+$ 2 bundles of 100's.

$7 \times 8 = 56$; adding all the bundles of 1,000's, result $= 5$ bundles of 10,000's $+$ 6 bundles of 1,000's.

Note that the work can be shortened by doing the "carrying" mentally.

8,639	$7 \times 9 = 63$. Write 3, carry 6 tens.
7	$7 \times 3 = 21 + 6 = 27$. Write 7 tens and carry the 2 hundreds.
60,473	$7 \times 6 = 42 + 2 = 44$. Write 4 hundreds and carry 4 thousands.
	$7 \times 8 = 56 + 4 = 60$. Write 60.

88. What is the procedure when the numbers to be multiplied contain more than one digit?

EXAMPLE: 698×457. It would not be convenient to set down 698 to be added 457 times.

698		
457		
4,886	(7×698)	Multiplier $(457 = 400 + 50 + 7)$
34,900	(50×698)	
279,200	(400×698)	
318,986	= Product	

Multiplying by 457 is therefore the same as multiplying by 7, by 50, and by 400, and adding the results.

(*a*) First multiply 698 by 7.

$7 \times 8 = 56$. Write 6, carry 5.
$7 \times 9 = 63 + 5 = 68$. Write 8, carry 6.
$7 \times 6 = 42 + 6 = 48$. Write 48.

(*b*) Then multiply by 50. Write 0 in units column and then multiply 698 by 5.

$5 \times 8 = 40$. Write zero, carry 4.
$5 \times 9 = 45 + 4 = 49$. Write 9, carry 4.
$5 \times 6 = 30 + 4 = 34$. Write 34.

(*c*) Then multiply 698 by 400. Write 00 and multiply 698 by 4.

$4 \times 8 = 32$. Write 2, carry 3.
$4 \times 9 = 36 + 3 = 39$. Write 9, carry 3.
$4 \times 6 = 24 + 3 = 27$. Write 27.

Now add the three results to get 318,986 = product. Of course, you may omit writing the zeros when you remember to move the product one place to the left when multiplying by the digit in the tens column, and two places to the left when multiplying by the digit in the hundreds column, etc.

```
     698
     457
   -----
   4,886
  34 90
 279 2
 -------
 318,986
```

89. How can the fact that either number may be used as the multiplier serve to provide a check on our multiplication?

EXAMPLE (as above): Reverse. Use 698 as the multiplier.

```
     457
     698
   -----
   3,656
  41 13
 274 2
 -------
 318,986 (= check of above)
```

90. How can we extend the multiplication table beyond 12 × 12 by making use of the smaller products by 2 or by 4?

EXAMPLES:

(*a*) $14 \times 13 = 2 \times 7 \times 13 = 91 \times 2 = 182$. Split 14 into 7×2
(*b*) $16 \times 13 = 2 \times 8 \times 13 = 104 \times 2 = 208$. Split 16 into 8×2
(*c*) $18 \times 13 = 2 \times 9 \times 13 = 117 \times 2 = 234$. Split 18 into 9×2
(*d*) $16 \times 16 = 4 \times 4 \times 16 = 4 \times 64 = 256$. Split 16 into 4×4

91. How can multiplication by two-digit numbers be simplified?

Convert one two-digit number into two one-digit numbers.

Ex. (*a*) $27 \times 16 = 27 \times 2 \times 8 = 54 \times 8 = 432$
(*b*) $27 \times 15 = 27 \times 3 \times 5 = 81 \times 5 = 405$

92. How can the multiplication of two 2-digit numbers having the same figure in the tens place be simplified?

(*a*) Multiply the units.
(*b*) Add the units and multiply the sum by the tens digit. *Annex* a zero.
(*c*) Multiply the tens. *Annex* 2 zeros.
(*d*) Add (*a*) + (*b*) + (*c*).

EXAMPLES:

(1) 78 × 75

Multiply units (8 × 5 =)	40	
Add units (8 + 5 = 13)		
Multiply by 7 (= 91)	910	←annex zero
Multiply tens (7 × 7 = 49)	4,900	←annex 2 zeros
Ans.	5,850	

(2) 96 × 92, quickly

6 × 2 =	12
(6 + 2) × 9 =	720
9 × 9 =	8,100
Ans.	8,832

(3) 89 × 83

9 × 3 =	27
(9 + 3) × 8 =	960
8 × 8 =	6,400
Ans.	7,387

93. How can multiplication be simplified by multiplying one factor and dividing the other factor by the same quantity?

Ex. (*a*) 35 × 16.

Multiply factor 35 by 2, getting 70
Divide factor 16 by 2, getting 8

70 × 8 = 560 = Product

The product is the same because

$$35 \times 16 \times \frac{2}{2} = (35 \times 2) \times \frac{16}{2} = 560.$$

Ex. (*b*) $27\frac{1}{2} \times 24 = 27\frac{1}{2} \times 24 \times \frac{2}{2} = 55 \times 12 = 660$ Ans.

This could also be done as $27\frac{1}{2} \times 24 \times \frac{4}{4} = 110 \times 6 = 660$.

94. What can be done when multiplication may simplify one of the factors, but when the other factor is not divisible by the same number?

If multiplication of one factor makes that factor simpler, use the result as the multiplier and divide the *product* by the same number used to simplify the multiplier.

Ex. (*a*) 45 × 29 Multiply factor 45 by 2 getting 90
Now 90 × 29 = 2,610
Divide this by 2, getting $\frac{2,610}{2} = 1,305$ Ans.

Ex. (*b*) 323 × 35

$35 \times 2 = 70 \qquad 70 \times 323 = 22{,}610 \qquad \frac{22{,}610}{2} = 11{,}305$ Ans.

Note this simplification applies to numbers ending in 5 up to 55 to give procedures within the range of the multiplication table.

Ex. (*c*) 271 × 55

$55 \times 2 = 110 \qquad 271 \times 110 = 29{,}810 \qquad \frac{29{,}810}{2} = 14{,}905$ Ans.

95. When the tens digits are alike and the units digits add up to 10, how is multiplication simplified?

Write the product of the units digits. Increase one of the tens digits by 1 and multiply by the other.

Ex. (*a*)
```
   43     Units 3 × 7 = 21
   47     Tens 4 + 1 = 5   5 × 4 = 20
 ─────
 2,021 Ans.
```

Ex. (*b*)
```
   69     Units 1 × 9 = 9, but write 09 when product
   61                      is less than 10.
 ─────
 4,209 Ans.  Tens 6 + 1 = 7   7 × 6 = 42
```

Ex. (*c*)
```
   75     Units 5 × 5 = 25
   75     Tens 7 + 1 = 8   8 × 7 = 56
 ─────
 5,625 Ans.
```

96. When the units digits are alike and the tens digits add up to 10, how is multiplication simplified?

Write the product of the units digits. Add units digit to product of tens digits.

Ex. (*a*)
```
        64  Units 4 × 4 = 16
        44  Tens 4 × 6 = 24 and add units 4, getting 28.
      ─────
Ans.  2,816
```

Ex. (*b*)
```
        79  Units 9 × 9 = 81
        39  Tens 3 × 7 = 21 + units 9 = 30
      ─────
Ans.  3,081
```

Ex. (*c*)
```
        83  Units 3 × 3 = 9; write 09 when product is less
        23                 than 10.
      ─────
Ans.  1,909  Tens 2 × 8 = 16 + 3 = 19
```

97. When neither of above combinations is applicable, how may so-called cross multiplication be applied to advantage?

Ex. (*a*) 53 Units 6 × 3 = 18; write 8, carry 1 tens.
46 Tens (6 × 5) + 1 = 31 tens
Ans. 2,438 Tens 4 × 3 = 12 tens
43 tens; write 3 tens, carry 4 hundreds.
Hundreds 4 × 5 = 20 + 4 = 24 hundreds.

Ex. (*b*) 81 Units 4 × 1 = 4
24 Tens 4 × 8 = 32 + (2 × 1) = 34; write 4, carry 3.
Ans. 1,944
Hundreds 2 × 8 = 16 + 3 = 19.

Ex. (*c*) 73 Units 9 × 3 = 27; write 7, carry 2 tens.
59 Tens 9 × 7 = 63 (+ 2) + (5 × 3) = 80 tens; write 0, carry 8 hundreds.
Ans. 4,307
Hundreds 5 × 7 = 35 + 8 = 43 hundreds.

98. When the units digits are 5 and the sum of the tens digits is even, how is multiplication simplified?

The product will end in 25. Multiply the tens digits and add half their sum.

Ex. (*a*) 95 Product ends in 25
55
Ans. 5,225 Tens $9 \times 5 = 45$ plus $\left(\frac{9+5}{2} = 7\right) = 52$

Ex. (*b*) 75 Product ends in 25
55
Ans. 4,125 Tens $5 \times 7 = 35$ plus $\left(\frac{5+7}{2} = 6\right) = 41$

99. When the units digits are 5 and the sum of the tens digits is odd, how is multiplication simplified?

The product will end in 75. Multiply tens digits and add half their sum, discarding fraction.

Ex. (*a*) 75 Product ends in 75
45
Ans. 3,375 Tens $4 \times 7 = 28$ plus $\frac{7+4}{2}$ [= 5 (discard fraction)] = 33

This method may be used when there are only two and not more than three digits in *either* multiplier or multiplicand. When dollars and cents are involved, the two end digits are cents and digits to the left are dollars.

Ex. (*b*) 155 × 75 = Ans. 11,625

Sum of 7 + 15 is even; ∴ product ends in 25.

$7 \times 15 = 105 \text{ plus } \frac{7 + 15}{2} [= 11] = 116$

Ex. (*c*) \$ 1.25 × 35 = Ans. \$ 43.75

Sum of 3 + 12 is odd; ∴ product ends in 75.

$3 \times 12 = 36 \text{ plus } \frac{12 + 3}{2} [= 7] = 43$

Ex. (*d*) \$ 1.55 × 65 = Ans. \$100.75

Sum of 15 + 6 is odd; ∴ product ends in 75.

$6 \times 15 = 90 \text{ plus } \frac{6 + 15}{2} [= 10] = 100$

100. What is meant by left-hand multiplication, or what is sometimes called inverted multiplication?

Multiply left-hand figures first and then the next, and add the products.

Ex. (*a*) 89 × 8	8 × 80 =	640	Multiply by left-hand figure;
	9 × 8 =	72	then the next.
	Ans.	712	

Ex. (*b*) 746 × 9	700 × 9 =	6,300	Multiply left-hand figure;
	40 × 9 =	360	then the next;
	6 × 9 =	54	then the next.
	Ans.	6,714	

101. What is meant by an aliquot (ăl′i-kwŏt) part of a number?

It is a quantity which can be a divisor of a number without leaving a remainder. It is therefore a factor of the number.

$$\frac{\text{Given number } (= \text{base})}{\text{A whole number}} = \text{aliquot part } (= \text{factor}).$$

Ex. (*a*) 5 is an aliquot part (or factor) of 20 or of 35. When 20 or 35 is divided by 5 there is no remainder. 5 is a factor of either number.

$$\frac{20}{4} = 5 = \text{aliquot part} \qquad \frac{35}{7} = 5 = \text{aliquot part.}$$

Ex. (*b*) $12\frac{1}{2}$, $16\frac{2}{3}$, $33\frac{1}{3}$, and 25 go into 100: 8, 6, 3, and 4 times respectively and are aliquot parts of 100 or factors of 100.

Ex. (*c*) $8\frac{1}{3}$¢, 10¢, and 25¢ are aliquot parts of \$1.00 since they are contained 12, 10, and 4 times respectively in \$1.00.

102. What is meant by a fractional equivalent of an aliquot part?

By definition

$$\frac{1}{\text{a whole number}} \times \text{given number } (= \text{base}) = \text{aliquot part.}$$

Then

$$\frac{1}{\text{a whole number}} = \text{fractional equivalent of the aliquot part.}$$

Ex. (*a*) $\frac{1}{8} \times 100 = 12\frac{1}{2}$ (= aliquot part of 100); 100 is the base. Then $\frac{1}{8}$ = fractional equivalent of the aliquot part of 100 (= $12\frac{1}{2}$).

Ex. (*b*) $\frac{1}{6} \times 100 = 16\frac{2}{3}$ (= aliquot part of 100). Then $\frac{1}{6}$ = fractional equivalent of $16\frac{2}{3}$ aliquot part of 100.

It is seen that the fractional equivalent has a numerator of 1 and a denominator which is the number of times that the aliquot part is contained in the given number.

103. When are some numbers useful while not aliquot parts themselves?

They are useful when they are convenient multiples of aliquot parts.

Ex. (*a*) $37\frac{1}{2}$ is not an aliquot part of 100 since it does not go into 100 a whole number of times, but $12\frac{1}{2}$ is an aliquot part of 100 and $37\frac{1}{2}$ is $3 \times 12\frac{1}{2}$. The fractional equivalent of $12\frac{1}{2}$ is $\frac{1}{8}$ of 100.

$$\therefore 37\tfrac{1}{2} \text{ is } \tfrac{3}{8} \text{ of } 100.$$

Ex. (*b*) $66\frac{2}{3}$ is $2 \times 33\frac{1}{3}$. The fractional equivalent of $33\frac{1}{3}$ is $\frac{1}{3}$.

$$\therefore 66\tfrac{2}{3} \text{ is } \tfrac{2}{3} \text{ of } 100.$$

Ex. (*c*) 75 is 3×25. The fractional equivalent of 25 is $\frac{1}{4}$.

$$\therefore 75 \text{ is } \tfrac{3}{4} \text{ of } 100.$$

104. What are some of the aliquot parts of 100 and their fractional equivalents?

We know that an aliquot part of 100 is a factor of 100.

Aliquot part of 100	Fractional Equivalent	Aliquot part	Fractional Equivalent	Aliquot part	Fractional Equivalent
2	$\frac{1}{50}$	$12\frac{1}{2}$	$\frac{1}{8}$	60	$\frac{3}{5}$
4	$\frac{1}{25}$	$13\frac{1}{3}$	$\frac{2}{15}$	$66\frac{2}{3}$	$\frac{2}{3}$
5	$\frac{1}{20}$	$16\frac{2}{3}$	$\frac{1}{6}$	75	$\frac{3}{4}$
$6\frac{1}{4}$	$\frac{1}{16}$	20	$\frac{1}{5}$	80	$\frac{4}{5}$
$6\frac{2}{3}$	$\frac{1}{15}$	25	$\frac{1}{4}$	$83\frac{1}{3}$	$\frac{5}{6}$
$8\frac{1}{3}$	$\frac{1}{12}$	$33\frac{1}{3}$	$\frac{1}{3}$	$87\frac{1}{2}$	$\frac{7}{8}$
		50	$\frac{1}{2}$		

105. How may aliquot parts of 100 be written as decimals?

An aliquot part of 100 means so many hundredths, and may be written as a decimal. The base is 100.

EXAMPLE: (as above)

Cipher in front of aliquot part:

.02, .04, .05, .0625, .0666..., .0833...

Decimal point in front of aliquot part:

.125, .1333..., .1666..., .20, .25.

106. Why are aliquot parts useful in calculations involving dollars?

Aliquot parts of 100 have 100 parts as their bases. As \$1.00 = 100 cents, then $\frac{1}{3}$ of a dollar = $33\frac{1}{3}$ cents, and $\frac{1}{5}$ of a dollar = 20 cents.

EXAMPLE: Find cost of 72 articles, when the price of one is $16\frac{2}{3}$¢.

$$72 \times \$\tfrac{1}{6} = \$12.00 \quad (16\tfrac{2}{3}¢ = \$\tfrac{1}{6})$$

If the price of an article were a dollar, the total cost would be \$72.00, but since the price is only $\frac{1}{6}$ of a dollar, the total cost is $\frac{1}{6} \times 72 = \$12.00$.

107. How may aliquot parts of 100 be used in multiplication?

(*a*) To multiply by 50 (= $\frac{1}{2}$ of 100). Multiply by 100 by annexing two zeros. Then divide by 2 to multiply by 50 (= $\frac{1}{2}$ of 100).

EXAMPLE: $$46 \times 50 = \frac{4{,}600}{2} = 2{,}300$$

(*b*) To multiply by 25 (= $\frac{1}{4}$ of 100). Annex two zeros to multiply by 100. Then divide by 4 to multiply by 25 (= $\frac{1}{4}$ of 100).

EXAMPLE: $$32 \times 25 = \frac{3{,}200}{4} = 800$$

(*c*) To multiply by 20 (= $\frac{1}{5}$ of 100). Annex two zeros to multiply by 100. Since 20 is $\frac{1}{5}$ of 100, divide by 5.

EXAMPLE: $$65 \times 20 = \frac{6{,}500}{5} = 1{,}300$$

In this case it would generally be easier to multiply directly.

(*d*) To multiply by 75 (= $\frac{3}{4}$ of 100). Annex two zeros to multiply by 100. Since 75 = $\frac{3}{4}$ of 100, multiply by $\frac{3}{4}$.

EXAMPLE: $$39 \times 75 = 3{,}900 \times \tfrac{3}{4} = \frac{11{,}700}{4} = 2{,}925$$

108. What is the practical use of aliquot parts in multiplication?

Aliquot parts enable us to dispense with fractions. For our use aliquot parts are applicable to bases of hundreds and other decimal numbers.

Ex. (*a*) What is the cost of 65 articles at \$2.50 each? The base here is 10 and $2\frac{1}{2}$ is $\frac{1}{4}$ of 10. Then add one zero and divide by 4:

$$\frac{650}{4} = \$162.50.$$

Ex. (*b*) How much will 49 items at $\$3.12\frac{1}{2}$ cost? Multiply 49 by $3 = \$147$, and add to it $\frac{49}{8} = \$6.125$ ($12\frac{1}{2}¢ = \frac{1}{8}$ of \$1.00).

$$\text{Cost} = \$147.00 + \$6.125 = \$153.12\tfrac{1}{2}.$$

Ex. (*c*) What is the cost of 38 articles at $62\frac{1}{2}$ cents? $62\frac{1}{2}$ cents = $\frac{5}{8}$ of \$1.00. But $\frac{5}{8} = \frac{4}{8} + \frac{1}{8} = \frac{1}{2} + \frac{1}{8}$. Then

$$\$\tfrac{38}{2} + \tfrac{38}{8} = \$19.00 + \$4.75 = \$23.75.$$

Ex. (*d*) What is the result of $37{,}519 \times 125$?

As 125 is $\frac{1}{8}$ of 1,000, annex three zeros and divide by 8. This multiplies the number first by 1,000 and then divides by 8 to find 125 as a multiplier:

$$\frac{37{,}519{,}000}{8} = 4{,}689{,}875 \text{ Ans.}$$

Also, since $125 = (100 + 25)$, then

$$37{,}519\ (100 + 25) = 3{,}751{,}900 + \frac{3{,}751{,}900}{4}$$

$$= 3{,}751{,}900 + 937{,}975 = 4{,}689{,}875 \text{ Ans.}$$

Ex. (*e*) What is the cost of each of the following?

$$\begin{aligned} 24 \text{ lb. at } 33\tfrac{1}{3}¢ &= 24 \times \tfrac{1}{3} = \$\ 8.00 \\ 42 \text{ lb. at } 25¢ &= 42 \times \tfrac{1}{4} = \$10.50 \\ 72 \text{ lb. at } 37\tfrac{1}{2}¢ &= 72 \times \tfrac{3}{8} = \$27.00 \\ 16 \text{ lb. at } 75¢ &= 16 \times \tfrac{3}{4} = \$12.00 \end{aligned}$$

109. May the number of articles and the price be interchanged as a means of simplifying a problem in aliquot parts?

Yes. Thus, $83\frac{1}{3}$ yards at \$3.15 can be changed to 315 yards at $83\frac{1}{3}¢$.

$$83\tfrac{1}{3}¢ = \tfrac{5}{6} \text{ of } \$1.00 \qquad \therefore \tfrac{5}{6} \times 315 = \frac{1{,}575}{6} = \$262.50$$

EXAMPLE: What is the cost of $16\frac{2}{3}$ yards of cloth at 69¢ a yard? This can be changed to 69 yards at $16\frac{2}{3}$¢ a yard.

At $1.00 per yard, 69 yards would cost $69.

But $16\frac{2}{3}$¢ $= \frac{1}{6}$ of $1.00 $\quad \therefore \frac{69}{6} =$ $11.50 Ans.

110. What is the cost of 1,780 lb. of feed at $15.00 a ton?

$$\$15 \text{ per ton} = \frac{1{,}500¢}{2{,}000 \text{ lb.}} = \tfrac{3}{4}¢ \text{ per lb. or } 75¢ \text{ per 100 lb.}$$

At 1¢ per lb. ($1.00 per 100 lb.), 1,780 lb. costs $17.80.

75¢ $= \frac{3}{4}$ of $1.00.

$\therefore \frac{3}{4} \times \$17.80 = \$13.35 =$ cost of 1,780 lb. at $15.00 per ton.

111. How can we simplify the multiplication by 24?

Multiply by 25 by annexing two zeros and dividing by 4. Subtract the original number from the result.

Ex. (*a*) 28×24. $\quad 28 \times 25 = \dfrac{2{,}800}{4} = 700$

$$28 \times 24 = 700 - 28 = 672$$

Ex. (*b*) A variation: 261×124. $\quad 124 = (100 + 24)$

Then

$$\begin{aligned} 261 \times 124 &= 261(100 + 24) \\ &= 26{,}100 + \left[\frac{26{,}100}{4} - 261\right] = 26{,}100 + 6{,}264 \\ &= 32{,}364 \end{aligned}$$

112. How can we simplify the multiplication by 26?

Multiply by 25 by annexing two zeros and dividing by 4. Add the original number to this.

Ex. (*a*) $261 \times 26 = \dfrac{26{,}100}{4} + 261 = 6{,}525 + 261 = 6{,}786$

Ex. (*b*) $261 \times 126 = 261(100 + 26) = 26{,}100 + \left[\dfrac{26{,}100}{4} + 261\right]$

$$= 26{,}100 + 6{,}786 = 32{,}886$$

113. How can we multiply a number by 9, using subtraction?

EXAMPLE: 66,492 × 9 = 598,428
66,492(10 − 1) = 664,920 − 66,492

```
 664,920   To do it in one line, imagine a zero annexed to the
− 66,492      number.  Subtract 2 from zero getting 8.
 598,428   Carry 1 to the 9 at left,        10 from 12 = 2,
              getting 10.                       carry 1.
           Add 1 to 4 at left,              5 from 9 = 4, no carry.
              getting 5.                 ┌--Add 1 to 6 (at left),
           6 from 14 = 8, carry 1 --------┘    getting 7.
           7 from 16 = 9, carry 1           1 from 6 = 5 = end
                                                number at left.
```

114. How can we multiply by 11 using addition?

EXAMPLE:

76,492 × 11 = 76,492(10 + 1) = 764,920 + 76,492
= 841,412

In one line:

Put down 2. Add the next figure 9 to the 2.
Put down 1, carry 1. Then 4 + 1 + 9 = 14.
Put down 4, carry 1. Then 6 + 1 + 4 = 11.
Put down 1, carry 1. Then 7 + 1 + 6 = 14.
Put down 4, carry 1. Then 7 + 1 = 8.

115. How can we multiply by 111 by using addition?

EXAMPLE:

```
76,492 × 111 = 76,492(100 + 10 + 1) = 7,649,200
                                        764,920
                                         76,492
                                      ---------
                                      8,490,612
```

In one line:

76,492 × 111	Put down 2.
Add 9 + 2 = 11	Put down 1, carry 1.
Add 4 + 9 + 2 + carry 1 = 16	Put down 6, carry 1.
Add 6 + 4 + 9 + 1 carry = 20	Put down 0, carry 2.
Add 7 + 6 + 4 + 2 carry = 19	Put down 9, carry 1.
Add 7 + 6 + 1 carry = 14	Put down 4, carry 1.
Add 7 + 1 carry = 8	Put down 8.

116. How can we simplify the multiplication by 8 and by 7?

To multiply by 8, annex a zero and subtract twice the number.

```
EXAMPLE: 579 × 8 = 5,790 − (2 × 579) =   5,790
                                        −1,158
                                         4,632  Ans.
```

To multiply by 7, annex a zero and subtract 3 times the number.

EXAMPLE: $579 \times 7 = 5{,}790 - (3 \times 579) =$

$$\begin{array}{r} 5{,}790 \\ -1{,}737 \\ \hline 4{,}053 \end{array} \text{ Ans.}$$

117. How do we multiply by 99, 98, 97 or by 999, 998, 997?

Annex the proper number of zeros and subtract the required number of times.

$579 \times 99 = 57{,}900 - 579 =$

$$\begin{array}{r} 57{,}900 \\ 579 \\ \hline 57{,}321 \end{array} \text{ Ans.}$$

$579 \times 98 = 57{,}900 - 2 \times 579 =$

$$\begin{array}{r} 57{,}900 \\ -\ 1{,}158 \\ \hline 56{,}742 \end{array} \text{ Ans.}$$

$579 \times 97 = 57{,}900 - 3 \times 579 =$

$$\begin{array}{r} 57{,}900 \\ -\ 1{,}737 \\ \hline 56{,}163 \end{array} \text{ Ans.}$$

118. What is meant by the complement of a number?

The difference between that number and the unit of a next higher *order*.

Ex. (*a*) Complement of 7 is 3 because the difference between 7 and 10 is 3; 10 is the next higher order of 7.

Ex. (*b*) Complement of 58 is 42 because 100 − 58 is 42. 100 is the next higher order of 58.

119. How is complement multiplication performed?

(*a*) Find the complement of each number.
(*b*) Multiply the complements together.
(*c*) Subtract one of the complements from the other number and multiply this by 100.
(*d*) Add (*b*) to (*c*).

Ex. (*a*) Multiply 92×96. $100 - 92 = 8 =$ complement
$100 - 96 = 4 =$ complement

$8 \times 4 = 32 =$ product of complements

Number 92 minus 4 (= complement of 96) = 88

$88 \times 100 = 8{,}800$

$8{,}800 + 32 = 8{,}832$ Ans.

Ex. (*b*) Multiply 86 × 93. Complements are 14 and 7.

14 × 7 = 98 = product of complements
86 − 7 = 79   79 × 100 = 7,900
7,900 + 98 = 7,998 Ans.

Ex. (*c*) Multiply 942 × 968. Complements are 58 and 32.

58 × 32 = 1,856 = product of complements
942 − 32 = 910. (Here multiply by 1,000.)
910 × 1,000 = 910,000
910,000 + 1,856 = 911,856 Ans.

It may not pay to use this method with three figures.

120. How can we multiply by a number between 12 and 20 using only one line in the product?

Multiply as usual by the *units* figure of the multiplier. Carry as usual, but also add the figure on the *right* of the figure multiplied. This latter addition takes care of the tens figure of the multiplier.

EXAMPLE:
```
48,623          48,623
  × 16            × 16
-------         -------
777,968         291,738
                486,23
                -------
                777,968
```

First 6 × 3 = 18. Put down 8, carry 1.
Next 6 × 2 = 12 + 1 + 3 (= figure to right of one multiplied)
= 16. Put down 6, carry 1.
6 × 6 = 36 + 1 + 2 (= figure to right of one multiplied)
= 39. Put down 9, carry 3.
6 × 8 = 48 + 3 + 6 (= figure to right of one multiplied)
= 57. Put down 7, carry 5.
6 × 4 = 24 + 5 + 8 (= figure to right of one multiplied)
= 39. Put down 7, carry 3.
3 + 4 (= figure to right of one multiplied)
= 7. Put down 7.

All the above can be done mentally of course. As you see by ordinary multiplication, the multiplication of the tens figure 1 of the multiplier moves the entire multiplicand one place to the left and accounts for the addition of the figure to the right of the one being multiplied in the one-line process.

121. What is meant by cross multiplication?

A method of multiplying by a number of more than one digit, without putting down the partial products. The partial products are kept in mind and only one line results as the answer. The secret is to start with the right-hand digit of the multiplier and continue to progress to each digit of the multiplier and as each is finished start another to the left. Get the units first, then add up the tens, hundreds, thousands, etc., using each digit of the multiplier or the multiplicand. Add the carry-over figure. Put each product in its proper place.

122. What is the result of 76 × 64 using cross multiplication?

Units: 4 × 6 = 24. Put down 4, carry 2
Tens: (4 × 7 = 28) + (6 × 6 = 36) + 2 carry = 66 Put down 6, carry 6.
Hundreds: (6 × 7 = 42) + 6 carry = 48. Put down 48.

```
   76
   64
-----
4,864  Ans.
```

123. What is the result of 847 × 76 using cross multiplication?

Units: 6 × 7 = 42. Put down 2, carry 4.
Tens: (6 × 4 + 4 carry) + (7 × 7) = 77. Put down 7, carry 7.
Hundreds: (6 × 8 + 7 carry) + (7 × 4) = 83. Put down 3, carry 8.
Thousands: 7 × 8 + 8 carry = 64. Put down 64.

```
   847
    76
------
64,372  Ans.
```

124. How can we check a multiplication by "casting out nines"?

(*a*) Get the remainder by adding digits of multiplicand.

(*b*) Get the remainder by adding digits of multiplier.

(*c*) Multiply remainders (*a*) and (*b*) together and get remainder of this product.

(*d*) Get remainder of the answer (or product).

If remainder of (*c*) and (*d*) are alike, the multiplication is in all probability correct.

All 9 digits or those which add up to 9 are discarded right away.

EXAMPLE:

```
         2 8·9·1·2   multiplicand, 2 + 2 = 4 = Remainder
           7 9·8 6   multiplier, 7 + 8 + 6 = 21, 2 + 1 = ③ = Rem.
-------------------
2 2 8,2·8·9;1·5·2·   Ans. (= product), 2 + 2 + 8 = 12, 1 + 2 = ③
```

Remainder of multiplicand (4) × remainder of multiplier (3) = 12
1 + 2 = ③ = same as remainder of answer or product.

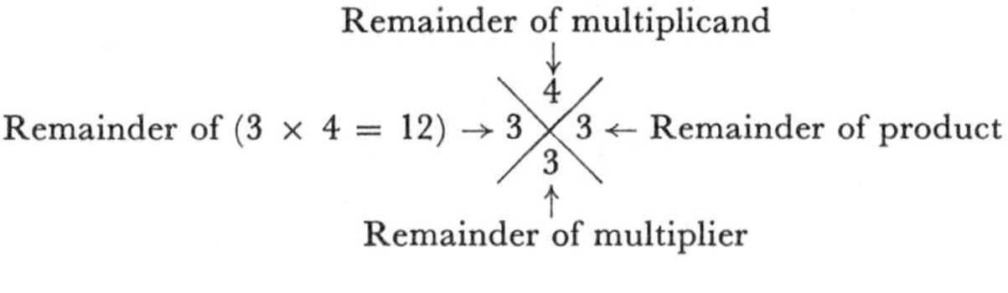

This is not an absolute proof but only a test of the correctness of the multiplication. The reversing of multiplier and multiplicand requires more time, but it is more accurate because it eliminates the possibility of transposed figures or of nines and zeros being added or omitted erroneously.

PROBLEMS

1. Multiply 54 by 10, by 100, by 1,000.

2. Multiply 820 by 10, by 100, by 1,000.

3. Multiply 1,762 by 10, by 100, by 1,000.

4. Multiply 631 by 60.

5. Multiply 45 by 40, by 400, by 4,000, by 400,000.

6. Multiply 4,700 by 4, by 40, by 400, by 4,000, by 40,000.

7. Multiply 6,390 by 300.

8. Multiply:
(*a*) 870 by 3,600 (*b*) 785,340 by 4,700 (*c*) 98,750 by 400
(*d*) 87,953 by 45,000 (*e*) 48,800 by 78,000 (*f*) 780,000 by 630
(*g*) 387,470 by 4,000

9. What is the product of:
(*a*) 4,738 multiplied by 6 (*b*) 892 by 8 (*c*) 953 by 67
(*d*) 628 by 86 (*e*) 438 by 99 (*f*) 673 by 83 (*g*) 768 by 57
(*h*) 4,174 by 647 (*i*) 587 by 756 (*j*) 9,046 by 839
(*k*) 3,490 by 874 (*l*) 5,947 by 638 (*m*) 6,084 by 519
(*n*) 7,493 by 349 (*o*) 9,486 by 305 (*p*) 9,385 by 3,005
(*q*) 3,795 by 803 (*r*) 9,476 by 8,007 (*s*) 2,583 by 7001
(*t*) 9,434 by 8,002 (*u*) 8,754 by 408 (*v*) 7,004 by 1,371
(*w*) 8,745 by 49 (*x*) 6,354 by 684 (*y*) 2,851 by 1,212
(*z*) 8,172 by 899?

10. Multiply:
(*a*) $38.85 by 375 (*b*) $731.40 by 457 (*c*) $872.34 by 741
(*d*) $400.10 by 856 (*e*) $1,340.35 by 704 (*f*) $4,650.20 by 708.

11. A mechanic earns $28.85 a day. What will his pay be for a five-day week? For a month of 22 days?

12. If 28 yards of carpet are required for a floor, what will be the cost at $9.25 a yard?

13. On October 1, John got a temporary job paying $82 a week. How much did he earn in 23 weeks?

14. If it costs $40.65 for labor and $36.29 for material to spray an acre of vineyard 5 times, what will be the cost to spray 8 acres 5 times?

15. There are 21,750 cubic feet in the first 6 inches of top soil of an acre of ground. How much will this soil weigh at 80 lb. per cubic foot?

16. A man bought 1,124 acres of land at $225 an acre. He spent $83,700 for improvements and then sold 8 acres at $450 an acre, 270 acres at \$535 an acre, 325 acres at \$380 an acre, 360 acres at \$660 an acre, and the rest at \$100 an acre. How much did he gain or lose?

17. If you bought \$15 worth of books a month for 28 months, how much would you have spent?

18. Joe drove a car 400 miles at 40 miles per hour for 20 days. How many miles did he cover?

19. What is: (*a*) 14 × 17 (*b*) 16 × 17 (*c*) 18 × 17 (*d*) 16 × 19? Make use of the smaller products by 2 or by 4.

20. What is: (*a*) 29 × 18 (*b*) 29 × 15 (*c*) 37 × 16 (*d*) 46 × 14? Convert one two-digit number into two one-digit numbers.

21. Multiply: (*a*) 85 × 87 (*b*) 48 × 49 (*c*) 58 × 53 (*d*) 37 × 32 (*e*) 65 × 67 (*f*) 99 × 94 (*g*) 74 × 72 (*h*) 26 × 28 (*i*) 17 × 18 by the method used when the tens figures are alike.

22. Multiply: (*a*) 45 × 16 (*b*) $23\frac{1}{2}$ × 22 (*c*) 32 × 18 (*d*) $22\frac{1}{2}$ × 24 (*e*) 18 × 18 (*f*) 15 × 16 (*g*) $33\frac{1}{2}$ × 14; by multiplying one factor and dividing the other factor by the same quantity.

23. Multiply: (*a*) 35 × 27 (*b*) 237 × 35 (*c*) 117 × 55 (*d*) 42 × 15 (*e*) 89 × 45; by multiplying the factor ending in 5 to simplify it and dividing the results by the same number.

24. Multiply: (*a*) 52 × 58 (*b*) 63 × 67 (*c*) 79 × 71 (*d*) 48 × 42 (*e*) 85 × 85 (*f*) 23 × 27 (*g*) 37 × 33 by the method used when units add up to 10 and tens digits are alike.

25. Multiply: (*a*) 63 × 43 (*b*) 75 × 35 (*c*) 94 × 14 (*d*) 47 × 67 (*e*) 58 × 58 (*f*) 84 × 24 (*g*) 26 × 86 by the method used when units digits are alike and tens digits add up to 10.

26. Multiply by cross multiplication method, getting answer in one line: (*a*) 63 × 54 (*b*) 82 × 23 (*c*) 72 × 48 (*d*) 52 × 43 (*e*) 48 × 69 (*f*) 91 × 18.

27. Multiply: (*a*) 95 × 45 (*b*) 75 × 65 (*c*) 65 × 85 (*d*) 35 × 55 (*e*) 95 × 35 (*f*) 75 × 55 (*g*) 35 × 35 (*h*) 85 × 75 (*i*) 145 × 65 (*j*) \$1.35 × 45 (*k*) \$1.56 × 75 (*l*) \$2.15 × 95 by simplified method.

28. Multiply: (*a*) 87 × 7 (*b*) 92 × 8 (*c*) 64 × 6 (*d*) 657 × 9 (*e*) 49 × 5 (*f*) 432 × 7 by left-hand multiplication.

29. What part of 100 is: (*a*) 50? (*b*) $12\frac{1}{2}$? (*c*) $37\frac{1}{2}$? (*d*) $62\frac{1}{2}$? (*e*) $58\frac{1}{3}$? (*f*) $16\frac{2}{3}$? (*g*) $41\frac{2}{3}$? (*h*) $33\frac{1}{3}$? (*i*) $8\frac{1}{3}$?

30. What part of 10 is: (*a*) 1.25? (*b*) $3\frac{3}{4}$? (*c*) $8\frac{1}{3}$? (*d*) $6\frac{2}{3}$? (*e*) 7.5? (*f*) $6\frac{1}{4}$? (*g*) $8\frac{3}{4}$? (*h*) $3\frac{1}{3}$?

31. What part of 1 is: (*a*) .25? (*b*) .375? (*c*) .625? (*d*) .125?

32. What part of 1,000 is: (*a*) 125? (*b*) 875? (*c*) 625? (*d*) 375?

33. What is the cost of 84 articles when the price of one is $16\frac{2}{3}$¢?

34. Multiply the following by the aliquot-part method:
(*a*) $41\frac{2}{3}$ × 64 (*b*) $87\frac{1}{2}$ × 72 (*c*) 25 × 5,744 (*d*) $62\frac{1}{2}$ × 68
(*e*) $83\frac{1}{3}$ × 54 (*f*) $16\frac{2}{3}$ × 72 (*g*) $8\frac{1}{3}$ × 240 (*h*) 75 × 48
(*i*) $66\frac{2}{3}$ × 636 (*j*) $12\frac{1}{2}$ × 144 (*k*) $37\frac{1}{2}$ × 136 (*l*) 20 × 85
(*m*) 58 × 50 (*n*) 48 × 25 (*o*) 2,840 × 75

35. What is the cost of:
(*a*) 85 articles at \$2.50 each, using aliquot-part method?
(*b*) 58 articles at \$2.12½ (*c*) 46 articles at 62½¢
(*d*) 36 lb. at 33⅓¢ per lb.? (*e*) 48 lb. at 25¢? (*f*) 56 lb. at 37½¢?
(*g*) 24 lb. at 75¢? (*h*) 66⅔ yd. at \$6.24 per yd.?
(*i*) 12½ yd. at 72¢?

36. What is the cost of 1,860 lb. of feed at \$12 a ton? Make use of aliquot-part method.

37. Find the cost of 72 lawn mowers at \$125 each, using aliquot part.

38. What is the cost of 48 radios at \$62.50 each? Use aliquot-part method.

39. Multiply: (*a*) 32 × 24 (*b*) 68 × 24 (*c*) 242 × 124 (*d*) 57 × 24, using simplified multiplication by 24.

40. Multiply: (*a*) 242 × 26 (*b*) 242 × 26 (*c*) 32 × 26 (*d*) 68 × 26 (*e*) 57 × 26 using simplified multiplication by 26.

41. Multiply: (*a*) 57,384 × 9 (*b*) 58,761 × 9 (*c*) 4,328 × 9 (*d*) 98,989 × 9 (*e*) 847,632 × 9, using subtraction method.

42. Multiply: (*a*) 87,583 × 11 (*b*) 9,898 × 11 (*c*) 57,384 × 11 (*d*) 58,761 × 11 (*e*) 4,328 × 11 (*f*) 847,632 × 11, using addition method.

43. Multiply: (*a*) 687 × 8 (*b*) 687 × 7 (*c*) 432 × 8 (*d*) 432 × 7 (*e*) 982 × 8 (*f*) 982 × 7, by annexing a zero and subtracting either twice or three times the number.

44. Multiply: (*a*) 687 × 99 (*b*) 687 × 98 (*c*) 687 × 97 (*d*) 982 × 99 (*e*) 982 × 98 (*f*) 982 × 97, by adding two zeros and subtracting the required number of times the number.

45. Multiply: (*a*) 84 × 98 (*b*) 94 × 96 (*c*) 86 × 93 (*d*) 79 × 95 (*e*) 82 × 88 (*f*) 982 × 978, by using complement multiplication.

46. Multiply: (*a*) 37,512 × 16 (*b*) 8,762 × 14 (*c*) 982 × 18 (*d*) 76,582 × 12 (*e*) 8,462 × 13 (*f*) 6,879 × 19, using only one line in the product as shown in text examples.

47. Multiply: (*a*) 84 × 76 (*b*) 758 × 84 (*c*) 68 × 47 (*d*) 832 × 59 (*e*) 54 × 132 (*f*) 38 × 78 (*g*) 176 × 42 (*h*) 872 × 74, using cross multiplication, and check results by "casting out nines."

CHAPTER IV

DIVISION

125. What is meant by division?

Division is the inverse of multiplication. As we have seen that multiplication is merely a simplified form of addition, we can conclude that its inverse, division, in its simplest form, is merely repeated subtraction.

Ex. (*a*) When we multiply 8 four times we get $8 \times 4 = 32$, which is simplified addition $8 + 8 + 8 + 8 = 32 =$ product. Now dividing the product 32 by 8 we get 4.

$$32 - 8 = 24 \qquad 24 - 8 = 16 \qquad 16 - 8 = 8 \qquad 8 - 8 = 0$$

We have subtracted 8 successively from 32 in four steps to get

$$\frac{32}{8} = 4 = \text{division.}$$

Ex. (*b*) Suppose you have 972 apples and you want to divide them equally among 324 men. How many apples will each man receive?

$$972 - 324 = 648 \qquad 648 - 324 = 324 \qquad 324 - 324 = 0$$

Count the number of subtractions, which is 3, and you get 3 apples for each man.

Ex. (*c*) How many 2's in 8? Subtract 2 from 8 as many times as possible, noting the number of times, 4, as the answer.

126. In what other ways may division be thought of?

(*a*) Division proper, a species of measurement, as finding how many times one number is contained in another.

(*b*) Partition, which is dividing a number into equal parts, the number of such parts being given. This is important with concrete numbers and is of no importance with abstract numbers.

Ex. (*a*) How many times is 7 contained in 35?

$\frac{35}{7} = 5$ times. You measure off 7 five times and you get 35.

Ex. (*b*) If 3 gallons of milk yield 21 ounces of butter, how many ounces will 1 gallon yield?

Think of 21 ounces as divided into 3 equal parts, which will result in 7 ounces in each part

$$\frac{21 \text{ oz.}}{3} = 7 \text{ oz.}$$

127. What are the terms of a division?

Dividend = The number to be divided, or separated into equal parts. Number in front of division sign.

Divisor = The number of equal parts into which dividend is to be separated, or the number by which dividend is to be divided. Number following division sign.

Quotient = Result obtained by division.

EXAMPLES:

(*a*) $42 \div 7 = 6$ or Dividend (= 42) ÷ Divisor (= 7) = Quotient (= 6).

(*b*) $\frac{42}{7} = 6$ or $\frac{\text{Dividend } (= 42)}{\text{Divisor } (= 7)} = \text{Quotient } (= 6).$

(*c*) $7\overline{)42}$ with quotient 6, or Divisor (= 7)$\overline{)\text{Dividend } (= 42)}$ with Quotient (= 6).

128. When the dividend is concrete and the divisor is abstract, what is the quotient?

The quotient is like the dividend.

EXAMPLE: If 3 gallons of milk yield 21 ounces of butter, we find the number of ounces contained in 1 gallon of milk by dividing 21 ounces by 3 (not by 3 gallons), getting 7 ounces. The divisor here (3) is an abstract number and the term 3 gallons serves only to indicate the number of groups into which 21 ounces is to be separated.

$$\frac{21 \textit{ ounces} \text{ (concrete)}}{3 \text{ (abstract)}} = 7 \textit{ ounces}$$

129. What is the result when both the dividend and divisor are concrete?

The dividend and divisor must be *alike* and the quotient will be *abstract.*

EXAMPLE: $\frac{21 \text{ ounces}}{7 \text{ ounces}} = 3 \qquad \frac{\text{Dividend (concrete)}}{\text{Divisor (concrete)}} = \text{abstract.}$

Seven ounces goes into 21 ounces three times.

130. What is meant by a remainder in division?

When division is not exact, the part of the dividend remaining is called the *remainder.*

EXAMPLE: $17 \div 2 = 8$ with 1 as a remainder.

$\frac{17}{2} = 8 + \frac{1}{2}$. The remainder is placed over the divisor as $\frac{1}{2}$ here.

131. Why may we think of division as the process of finding one factor when the product and the other factor are given?

EXAMPLE: In $7 \times 3 = 21$ we have multiplication.

$$\text{Factor } (= 7) \times \text{Factor } (= 3) = \text{Product } (= 21)$$

In $\frac{21}{3} = 7$, we have division.

$$\text{Given } \frac{\text{Product } (= 21)}{\text{One factor } (= 3)} = \text{Quotient } (= \text{other factor} = 7)$$

132. How can we make use of the fact that division is the opposite of multiplication?

EXAMPLE: What number multiplied by 324 would give 972?

We know that $324 = 300 + 20 + 4$

$972 = 900 + 70 + 2$

We can readily see that

$$\begin{aligned} 3 \times 300 &= 900 \\ 3 \times 20 &= 60 \\ 3 \times 4 &= 12 \\ & \overline{972} \end{aligned}$$

or $3 \times 324 = 972$ ($= 900 + 60 + 12$)

The reverse is also true: $\frac{972}{3} = 324$

$$324\overline{)972}^{\;3} \qquad 3\overline{)972}^{\;324}$$

133. If we wanted to divide 3,492 men into 4 groups, how would we proceed?

(*a*) $8 \times 4 = 32$ or *800* complete 4's = 3,200

(*b*) Subtract 3,200 from 3,492

(*c*) $7 \times 4 = 28$ or $70 \times 4 = 280$

(*d*) Subtract 280 from 292

(*e*) $3 \times 4 = 12$

(*f*) Adding the quotients we get $800 + 70 + 3 = 873$

```
       873  (= 800 + 70 + 3)
4)   3,492
    -3,200  (= 4 × 800)
       292  leaves 292 men
            still to be counted
      -280  (= 4 × 70)
        12  leaves 12 men
            still to be counted
       -12  (= 4 × 3)
```

This process can be shortened by omitting the zeros as is done in multiplication. Bring down only the number or numbers to be used in the next part of the example. Be careful in placing the numbers directly under the columns in which they first appeared.

When dividing with only one digit, we may shorten the step still further by "thinking" the subtractions and carrying the remainders.

$$\begin{array}{rrl} & 873 & (= \text{Quotient}) \\ (\text{Divisor} =)\ 4\overline{)} & 3{,}492 & (= \text{Dividend}) \\ & -3\,2\downarrow & (= 4 \times 8) \\ & 29 & \\ & -28\downarrow & (= 4 \times 7) \\ & 12 & \\ & -12 & (= 4 \times 3) \\ & 0 & \end{array}$$

"Think" subtract 8 × 4 = 32 from 34, carry 2 to the 9, making 29.

$$\begin{array}{r} 8\ 7\ 3 \\ 4\overline{)3{,}4^{2}9^{1}2} \end{array}$$

"Think" subtract 7 × 4 = 28 from 29, carry 1 to 2, making 12.

"Think" subtract 3 × 4 = 12 from 12, getting 0, which is zero remainder.

134. What is meant by "short division" and what is the process in simple form?

When the divisor is so small that the work can be performed mentally, the process is called short division.

EXAMPLE: Divide 9,712 by 4. Write as

$$\begin{array}{r} 4\overline{)9{,}712} \\ 2{,}428 \end{array}$$

(*a*) Begin at left. Find how many times divisor 4 is contained in the first figure of the dividend.

4 is contained in 9 two times with a remainder 1.

(*b*) Reduce the 1 to the next lower order, making 10, which with 7 makes 17.

4 is contained in 17 four times with a remainder 1.

(*c*) Reduce this 1 to the next lower order, making 10, which with 1 makes 11.

4 is contained in 11 two times with a remainder of 3.

(*d*) Reduce this 3 to the next lower order, making 30, which with 2 makes 32.

4 is contained in 32 eight times with no remainder.

135. How do we divide 3,762 by 7 using short division?

$$\begin{array}{rl} 7\overline{)3{,}7^{2}6^{5}2} & (= \text{Dividend}) \\ (= \text{Quotient})\ 5\ 3\ 7\ \ \tfrac{3}{7} & (= \text{Remainder}) \end{array}$$

(*a*) 7 is not contained in the first figure of the dividend 3 and 3 must be reduced to the next lower order, making 30, which with 7 makes 37.

(*b*) 7 is contained in 37 five times with 2 remainder. Reduce 2 to next lower order, making 20, which with 6 makes 26.

(*c*) 7 is contained in 26 three times with 5 remainder. Reduce 5 to next lower order, making 50 which with 2 makes 52.

(*d*) 7 is contained in 52 seven times with 3 remainder, which is written $\frac{3}{7}$.

136. How do we proceed with long division?

EXAMPLE: To divide 73,158 (= Dividend) by 534 (= Divisor).

	137	(= Quotient)
(Divisor =) 534)	73,158	(= Dividend)
	−53 4↓	(= 534 × 1)
	19 75	(= Partial dividend)
	16 02↓	(= 534 × 3)
	3 738	(= Partial dividend)
	3 738	(= 534 × 7)
	0	(= Remainder)

(*a*) Since divisor has 3 digits, take the first 3 digits of the dividend and ask how many times divisor 534 is contained in 731 (= first 3 digits of dividend). (Usually a clue is given by trial of the first figure of divisor, which here is 5, and finding how many times it is contained in first figures of dividend, here 7.) Divide 5 into 7, or 1.

(*b*) Write partial quotient 1 over the last figure of 731. Here 1 goes over the 1 of 731.

(*c*) Subtract 1 × 534 from 731, getting 197, and bring down the 5 which is the next digit of the dividend. This results in the partial dividend 1,975.

(*d*) Divide first figure 5 of divisor into 19 (= the first two figures of partial dividend). Write partial quotient 3 over 5 of the original dividend.

(*e*) Subtract 3 × 534 = 1,602 from 1,975, getting 373, and bring down the 8 which is the next digit of the dividend. This results in the partial dividend 3,738.

(*f*) Divide 5 of divisor into 37 of partial dividend. Write partial quotient 7 over 8 of original dividend.

(*g*) Subtract 7 × 534 = 3,738 from 3,738 of partial dividend, getting zero remainder. Quotient is therefore 137 exact.

137. What do we do when the last subtraction is not zero?

EXAMPLE: Divide 73,170 by 534.

The remainder 12 is expressed as 12 over the divisor, or $\frac{12}{534}$ here. The quotient is $137\frac{12}{534}$.

Sometimes we place a decimal point after the last digit of the dividend, add zeros, and continue the process of division to express the remainder as a decimal.

```
                   137 12/534
(Divisor =) 534)73,170  (= Dividend)
                53 4↓
                19 77
                16 02↓
                 3 750
                 3 738
                    12  (= Remainder)
```

138. What is the principle of the trial divisor in long division?

EXAMPLE: Divide 236,987 by 863.

(*a*) Ordinarily try first left-hand digit of divisor into the first two digits of dividend, as 8 of divisor into 23 of dividend.

(*b*) But when the second digit of divisor is number 5 or greater (6 in this case), then increase the first digit of divisor by 1 and try in dividend. Here try 9 into 23, getting 2 as quotient.

```
                    274 525/863  (= Quotient)
(Divisor =) 863)236,987  (= Dividend)
                172 6↓   (= 863 × 2)
                 64 38   (= Partial dividend)
                 60 41↓  (= 863 × 7)
                  3 977  (= Partial dividend)
                  3 452  (= 863 × 4)
                    525  (= Remainder)
```

(*c*) In the next partial dividend try 9 into 64, getting 7 as quotient.

(*d*) In the following partial dividend try 9 into 39, getting 4 as quotient.

(*e*) Remainder here is $\frac{525}{863}$. Quotient is $274\frac{525}{863}$.

139. What is the rule for long division?

(*a*) Write divisor at left of dividend with a curved line between them. Take the fewest number of digits at left of dividend that will contain divisor and write this partial quotient on top over the right-hand digit of the partial dividend.

(*b*) Multiply entire divisor by this partial quotient and write the product under the partial dividend used.

(*c*) Subtract this product and to remainder annex (bring down) the next figure of dividend for the second partial dividend.

(*d*) Divide as before, and continue process until all digits of dividend have been used to make partial dividends.

(*e*) When there is a remainder, write it with the quotient.

140. What is a pure proof of any division?

Multiply divisor by quotient and to this product add the remainder, if any. The result should equal the dividend.

EXAMPLE:

$$274\tfrac{525}{863} \quad (= \text{Quotient})$$
$$(\text{Divisor} =)\ 863\overline{)236{,}987} \quad (= \text{Dividend})$$

863 × 274	= 236,462
Add Remainder	525
Result	236,987 (= Dividend)

141. What is the procedure for division with United States money?

Divide as in integral numbers, writing the first digit of the quotient over the right-hand digit of the first partial dividend. (Place the decimal point in the quotient directly over the decimal point in the dividend.)

EXAMPLE: Divide \$829.11 by 87.

$$
\begin{array}{rrl}
 & \$9.53 & (= \text{Quotient}) \\
(\text{Divisor} =)\ 87\overline{)} & \$829.11 & (= \text{Dividend}) \\
 & -783 & (= 9 \times 87) \\
 & 461 & (= \text{Partial dividend}) \\
 & -435 & (= 5 \times 87) \\
 & 261 & (= 3 \times 87) \\
 & -261 & \\
 & 0 &
\end{array}
$$

142. What is the quotient of the division of \$45.36 by \$0.27?

Change the dividend and divisor to cents, which gives 4,536 cents divided by 27 cents. The quotient is 168, which is an abstract number showing the number of times 27¢ goes into 4,536¢.

$$
\begin{array}{rrl}
 & 168 & (= \text{Quotient}) \\
(\text{Divisor} =)\ 27\overline{)} & 4{,}536 & (= \text{Dividend}) \\
 & -2\,7 & \\
 & 1\,83 & \\
 & -1\,62 & \\
 & 216 & \\
 & -216 & \\
 & 0 &
\end{array}
$$

143. How can factoring of the divisor be used to reduce a problem of long division to a series of short divisions?

EXAMPLE: Divide 27,216 by 432. Here the divisor 432 can be factored down far enough to give a series of short divisions by the factors, which procedure is substituted for the long division.

$$
\begin{array}{rl}
12\overline{)27{,}216} & \\
12\overline{)2{,}268} & \\
3\overline{)189} & \\
63 & (= \text{Quotient) with no Remainder.}
\end{array}
$$

$$432 = \text{Divisor} = 12 \times 12 \times 3$$

144. What is the procedure for the above when there is a remainder?

EXAMPLE: Divide 47,897 by 18.

Factor divisor 18 as $2 \times 3 \times 3 = 18$. Divide by 2, then the quotient of this by 3, and the quotient of this by 3:

2)47,897			
3)23,948	+1 =	Remainder	1
3)7,982	+2 =	Remainder	5
2,660	+2 =	Remainder	17

Quotient is $2{,}660\frac{17}{18}$.

The first remainder 1 remains unchanged.

The second division has a remainder 2. As this division is of one half the number by 3, you multiply the remainder 2 by 2, getting 4, and adding this to the previous remainder, getting $4 + 1 = 5$.

The next division is of one sixth of the number by 3. You then multiply this remainder by 6, getting 12, and add this to the previous 5, getting 17, which is the final remainder.

It is seen that each remainder except the first is multiplied by the factors of the divisions preceding its own and the sum of the products is the total remainder.

145. What is the quotient of 65,349 by 126 using the factoring-of-the-divisor method?

Divisor $126 = 2 \times 3 \times 3 \times 7$.

2)65,349			
3)32,674	+1	...	1
3)10,891	+1	...	$3 = (2 \times 1 + 1)$
7)3,630	+1	...	$9 = (2 \times 3 \times 1 + 3)$
518	+4	...	$81 = (2 \times 3 \times 3 \times 4 + 9)$

Quotient $= 518\frac{81}{126}$.

146. What is the procedure for dividing by 10, 100, 1,000, etc.?

Set off as many figures at the right of the dividend as there are ciphers in the divisor. The figures thus set off are the remainder. The other figures are the quotient.

Ex. (*a*) $65 \div 10 = 6$, with 5 as remainder or $\frac{65}{10} = 6\frac{5}{10} = 6.5$. (One cipher in divisor. Set off 1 figure at right of dividend.)

Ex. (*b*) $579 \div 100 = 5$ with 79 as remainder or $\frac{579}{100} = 5\frac{79}{100} = 5.79$. (Two ciphers in divisor. Set off 2 figures at right of dividend.)

Ex. (*c*) $5{,}670 \div 200 = 200\overline{)5{,}670} = 2\overline{)56.70}$

28.35 Ans.

Dividing 200 by 100 we get 2.

Dividing 5,670 by 100 we get 56.70.

Now dividing 56.70 by 2 we get 28.35 (Ans.).

Ex. (*d*) 89,644 ÷ 4,000 = 4,000)89,644 = 4)89.644
22.411 Ans.

When the divisor ends in one or more ciphers, cut these off, and also cut off an equal number of figures from the right of the dividend. Then divide by the figures remaining.

Ex. (*e*) 8.743 ÷ 700 = 0.08743 ÷ 7 = 0.01249.

147. How do we apply the excess-of-nines method to prove the correctness of a division?

(*a*) Get excess of 9's in *divisor.*
(*b*) Get excess of 9's in *quotient.*
(*c*) Multiply these two excesses and get excess of 9's of the *product.*
(*d*) *Add* to this the excess of 9's in *remainder.* Get excess of sum.
(*e*) Get excess of 9's in *dividend* and compare.

	(Dividend)		(Divisor)		(Quotient)	(Remainder)
EXAMPLE:	8,974	÷	173	=	51	$\frac{151}{173}$

(*a*) Divisor 173 excess of 9's = 2
(*b*) Quotient 51 excess of 9's = 6
(*c*) Product of excesses 2 × 6 = 12 = 3
(*d*) Remainder 151 excess of 9's = 7
Sum = 10 excess of 9's = 1
(*e*) Dividend 8,974 excess of 9's = 1 ←——check——↑

A quotient may be incorrect even though the excess-of-nines might check, but this happens rarely.

148. What is meant by an even number?

A number divisible by 2 is called an even number. An even number may end in 2, 4, 6, 8, or in a zero.

EXAMPLES: 42, 54, 76, 68, 970 are even numbers. Each divided by 2 results in 21, 27, 38, 34, 485.

149. How can we know when a number is divisible by 3?

When the sum of its digits is divisible by 3, the number itself is divisible by 3.

Ex. (*a*) Number = 213. Add digits 2 + 1 + 3 = 6. Now sum 6 is divisible by 3. Therefore number 213 is divisible by 3. Ans. = 71.

Ex. (*b*) Number = 531. Add digits 5 + 3 + 1 = 9. Sum 9 is divisible by 3. Therefore number 531 is divisible by 3. Ans. = 177.

150. If we have an even number and it is divisible by 3, by what other number is it also divisible?

The number is also divisible by 6 because an even number is divisible by 2, and $2 \times 3 = 6$.

EXAMPLE: Given number = 162 which is an even number. Add digits $1 + 6 + 2 = 9$, which is divisible by 3.

$$\therefore \text{ 162 is divisible by 6 or } \frac{162}{6} = 27 \text{ Ans.}$$

151. When is a number divisible by 4?

When its last two digits are divisible by 4.

EXAMPLE: 7,624. Last two digits 24 are divisible by 4.

$$\therefore \frac{7{,}624}{4} = 1{,}906 \text{ Ans.}$$

152. When is a number divisible by 5?

When it ends in 5 or zero.

Ex (*a*) $\frac{7{,}865}{5} = 1{,}573$ Ex. (*b*) $\frac{7{,}860}{5} = 1{,}572$

153. What number, or any multiples of it, can be divided by 7, 11, or 13?

Number 1,001, or any of its multiples, can be divided by 7, 11, or 13.

Ex. (*a*) $\frac{4{,}004}{7} = 572 \quad \frac{4{,}004}{11} = 364 \quad \frac{4{,}004}{13} = 308$

Ex. (*b*) $\frac{12{,}012}{7} = 1{,}716 \quad \frac{12{,}012}{11} = 1{,}092 \quad \frac{12{,}012}{13} = 924$

154. When is a number divisible by 8?

When the number ends in three zeros or when the last three digits are divisible by 8.

Ex. (*a*) $\frac{1{,}000}{8} = 125 \quad \frac{125{,}000}{8} = 15{,}625$

Because 1,000 is divisible by 8, whatever precedes the last three figures merely adds that many thousands and does not affect the divisibility by 8.

Ex. (*b*) $\frac{136}{8} = 17$. Now add 1,000, getting 1,136. Then

$$\frac{1{,}136}{8} = \frac{1{,}000}{8} + \frac{136}{8} = 125 + 17 = 142$$

Again add 1,000 getting 2,136. Then

$$\frac{2{,}136}{8} = \frac{2{,}000}{8} + \frac{136}{8} = 250 + 17 = 267$$

Ex. (*c*) $\frac{8{,}654{,}136}{8} = 1{,}081{,}767$

No matter how many figures are placed in front of the original 136, the number is divisible by 8.

Ex. (*d*) 29,632. Consider $\frac{632}{8} = 79$ = divisible by 8

$\therefore$ 29,632 is divisible by 8, getting 3,704. Ans.

155. When is a number divisible by 9?

When the sum of its digits is divisible by 9.

Ex. (*a*) Number is 8,028. Add digits $8 + 0 + 2 + 8 = 18$. $\frac{18}{9} = 2$.

$\therefore$ 8,028 is divisible by 9. $\frac{8{,}028}{9} = 892$.

Ex. (*b*) Number $\frac{38{,}934}{9} = 4{,}326$. Add digits $3 + 8 + 9 + 3 + 4 = 27$ and $\frac{27}{9} = 3$.

156. When is a number divisible by 25?

When it ends in two zeros or in two digits forming a multiple of 25.

Ex. (*a*) $\frac{8{,}9\mathit{00}}{25} = 356$ Ex. (*b*) $\frac{8{,}6\mathit{75}}{25} = 347$

157. When is a number divisible by 125?

When it ends in three zeros or in three digits forming a multiple of 125.

Ex. (*a*) $\frac{89{,}\mathit{000}}{125} = 712$ Ex. (*b*) $\frac{89{,}\mathit{625}}{125} = 717$

158. What is the criterion for a number divisible by 11?

(*a*) When the sum of even-placed digits equals the sum of odd-placed digits.

Ex. (*a*) 2,649,867

. . . . →. Add odd-placed digits 7 + 8 + 4 + 2 = *21.*

x x x →x Add even-placed digits 6 + 9 + 6 = *21.*

$\therefore \frac{2{,}649{,}867}{11} = 240{,}897$ (divisible by 11) (*b*) When the difference between the sum of the odd- and even-placed digits is divisible by 11 the number itself is divisible by 11.

Ex. (*b*) 65,835 — odd-placed (marked ·) $5 + 8 + 6 = 19$; even-placed (marked ×) $3 + 5 = 8$; $19 - 8 = 11$

$$\therefore \frac{65{,}835}{11} = 5{,}985$$

Ex. (*c*) 29,271 — odd-placed (marked ·) $1 + 2 + 2 = 5$; even-placed (marked ×) $7 + 9 = 16$; $16 - 5 = 11$

$$\therefore \frac{29{,}271}{11} = 2{,}661$$

159. How can we tell in advance what the remainder will be, when the divisor is 9?

Add the digits and then add the digits of this sum. This last is the remainder.

Ex. (*a*) 867. Add digits $8 + 6 + 7 = 21$ (= Sum).
Add digits of sum $2 + 1 = 3$ (= Remainder).

$$\frac{867}{9} = 96 + \frac{3}{9} \text{ (= Remainder)}$$

Ex. (*b*) 973,285.
Add digits $9 + 7 + 3 + 2 + 8 + 5 = 34$ (= Sum).
Add digits of sum $3 + 4 = 7$ (= Remainder).

$$\frac{973{,}285}{9} = 108{,}142 + \frac{7}{9} \text{ (= Remainder)}$$

160. What is a short-cut way of dividing by 5?

Multiply by 2 and point off one decimal place to the left.

Ex. (*a*) $23 \div 5$. $23 \times 2 = 46$. Point off one place to left, 4.6.Ans. To point off one decimal place means dividing by 10.

$$\frac{23}{5} \text{ is the same as } \frac{23}{\cancel{5}_{1}} \times \frac{\cancel{10}^{2}}{10} = \frac{23 \times 2}{10} = 4.6.$$

Ex. (*b*) $\frac{832}{5}$. $832 \times 2 = 1{,}664$. Point off one place, 166.4.Ans.

161. What is a simple way of dividing by 25?

Multiply by 4 and point off two places to the left.

Ex. (*a*) $1{,}394 \div 25$. $1{,}394 \times 4 = 5{,}576$. Point off two places, 55.76.

To point off two places to the left means dividing by 100.

$$\frac{1{,}394}{25} \text{ is the same as } \frac{1{,}394}{\underset{1}{\cancel{25}}} \times \frac{\overset{4}{\cancel{100}}}{100} = \frac{1{,}394 \times 4}{100} = 55.76 \text{ Ans.}$$

$$\text{Ex. } (b)\ \frac{0.176}{25} = \frac{0.176 \times 4}{100} = \frac{0.704}{100} = 0.00704 \text{ Ans.}$$

162. What is a simple way to divide by 125?

Multiply by 8 and point off three places to the left.

Example: $7{,}856 \div 125$. $7{,}856 \times 8 = 62{,}848$. Set off three places to left, getting 62.848.

Dividing by 1,000 means setting off three places to the left.

$$\frac{7{,}856}{125} \text{ is the same as } \frac{7{,}856}{\underset{1}{\cancel{125}}} \times \frac{\overset{8}{\cancel{1{,}000}}}{1{,}000} = \frac{7{,}856 \times 8}{1{,}000} = 62.848 \text{ Ans.}$$

163. What is the short-cut way of dividing by any aliquot part of 100?

Multiply by the *inverted* fraction represented by the aliquot and point off two places to the left.

Ex. (*a*) $875 \div 33\frac{1}{3}$ ($33\frac{1}{3}$ is $\frac{1}{3}$ of 100). *Invert* $\frac{1}{3}$, getting $\frac{3}{1}$.
$875 \times 3 = 2{,}625$. Point off two places to left, getting 26.25.Ans.

Ex. (*b*) $6{,}000 \div 66\frac{2}{3}$ ($66\frac{2}{3} = \frac{2}{3}$ of 100). *Invert* $\frac{2}{3}$, getting $\frac{3}{2}$.
$6{,}000 \times \frac{3}{2} = 9{,}000$. Point off two places to left, getting 90.Ans.

Ex. (*c*) $12{,}367 \div 16\frac{2}{3}$ ($16\frac{2}{3} = \frac{1}{6}$ of 100). *Invert* $\frac{1}{6}$, getting $\frac{6}{1}$.
$12{,}367 \times 6 = 74{,}202$. Point off two places to left, getting 742.02.Ans.

164. What is a simple way of dividing by 99?

(*a*) Add the two right-hand digits to the rest of the number. Put this sum down under the original number.

(*b*) Add the two right-hand digits of this to the rest of its number and put this down under the other two.

(*c*) Keep up this process until 99 or a quantity less than 99 is left. This is the remainder.

(*d*) Add up the digits to the left excluding the two right-hand digits of each number.

Ex. (1) 7812 09 ÷ 99

(*a*) Add 09 to 7812, getting 78 21

(*b*) Add 21 to 78, getting 99

	7812	09
	+78	21
		99 (= Remainder)
Quotient =	7890	

Remainder $\frac{99}{99} = 1$. Add 1 to quotient 7,890, getting 7,891.

Ex. (2) 9785 43 ÷ 99

(*a*) Add 43 to 9,785, getting 98 28

(*b*) Add 28 to 98, getting 1 26

(*c*) Add 1 to 26, getting 27

	9785	43
	+98	28
	+1	26
		27 (= Remainder)
Quotient =	9884	$\frac{27}{99}$

165. How can we make a number divisible by 3?

(*a*) Add the digits. Divide this sum by 3 and get remainder.

(*b*) Subtract remainder from original number.

EXAMPLE: 13,477. Add digits $1 + 3 + 4 + 7 + 7 = 22$.

$$\text{Divide by 3: } \frac{22}{3} = 7 + \frac{1}{3} \text{ (= Remainder)}$$

Subtract remainder 1 from 13,477, getting 13,476.Ans. Now

$$\frac{13{,}476}{3} = 4{,}492.$$

166. How can we make a number divisible by 9?

(*a*) Add the digits. Divide this sum by 9. Get remainder.

(*b*) Subtract remainder from original number.

EXAMPLE: 13,477. Add digits $1 + 3 + 4 + 7 + 7 = 22$.

$$\frac{22}{9} = 2 + \frac{4}{9}$$

Subtract remainder 4 from 13,477, getting 13,473.

$$\frac{13{,}473}{9} = 1{,}497 \text{ Ans.}$$

167. How do we obtain an average of a number of items?

Divide the *sum* of the items *by* the *number* of *items* added.

EXAMPLE: Find the average sales made by a salesman during the week when his daily sales are:

Monday	\$268.60
Tuesday	\$329.85
Wednesday	\$97.45
Thursday	\$239.90
Friday	\$296.70

Number of items = 5) \$1,232.50 (= Sum)

\$246.50 (= Average)

We see that the sales for Monday, Tuesday, and Friday were above the average, while for Wednesday and Thursday they were below average.

168. What is the rule for finding the value of one of anything?

Always *divide by that* of which you want to find the value of *one*.

Ex. (*a*) If 75 books cost \$300, what is the cost of 1 book?

You want the cost of *1 book*, so divide by the *number* of *books*.

$$\frac{\$300}{75 \text{ books}} = \$4.00 \text{ for 1 book}$$

Ex. (*b*) If a dozen hats cost \$72, what is the cost of 1 hat?

You want the cost of *1 hat*, so divide by number of *hats*.

$$\frac{\$72}{12 \text{ hats}} = \$6 \text{ for 1 hat}$$

Ex. (*c*) If a pound of coffee costs 80¢, how many ounces can you get for 10¢?

First you want the number of ounces for 1 cent, so divide by cents

$$\frac{16 \text{ ounces}}{80¢} = \tfrac{1}{5} \text{ ounce for 1 cent}$$

For 10¢

$$10 \times \tfrac{1}{5} = 2 \text{ ounces}$$

Ex. (*d*) If a jeep used 16 gallons of gasoline in driving 288 miles, how much does it consume on a 486-mile trip?

You want first the number of miles for *1 gallon*, so divide by gallons

$$\frac{288 \text{ miles}}{16 \text{ gallons}} = 18 \text{ miles for 1 gallon}$$

then

$$\frac{488 \text{ miles}}{18 \text{ miles for 1 gallon}} = 27 \text{ gallons}$$

Ex. (*e*) If it takes 8 minutes for a pipe to fill a tank, how much of the tank will be filled in 1 minute?

You want the amount for *1 minute*, so divide by minutes:

$$\frac{1 \text{ tank}}{8 \text{ minutes}} = \tfrac{1}{8} \text{ tank in } \textit{1 minute}$$

PROBLEMS

1. How many times is 8 contained in 56?

2. If 3 gallons of milk yield 18 ounces of butter, how many ounces will 1 gallon yield?

3. If you have 1,048 pads of writing paper and you want to divide them equally among 262 employees, how many pads will each one receive?

4. Find the quotients of the following and prove each by multiplying the factors together.

(*a*)	$6 \div 2 =$	because $2 \times$	$= 6$
(*b*)	$9 \div 3 =$	because $3 \times$	$= 9$
(*c*)	$12 \div 4 =$	because $4 \times$	$= 12$
(*d*)	$18 \div 9 =$	because $9 \times$	$= 18$
(*e*)	$28 \div 7 =$	because $7 \times$	$= 28$
(*f*)	$42 \div 6 =$	because $6 \times$	$= 42$
(*g*)	$48 \div 8 =$	because $8 \times$	$= 48$
(*h*)	$66 \div 6 =$	because $6 \times$	$= 66$
(*i*)	$72 \div 9 =$	because $9 \times$	$= 72$
(*j*)	$84 \div 7 =$	because $7 \times$	$= 84$

5. If $1 is changed to five-cent pieces, how many are there?

6. If a man earns $16 while a boy earns $6, how much will the boy earn while the man earns $96?

7. If a man can pick twice as much fruit as a boy, and 4 boys and 3 men pick 5 acres of orchard in a day, what amount of ground does each cover?

8. If a man eats 380 grams of carbohydrates, 130 grams of protein, and 60 grams of fats each day, how much does he average per meal?

9. Divide:

(*a*) $3\overline{)963}$	(*b*) $4\overline{)884}$	(*c*) $2\overline{)462}$	(*d*) $2\overline{)432}$
(*e*) $6\overline{)432}$	(*f*) $8\overline{)512}$	(*g*) $7\overline{)637}$	(*h*) $5\overline{)475}$
(*i*) $9\overline{)468}$	(*j*) $6\overline{)756}$	(*k*) $4\overline{)548}$	(*l*) $7\overline{)239}$
(*m*) $3\overline{)5,472}$	(*n*) $5\overline{)5,385}$	(*o*) $9\overline{)80,208}$	(*p*) $8\overline{)40,576}$
(*q*) $6\overline{)63,516}$	(*r*) $9\overline{)94,806}$	(*s*) $7\overline{)47,607}$	(*t*) $8\overline{)21,176}$
(*u*) $12\overline{)56,216}$	(*v*) $15\overline{)83,542}$	(*w*) $11\overline{)73,236}$	(*x*) $9\overline{)87,585}$
(*y*) $12\overline{)83,691}$	(*z*) $15\overline{)98,254}$		

10. Find the quotient of:

(*a*) 1,607 ÷ 19 (*b*) 6,548 ÷ 89 (*c*) 3,402 ÷ 81
(*d*) 3,485 ÷ 873 (*e*) 54,963 ÷ 863 (*f*) 861,618 ÷ 843
(*g*) 879,384 ÷ 508 (*h*) 938,764 ÷ 879 (*i*) 42,896 ÷ 269
(*j*) 98,641 ÷ 679 (*k*) 3,862,847 ÷ 76,298 (*l*) $47\overline{)9{,}541}$
(*m*) $78\overline{)5{,}674}$ (*n*) $619\overline{)68{,}435}$ (*o*) $428\overline{)50{,}002}$

11. If there are 266 pages in a book and you can read 38 pages in an hour, how long will it take you to read it?

12. Find the quotient of:

(*a*) $18.36 ÷ 12 (*b*) 967.50 ÷ 43 (*c*) $967.50 ÷ $43
(*d*) $438.90 ÷ $21 (*e*) $438.90 ÷ 21¢

13. Divide by factoring method:

(*a*) 23,112 ÷ 108 (*b*) 39,798 ÷ 99 (*c*) 35,952 ÷ 84

14. Divide:

(*a*) 490 ÷ 10 (*b*) 48.7 ÷ 10 (*c*) 5,300 ÷ 100 (*d*) 1,587.4 ÷ 100
(*e*) 3.85 ÷ 10 (*f*) 8.745 ÷ 100 (*g*) 490 ÷ 20 (*h*) 48.7 ÷ 30
(*i*) 5,300 ÷ 400 (*j*) 3.85 ÷ 20 (*k*) 8.745 ÷ 700
(*l*) 697 ÷ 1,000 (*m*) 16,720 ÷ 800

15. Apply excess-of-nines method to prove the correctness of the divisions of problem 10.

16.

(*a*) Is 7,893 divisible by 3? (Using sum-of-digits method.)
(*b*) Is 3,876 divisible by 6? (Using short-cut method.)
(*c*) Is 3,876 divisible by 4? (Using last-two-digits method.)
(*d*) Is 8,695 divisible by 5? (Using criterion.)
(*e*) Is 14,014 divisible by 7, 11, or 13? (Using criterion.)
(*f*) Is 7,462,768 divisible by 8? (Using divisibility-of-last-3-digits method.)
(*g*) Is 8,658 divisible by 9? (Using sum-of-digits method.)
(*h*) Are 7,800 and 9,864,175 divisible by 25? (Use criterion.)
(*i*) Are 7,860,000 and 76,375 divisible by 125? (Use criterion.)
(*j*) Are 3,657,654 and 78,947 divisible by 11? (Use criterion.)

17. (*a*) What will be the remainder of 948 ÷ 9 (without dividing first)?
(*b*) Can you tell in advance the remainder of 864,893 ÷ 9?

18. (*a*) Divide 39 by 5 at once by short-cut method.
(*b*) Divide 482 by 25 by short-cut method.
(*c*) Divide 6,743 by 125 by short-cut method.

19. Divide the following by use of aliquot parts of 100:

(*a*) 986 ÷ $33\frac{1}{3}$ (*b*) 4,000 ÷ $66\frac{2}{3}$ (*c*) 13,478 ÷ $16\frac{2}{3}$
(*d*) 32,000 ÷ $87\frac{1}{2}$ (*e*) 630 ÷ $37\frac{1}{2}$ (*f*) 34,560 ÷ 5
(*g*) 4,500 ÷ $62\frac{1}{2}$ (*h*) 3,500 ÷ $83\frac{1}{3}$ (*i*) 3,475 ÷ 25
(*j*) 2,700 ÷ 75 (*k*) 1,400 ÷ 125

20. Divide: (*a*) 872,317 ÷ 99, (*b*) 867,432 ÷ 99 by simple method shown in text.

21. Make: (*a*) 25,694, (*b*) 85,642 divisible by 3 by method shown in text.

22. Make: (*a*) 25,694, (*b*) 85,642 divisible by 9 by method shown in text.

23. If six ranking candidates on an examination had marks of 92.1, 87.3, 85.6, 80.7, 80.2, and 79.1 respectively, what is the average mark?

24. Sixteen students in a class in arithmetic made the following grades on a test: 84, 96, 74, 93, 88, 86, 81, 77, 81, 94, 99, 86, 71, 69, 76, and 84. What was the average grade of the class?

25. An author received royalties from his publisher during a six-year period as follows: $897.65, $917.59, $893.25, $997.75, $1,146.79, and $1,238.32. What is the average yearly royalty?

26. If you have an apple orchard of 2,000 trees, if you use 4 gallons of spraying mixture for each tree, and you mix 1 lb. of Paris green at 80¢ per lb. with 150 gal. of water, what would be the cost of the Paris green for 2 sprayings? What would be the cost per tree?

CHAPTER V

FACTORS — MULTIPLES — CANCELLATION

169. What is a prime number?

A number divisible only by 1 and itself.

EXAMPLES: 1, 2, 3, 5, 7, 11, 13, 17, 19, 23, 29, 31, 37, etc., are prime numbers. Each is divisible only by 1 and itself.

170. What is a composite number?

One that is divisible by other numbers in addition to 1 and itself.

EXAMPLES: 4, 6, 8, 10, 12, 14, 16, 18, 36, 44, etc., are composite numbers.

171. What is a factor of a number?

An exact divisor of the number.

Ex. (*a*) 2 is a factor of 6 because 2 is an exact divisor of 6.

Ex. (*b*) 2, 3, 4, 6 are factors of 12 because each is an exact divisor of 12. If 3 is one factor of 12, then 4 is the other factor.

172. What is meant by factoring?

The process of separating a number into its factors.

173. What is a prime factor?

A factor which is a prime number.

Ex. (*a*) 2, 2, and 3 are prime factors of 12.
(*b*) 2, 2, 2, and 3 are the prime factors of 24.

Of course, 4, 6, 8, 12 are also factors of 24 but these are not prime factors.

174. What do we call a number that has the factor 2?

An *even number*. Numbers not divisible by 2 are called *odd numbers*.

175. What is meant by a common divisor or factor?

One that is common to two or more numbers.

EXAMPLE: 4 is a factor common to 12 and 36.
3 is a factor common to 12 and 36.
12, 6, 2 are factors common to 12 and 36.

Numbers that have no common factors are said to be *prime to each other.*

176. What facts regarding the divisibility of numbers are of assistance in factoring?

(*a*) 2 is a factor of all even numbers.
(*b*) 3 is a factor when the *sum of the digits* is divisible by 3.
(*c*) 4 is a factor when the two digits at the right are zeros or a number divisible by 4.
(*d*) 5 is a factor when the units figure is 5 or zero.
(*e*) 6 is a factor of all even numbers that are divisible by 3.
(*f*) 8 is a factor when the three digits at the right are zeros or a number divisible by 8.
(*g*) 9 is a factor when the sum of the digits is divisible by 9.
(*h*) 11 is a factor when the sum of the digits in the even places equals the sum of the digits in the odd places, or when the difference between these two sums is 11 or some multiple of 11.

177. How do we find the prime factors of a number?

Divide by a prime factor and continue to divide by a prime factor until the last quotient is a prime number.

Ex. (*a*) What are the prime factors of 720?

$$\begin{array}{r} 2\overline{)720} \\ 2\overline{)360} \\ 2\overline{)180} \\ 2\overline{)90} \\ 5\overline{)45} \\ 3\overline{)9} \\ 3 \end{array}$$

Thus, the prime factors are 2, 2, 2, 2, 5, 3, and 3.

Ex. (*b*) Find the prime factors of 7,644.

$$\begin{array}{r} 2\overline{)7{,}644} \\ 2\overline{)3{,}822} \\ 3\overline{)1{,}911} \\ 7\overline{)637} \\ 91 \end{array}$$

Thus, the prime factors are 2, 2, 3, 7, and 91.

178. What is meant by the greatest common divisor or factor, abbreviated G.C.D. or g.c.d.?

The largest divisor or factor common to two or more given numbers is the G.C.D.

Ex. (*a*) 6 is the greatest common divisor of 24 and 30.

Ex. (*b*) 8 is the greatest common divisor of 16, 24, and 32 because it is the largest number that will exactly divide each of the numbers.

179. What is the rule for finding the G.C.D. of two or more numbers?

Separate the numbers into their prime factors and get the product of the prime factors that are common to all the numbers.

Common (↓ 2, ↓ 3)

Ex. (*a*) $24 = 2 \times 2 \times 2 \times 3$
$30 = 2 \times 3 \times 5$

Factors 2 and 3 are common to both 24 and 30

$$\therefore 2 \times 3 = 6 = \text{G.C.D.}$$

Common (↓ 2, ↓ 2, ↓ 2)

Ex. (*b*) $16 = 2 \times 2 \times 2 \times 2$
$24 = 2 \times 2 \times 2 \times 3$
$32 = 2 \times 2 \times 2 \times 2 \times 2$

Factors 2, 2, 2 are common to all three numbers

$$\therefore 2 \times 2 \times 2 = 8 = \text{G.C.D.}$$

180. What is a more convenient method of finding G.C.D.?

Arrange the numbers as below and divide by some number which will exactly divide each of them. Continue doing this until no divisor can be found to divide each last quotient. Multiply all the common factors.

Common factor →	2)	48	72	96
Common factor →	2)	24	36	48
Common factor →	2)	12	18	24
Common factor →	3)	6	9	12
		2	3	4

Common factors → $2 \times 2 \times 2 \times 3 = 24 = \text{G.C.D.}$

181. What is meant by a multiple of a number?

It is the product of that number multiplied by an integer.

Ex. (*a*) 24 is a multiple of number 12 because 12 multiplied by an integer 2 = 24.

Ex. (*b*) What numbers are multiples of 8?

$2 \times 8 = 16 \qquad 3 \times 8 = 24 \qquad 4 \times 8 = 32$, etc.

Thus 16, 24, 32, etc., are multiples of 8.

182. What is meant by a common multiple of two or more numbers?

A number that is a multiple of each.

Ex. (*a*) 16 is a common multiple of 4 and 8 because either of them multiplied by an integer = 16.

Ex. (*b*) 18 is a common multiple of 2, 3, 6, and 9 because any of these multiplied by an integer = 18.

183. What is meant by the least common multiple (L.C.M.) of two or more numbers?

The least number that is a multiple of each.

Ex. (*a*) 18 is a common multiple of 3 and 6, but 12 is the least common multiple of 3 and 6 because 12 is the smallest number which contains each without a remainder.

Ex. (*b*) 72 is a common multiple of 6, 9, and 12, but 36 is the L.C.M. because it is the smallest number which contains each without a remainder.

184. What is a method of finding the least common multiple (L.C.M.) of 18, 28, and 36?

Separate each number into its prime factors. Multiply the factors using each factor the greatest number of times it occurs in any of the given numbers that are factored.

$$18 = 2 \times 3 \times 3$$
$$28 = 2 \times 2 \times 7$$
$$36 = 2 \times 2 \times 3 \times 3$$

2 does not appear as a factor more than twice in any number.
3 does not appear as a factor more than twice in any number.
7 appears once.

$\therefore\ 2 \times 2 \times 3 \times 3 \times 7 = 252$ = L.C.M. that will contain 18, 28, and 36 without a remainder.

185. What is another method of getting the L.C.M. of 18, 28, and 36?

Divide the numbers by any prime number that will exactly divide two or more of them. Any number not so divisible is brought down intact. Continue this process until no further division can be made. Multiply all divisors and the quotients remaining to get the L.C.M.

$$\begin{array}{r|rcrcr}
2 & 18 & - & 28 & - & 36 \\ \hline
2 & 9 & - & 14 & - & 18 \\ \hline
3 & 9 & - & 7 & - & 9 \\ \hline
3 & 3 & - & 7 & - & 3 \\ \hline
 & 1 & - & 7 & - & 1
\end{array}$$

$$\therefore\ 2 \times 2 \times 3 \times 3 \times 7 = 252 = \text{L.C.M.}$$

186. What is meant by cancellation?

Elimination of factors in the dividend and divisor before dividing. The quotient is not affected by elimination of factors which are common to both dividend and divisor.

Ex. (*a*) Divide 4,368 by 156 by factoring and cancelling.

$$\frac{4{,}368 = \cancel{4} \times 4 \times 7 \times \cancel{3} \times \cancel{13}}{156 = \cancel{4} \times \cancel{3} \times \cancel{13}} = 4 \times 7 = 28 \text{ Ans.}$$

The same answer can be obtained by long division. It is not necessary to separate the numbers into their prime factors. The criteria for divisibility of numbers may be used as shown in question 176.

Ex. (*b*) Compute $\dfrac{\cancel{26} \times \cancel{36} \times \overset{3}{\cancel{54}}}{\cancel{18} \times \underset{2}{\cancel{52}} \times \underset{2}{\cancel{72}}}$ by means of cancellation.

$$\text{Ans.} = \frac{3}{2 \times 2} = \frac{3}{4}$$

Ex. (*c*) Compute by cancellation: $\dfrac{\overset{5}{\cancel{65}} \times \overset{35}{\cancel{105}}}{\underset{\cancel{3}}{\cancel{39}}} = 175$ Ans.

13 is a factor of 39 and 65 three and five times respectively.
Then 3 is contained in 105 thirty-five times.
The product of the remaining factors 5 × 35 = 175 Ans.

Ex. (*d*) Compute by cancellation:

$$\frac{\cancel{38} \times \overset{\overset{9}{\cancel{27}}}{\cancel{54}} \times \overset{\overset{\cancel{3}}{\cancel{12}}}{\cancel{48}} \times \overset{2}{\cancel{60}}}{\underset{17}{\cancel{68}} \times \underset{\cancel{2}}{\cancel{76}} \times \underset{\underset{7}{\cancel{28}}}{\cancel{84}} \times \underset{\cancel{3}}{\cancel{90}}} = \frac{9 \times 2}{17 \times 7} = \frac{18}{119} \text{ Ans.}$$

Find factors common to numbers above the line and numbers below the line and cancel them.

PROBLEMS

1. Name two factors of 18, 30, 36, 81, 120.

2. Name three factors of 18, 32, 45, 50, 66.

3. Name a factor common to 12 and 36.

4. Name all the factors or exact divisors of 3, 7, 17.

5. Make a list of all prime numbers below 100.

6. Make a list of all odd numbers below 50.

7. Separate into prime factors: 4, 5, 7, 8, 10, 12, 13, 14, 16, 18, 21, 24, 25, 30, 34.

8. Separate the prime, composite, even, and odd numbers in the following: 1, 6, 7, 10, 11, 12, 14, 19, 20, 21, 24, 25, 27, 33, 34.

9. Give the prime factors of:

(*a*) 310 (*b*) 297 (*c*) 670 (*d*) 741 (*e*) 981 (*f*) 385 (*g*) 2,650
(*h*) 1,215 (*i*) 321 (*j*) 1,575 (*k*) 10,935 (*l*) 420 (*m*) 497
(*n*) 378 (*o*) 462 (*p*) 2,430 (*q*) 25,344 (*r*) 73,260 (*s*) 599,676
(*t*) 273,564 (*u*) 15,625 (*v*) 10,675 (*w*) 12,625 (*x*) 976
(*y*) 8,050 (*z*) 3,848

10. Find the G.C.D. (greatest common divisor) of:

(*a*) 68, 112, 240 (*b*) 21, 84, 126, 147 (*c*) 21, 28, 77
(*d*) 45, 72, 81 (*e*) 44, 77, 121 (*f*) 14, 98, 112 (*g*) 24, 80, 96
(*h*) 28, 42, 36 (*i*) 45, 72, 81 (*j*) 315, 2,267, 9,012
(*k*) 144, 576 (*l*) 820, 697 (*m*) 125, 175, 1,792 (*n*) 6,004, 3,318
(*o*) 1,254, 2,361, 8,163 (*p*) 1,086, 905

11. Give two multiples of:

(*a*) 9 and 3 (*b*) 7 and 5 (*c*) 9 and 2 (*d*) 3 and 7 (*e*) 8 and 5
(*f*) 6 and 3 (*g*) 8 and 2 (*h*) 9, 2, and 8 (*i*) 3, 6, and 9
(*j*) 8, 6, and 4

12. Find the L.C.M. (least common multiple) of:

(*a*) 9 and 12 (*b*) 21 and 36 (*c*) 5 and 15 (*d*) 12, 15, and 18
(*e*) 36, 42, and 48 (*f*) 3, 9, 18, and 27 (*g*) 5, 15, 25, and 35
(*h*) 14, 81, 35, and 15 (*i*) 32, 48, 35, and 70
(*j*) 72, 88, 96, and 124 (*k*) 11, 22, 55, and 110
(*l*) 21, 24, 26, and 28 (*m*) 9, 21, 42, and 63
(*n*) 36, 75, 48, and 24 (*o*) 7, 14, 56, and 84
(*p*) 24, 72, 128, and 240

13. Divide by cancellation method of factors and prove by long division:

(*a*) 38,367 ÷ 1,827 (*b*) 52,800 ÷ 3,520 (*c*) 90,384 ÷ 3,228
(*d*) 88,368 ÷ 3,682 (*e*) 32,768 ÷ 2,048

14. Solve by cancellation:

(*a*) $3 \times 27 \times 48 \times 81 \div 6 \times 9 \times 54 \times 210$
(*b*) $81 \times 16 \times 10 \times 12 \div 9 \times 27 \times 2 \times 5$
(*c*) $8 \times 12 \times 18 \times 32 \div 4 \times 6 \times 9 \times 16$
(*d*) $42 \times 36 \times 77 \times 22 \div 11 \times 6 \times 24 \times 21$
(*e*) $5 \times 30 \times 65 \times 125 \div 15 \times 75 \times 95$

15. How many lb. of butter at 55¢ a lb. can be exchanged for 30 doz. eggs at 66¢ a doz.? (By cancellation.)

16. How many days of 8 hr. each would one need to work at $2.30 an hour to pay for 8 tons of coal at $27.60 a ton? (Solve by cancellation.)

17. If 14 men earn $7,257.60 working 27 days of 8 hours each at $2.40 an hour, how long will it take 21 men working 8 hours a day at the same rate to earn the same amount? (Solve by cancellation.)

18. If you drove 20,000 miles on new tires before replacement and you paid $120 for the 4 new tires, what was the tire cost for each 100 miles? (Solve by cancellation.)

CHAPTER VI

COMMON FRACTIONS

187. What does a fraction mean?

The Latin *frangere* means "to break." The Latin *fractus* means "broken." Thus a fraction is a broken unit, or a *part* of a unit. Also "fraction" comes from the same Latin root as the word "fragment," meaning, "a part." Actually a fraction is any quantity numerically less than a unit.

188. What are the terms of a fraction?

Every fraction has a *numerator* placed above a horizontal line and a *denominator* placed below the line. The denominator is the *divisor* of the numerator.

EXAMPLE:

$$\frac{5 = \text{numerator}}{7 = \text{denominator}} = \text{fraction} = 5/7 \text{ (when line is inclined).}$$

189. What is assumed in expressing fractional division?

It is assumed that all of the parts into which an object has been divided are of exactly equal size.

190. What is meant when we say that a thing is divided equally into two parts, and how is the fraction expressed?

The object is said to be divided into halves. The object is divided into two parts. The object or unit to be divided is placed as the numerator of the fraction; the number of divisions is the denominator.

Thus, $\frac{1}{2}$ = one divided into two parts
= a fraction.

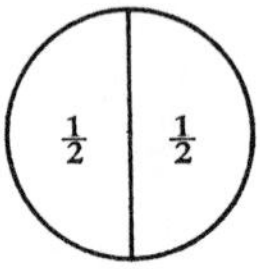

191. What is meant by $\frac{1}{3}, \frac{1}{4}, \frac{1}{5}, \frac{1}{6}$?

(*a*) $\frac{1}{3}$ = one (object or unit)
divided into 3 equal parts
= fraction (object divided into thirds)

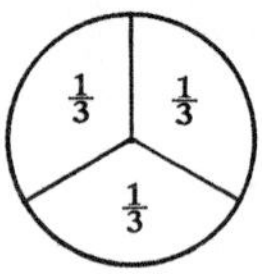

(*b*) $\frac{1}{4}$ = one divided into 4 equal parts = fraction (object divided into fourths)

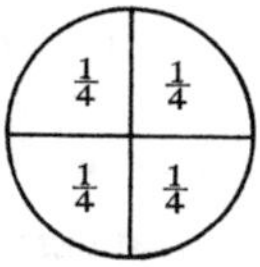

(*c*) $\frac{1}{5}$ = one divided into 5 equal parts = fraction (object divided into fifths)

(*d*) $\frac{1}{6}$ = one divided into 6 equal parts = fraction (object divided into sixths)

192. What is meant by a unit fraction?

When the numerator of a fraction is 1, it is called a unit fraction, as $\frac{1}{3}$, $\frac{1}{5}$, $\frac{1}{8}$.

193. What is a vulgar fraction and how is it classified?

A vulgar fraction is one expressed as a division.
The *divisor* classifies the fraction:

Ex. (*a*) $\frac{1}{3}$ is classified as *thirds* from its divisor 3.

Ex. (*b*) $\frac{11}{25}$ is classified as *twenty-fifths* from its divisor 25.

194. What are the parts of a vulgar fraction and how is it written?

The numerator is the dividend; the denominator is the divisor. It is written as a numerator above and denominator below a short horizontal or diagonal line or bar.

Ex. (*a*) $\frac{1}{3}$. Numerator tells us that only 1 of its class is considered.

Ex. (*b*) $\frac{11}{25}$. Numerator tells us that 11 of its class are taken.

195. What other meaning has the bar in a fraction?

The bar means "division" in the same way as the sign [÷].

Ex. (*a*) $\frac{5}{9}$ = five ninths, or five divided by nine = $5 \div 9$.

Ex. (*b*) $\frac{100}{4} = 25 = 100 \div 4$. Both expressions mean the same thing.

Ex. (*c*) $\frac{5}{16} = 5 \div 16$ or $16\overline{)5.0000}$ giving 0.3125

196. What are the three ways in which a fraction may be interpreted?

The fraction $\frac{3}{2}$, for example, may be thought of as:

(*a*) 3 units divided into 2 equal parts.

(*b*) 1 unit divided into 2 equal parts, with 3 of these parts taken, as: 3 times $\frac{1}{2} = \frac{3}{2}$.

(*c*) As an indicated division not yet performed.

EXAMPLES: Assume 1 or unity is a line 1 inch long:

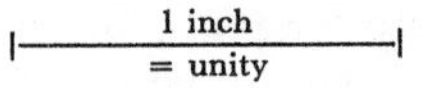

(*a*)

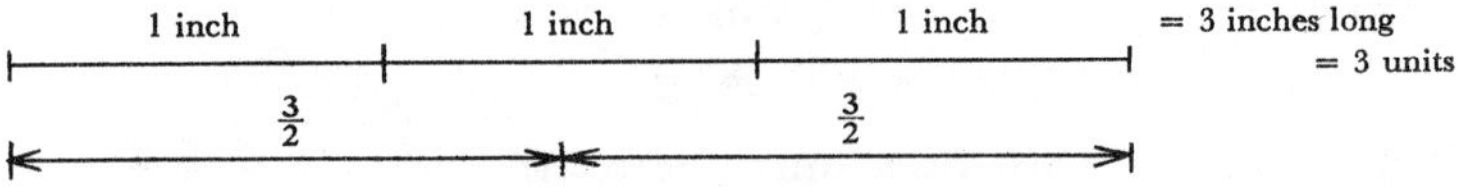

Three units divided into two equal parts. Each part = $\frac{3}{2}$.

(*b*)

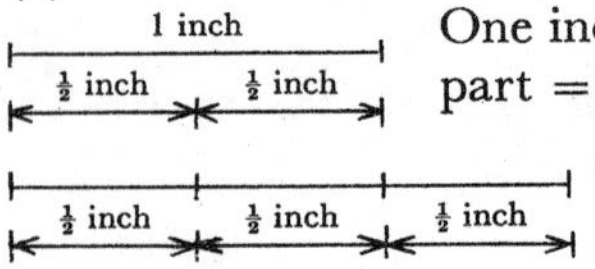

One inch divided into two equal parts. Each part = $\frac{1}{2}$.

= $\frac{3}{2}$ inches. Three such parts are taken.

(*c*) $\frac{3}{2}$ can be thought of as a division not yet performed.

197. When we add up all the fractional parts of a unit what do we get as a result?

We get the whole unit.

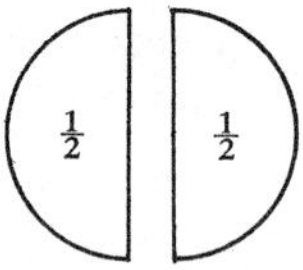

Ex. (*a*) $\frac{1}{2} + \frac{1}{2} = \frac{1+1}{2} = \frac{2}{2} = 1$ unit

= a whole unit.

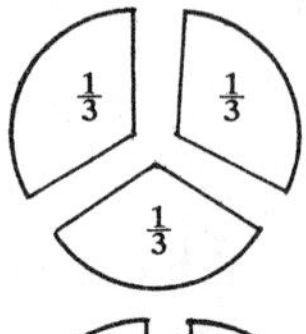

Ex. (*b*) $\frac{1}{3} + \frac{1}{3} + \frac{1}{3} = \frac{1+1+1}{3} = \frac{3}{3} = 1$ unit.

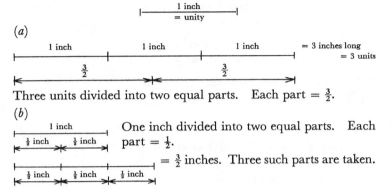

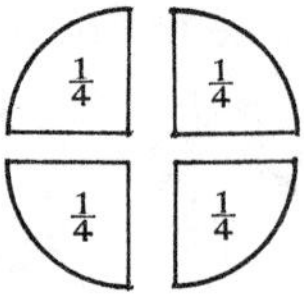

Ex. (*c*) $\frac{1}{4} + \frac{1}{4} + \frac{1}{4} + \frac{1}{4} = \frac{1+1+1+1}{4}$

$= \frac{4}{4} = 1$ unit.

Or, any fractional expression of a number divided by itself = 1 = unity, as $\frac{25}{25} = 1$.

198. What is a simple fraction?

One whose numerator and denominator are whole numbers.

EXAMPLE: $\frac{9}{13}$ and $\frac{5}{8}$ are simple fractions.

199. What is a compound fraction?

It is a fraction of a fraction.

EXAMPLE: $\frac{3}{8}$ of $\frac{5}{6}$ and $\frac{3}{4}$ of $\frac{5}{9}$ are compound fractions.

200. What is a complex fraction?

One in which either the numerator or denominator or both are not whole numbers.

Ex. (*a*) $\frac{6\frac{1}{8}}{15}$. Numerator is not a whole number.

Ex. (*b*) $\frac{7}{9\frac{1}{2}}$. Denominator is not a whole number.

Ex. (*c*) $\frac{4\frac{5}{6}}{6\frac{3}{4}}$. Both numerator and denominator are not whole numbers.

All the above are complex fractions.

201. What is a proper fraction?

One in which the numerator is less than the denominator.

EXAMPLE: $\frac{2}{5}$, $\frac{5}{8}$, $\frac{3}{4}$, $\frac{6}{7}$, $\frac{13}{21}$ are proper fractions. Each has a value less than a unit. Note that the numerator does not have to be 1.

202. What is an improper fraction?

One in which the numerator equals or exceeds the denominator. The fraction is thus equal to or greater than 1 unit.

Ex. (*a*) $\frac{4}{3} = \frac{3}{3} + \frac{1}{3} = 1 + \frac{1}{3} = 1\frac{1}{3}$.

Ex. (*b*) $\frac{48}{13} = \frac{39}{13} + \frac{9}{13} = \frac{13}{13} + \frac{13}{13} + \frac{13}{13} + \frac{9}{13} = 3 + \frac{9}{13} = 3\frac{9}{13}$.

203. What is a mixed number?

A whole number and a fraction taken together.

EXAMPLE: $1\frac{1}{3}$, $3\frac{9}{13}$, $1\frac{12}{17}$, $4\frac{3}{8}$, $24\frac{3}{5}$ are mixed numbers.

204. How may we shorten the process of finding the value of an improper fraction?

Divide the numerator by the denominator. Write the quotient as a whole number followed by a fraction in which the remainder is expressed as a numerator over the same denominator.

Ex. (*a*) $\frac{48}{13} = 3\frac{9}{13}$. Thirteen goes into 48 three times with a remainder of 9. $3\frac{9}{13}$ is a mixed number.

Ex. (*b*) $\frac{76}{23} = 3\frac{7}{23}$ = a mixed number.

205. How do we change a mixed number into an improper fraction?

Multiply the whole number by the denominator, add the numerator, and place this sum over the denominator.

Ex. (*a*)

$$8\tfrac{2}{5} = \frac{(\text{Whole number 8 times denominator 5}) + \text{numerator 2}}{\text{Denominator 5}}$$

$$= \frac{(8 \times 5 + 2)}{5} = \frac{42}{5} = \text{improper fraction.}$$

Ex. (*b*) $15\tfrac{1}{3} = \dfrac{(15 \times 3 + 1)}{3} = \dfrac{46}{3} = \text{improper fraction.}$

Ex. (*c*) $25\tfrac{4}{5} = \dfrac{(25 \times 5 + 4)}{5} = \dfrac{129}{5} = \text{improper fraction.}$

The reasoning is:

$$\text{If } \tfrac{5}{5} = 1, \text{ then } 25 \times 1 = 25 \times \tfrac{5}{5} = \tfrac{125}{5}.$$

Then $25\tfrac{4}{5} = \tfrac{125}{5} + \tfrac{4}{5} = \tfrac{129}{5}$. This is why we multiply the whole number by the denominator and add the numerator to get the total number of fifths in this case.

206. What happens to the value of a fraction when we multiply or divide both the numerator and the denominator by the same number?

The value of the fraction is unchanged.

Ex. (*a*)

$$\frac{1}{4} = \frac{\boxed{4} \times 1}{\boxed{4} \times 4} = \frac{4}{16} = \frac{\boxed{25} \times 1}{\boxed{25} \times 4} = \frac{25}{100} = \frac{\frac{1}{25}}{\frac{4}{25}} = \frac{1}{25} \times \frac{25}{4} = \frac{1}{4}$$

Ex. (*b*) $\dfrac{16}{24} = \dfrac{\frac{16}{8}}{\frac{24}{8}} = \dfrac{2}{3}$ (dividing numerator and denominator by 8).

207. When is a fraction said to be reduced to its lowest terms?

When the terms are prime to each other.

Ex. (*a*) $\tfrac{5}{6}$ is expressed in its lowest terms because 5 and 6 are prime to each other.

Ex. (*b*) $\tfrac{10}{12}$ is not expressed in its lowest terms because 2 is a factor common to both numerator and denominator.

208. How do we reduce a fraction to its lowest terms?

Divide both numerator and denominator by a common divisor and continue to divide until all common divisors are eliminated. This is done by cancelling the common factors.

Ex. (*a*) $\frac{15}{25} = \frac{3 \times \cancel{5}}{5 \times \cancel{5}} = \frac{3}{5}.$ The value of the fraction is not changed by reducing it to its lowest terms.

Ex. (*b*) $\frac{16}{48} = \frac{\cancel{4} \times \cancel{4}}{\cancel{4} \times \cancel{4} \times 3} = \frac{1}{3}.$ All like factors are cancelled out.

209. How can we change a fraction to higher terms?

Multiply both numerator and denominator by the same number.

Ex. (*a*) Change $\frac{3}{4}$ to twenty-fourths.

$$\frac{3 \times 6}{4 \times 6} = \frac{18}{24}$$

Multiply both numerator and denominator by 6.

Ex. (*b*) Change $\frac{9}{20}$ to hundredths.

$$\frac{9 \times 5}{20 \times 5} = \frac{45}{100}$$

Multiply both numerator and denominator by 5.

210. What must be done to fractions in giving the answer to a problem?

(*a*) Reduce fractions to lowest terms.

Example: $\frac{4}{12} = \frac{1}{3}$ Ans.

(*b*) Reduce improper fractions to mixed numbers.

Example: $\frac{8}{5} = 1\frac{3}{5}$ Ans.

211. How can we increase the value of a fraction?

(*a*) By multiplying the numerator by a number greater than 1.

Example: $\frac{3}{4}$ is increased to

$$\frac{3 \times 2}{4} = \frac{6}{4} = \frac{3}{2}$$

by multiplying numerator by 2, for example.

(*b*) By dividing the denominator by a number greater than 1.

Example: $\frac{3}{4}$ is increased to

$$\frac{3}{\left(\frac{4}{2}\right)} = \frac{3}{2}$$

by dividing denominator by 2, for example.

The *value* of the fraction has been doubled in each case.

EXAMPLE: Increase the value of $\frac{3}{4}$ three times.

$$\frac{3 \times 3}{4} = \frac{9}{4}. \quad \text{Multiply numerator by 3.}$$

$$\frac{3}{\frac{4}{3}} = 3 \times \frac{3}{4} = \frac{9}{4}. \quad \text{Divide denominator by 3.}$$

212. How can we decrease the value of a fraction?

(*a*) By dividing the numerator by a number greater than 1.

EXAMPLE: $\frac{4}{3}$ is decreased to

$$\frac{\frac{4}{2}}{3} = \frac{2}{3}$$

by dividing numerator by 2, for example.

(*b*) By multiplying the denominator by a number greater than 1.

EXAMPLE: $\frac{4}{3}$ is decreased to

$$\frac{4}{3 \times 2} = \frac{4}{6} = \frac{2}{3}$$

by multiplying the denominator by 2, for example.

The value of the fraction is reduced one-half in each case.

EXAMPLE: Decrease $\frac{24}{25}$ to one-sixth of its value.

$$\frac{\frac{24}{6}}{25} = \frac{4}{25} \qquad \frac{24}{25 \times 6} = \frac{4 \times \not{6}}{25 \times \not{6}} = \frac{4}{25}$$

213. How do we change a compound fraction to a simple fraction?

Place the product of the numerators over the product of the denominators.

Ex. (*a*) $\frac{3}{8}$ of $\frac{5}{6} = \frac{3}{8} \times \frac{5}{6} = \frac{15}{48}$ = simple fraction $= \frac{5}{16}$.

Ex. (*b*) $\frac{5}{6}$ of $\frac{7}{8} = \frac{5}{6} \times \frac{7}{8} = \frac{35}{48}$ = simple fraction.

214. How do we change a complex fraction to a simple fraction?

Divide the numerator by the denominator.

Ex. (*a*) $\frac{6\frac{1}{8}}{15} = 6\frac{1}{8} \div 15 = \frac{49}{8} \div 15 = \frac{49}{8} \times \frac{1}{15} = \frac{49}{120}$ Ans.

Ex. (*b*) $\frac{4\frac{5}{6}}{6\frac{3}{4}} = 4\frac{5}{6} \div 6\frac{3}{4} = \frac{29}{6} \div \frac{27}{4} = \frac{29}{\underset{3}{\not{6}}} \times \frac{\overset{2}{\not{4}}}{27} = \frac{58}{81}$ Ans.

215. What is another method of simplifying a complex fraction?

Multiply both numerator and denominator by a number that does not change the value of the fraction.

EXAMPLE: $\dfrac{4\frac{5}{6}}{6\frac{3}{4}} = \dfrac{4\frac{5}{6}}{6\frac{3}{4}} \times \dfrac{12}{12} = \dfrac{58}{81}$ Ans.

216. What is the condition for adding or subtracting of fractions?

The fractions must all be of the same class, which means the denominators must all be the same.

Add the numerators and place over the common denominator.

Ex. (*a*) Add $\frac{3}{5}$, $\frac{2}{5}$, and $\frac{4}{5}$.

$$\frac{3+2+4}{5} = \frac{9}{5} = 1\tfrac{4}{5} \text{ Ans.}$$

Ex. (*b*) If there are any whole numbers, add them also.

Add $1\frac{3}{5} + 3\frac{2}{5} + 12\frac{4}{5}$

Add whole numbers $1 + 3 + 12 = 16$

Add fractions $\frac{3}{5} + \frac{2}{5} + \frac{4}{5} = \dfrac{3+2+4}{5} = \dfrac{9}{5} = 1\frac{4}{5}$

Then $16 + 1\frac{4}{5} = 17\frac{4}{5}$ Ans.

217. What is the procedure when the denominators are not the same?

Find the "lowest common denominator," which is the smallest denominator into which all will divide evenly. This is the same as the L.C.M. previously studied.

Ex. (*a*) $\frac{1}{2} + \frac{2}{3} + \frac{5}{6}$. The lowest common denominator (L.C.D.) of 2, 3, and 6 is 6. All the denominators divide into 6 evenly.

Now $\frac{1}{2} = \frac{3}{6}$ $\frac{2}{3} = \frac{4}{6}$

$$\therefore \frac{1}{2} + \frac{2}{3} + \frac{5}{6} = \frac{3}{6} + \frac{4}{6} + \frac{5}{6} = \frac{3+4+5}{6} = \frac{12}{6} = 2$$

Ex. (*b*) Add $\frac{1}{2}$, $\frac{3}{5}$, $\frac{7}{10}$, $\frac{3}{4}$. (L.C.D. = 20.)

$$\frac{1}{2} + \frac{3}{5} + \frac{7}{10} + \frac{3}{4} = \frac{10}{20} + \frac{12}{20} + \frac{14}{20} + \frac{15}{20}$$

$$= \frac{10+12+14+15}{20} = \frac{51}{20} = 2\frac{11}{20}$$

Ex. (*c*) Add $5\frac{2}{5}$, $36\frac{1}{4}$, $16\frac{9}{10}$. (L.C.D. = 20.) Multiply each numerator by as many times as the denominator goes into the L.C.D.

$$5\tfrac{8}{20} + 36\tfrac{5}{20} + 16\tfrac{18}{20}$$

$$\frac{8+5+18}{20} = \frac{31}{20} = 1\,\frac{11}{20} \text{ (adding numerators)}$$

$$5 + 36 + 16 = 57 \text{ (adding whole numbers)}$$

$$57 + 1\tfrac{11}{20} = 58\tfrac{11}{20} \text{ Ans.}$$

218. What is the procedure for subtraction of fractions?

(*a*) Work with only two terms at a time.

(*b*) Change a mixed number first to an improper fraction when the mixed number is small.

(*c*) Find the L.C.D. (same as L.C.M.).

(*d*) Subtract smaller numerator from larger. Place result over L.C.D.

(*e*) Reduce to lowest terms.

Ex. (*a*) Subtract $\frac{3}{10}$ from $\frac{4}{5}$. (L.C.D. = 10.)

$$\tfrac{4}{5} - \tfrac{3}{10} = \tfrac{8}{10} - \tfrac{3}{10} = \tfrac{5}{10} = \tfrac{1}{2} \text{ Ans.}$$

Ex. (*b*) Subtract $4\frac{2}{5}$ from $6\frac{3}{8}$.

$$6\tfrac{3}{8} - 4\tfrac{2}{5} = \tfrac{51}{8} - \tfrac{22}{5} \quad \text{(L.C.D. = 40)}$$

$$\tfrac{255}{40} - \tfrac{176}{40} = \tfrac{79}{40} = 1\tfrac{39}{40} \text{ Ans.}$$

219. How do we subtract mixed numbers when they are large?

(*a*) Find the difference between the two fractions and then find the difference between the whole numbers. Borrow 1 from the minuend to increase its fraction when necessary.

Ex. (*a*)

$$\begin{array}{rrl} 48\tfrac{5}{8} & \tfrac{15}{24} & \text{(L.C.D. = 24)} \\ -24\tfrac{7}{12} & \tfrac{14}{24} & \\ \hline 24 & \tfrac{1}{24} & \text{Ans.} = 24\tfrac{1}{24} \end{array}$$

Ex. (*b*) From $148\frac{6}{27}$ take $89\frac{4}{9}$. Before $\frac{4}{9}$ or $\frac{12}{27}$ can be taken from $\frac{6}{27}$, you must borrow 1, or $\frac{27}{27}$, from the minuend, to make the fraction $\frac{33}{27}$. The minuend then becomes $147\frac{33}{27}$.

$$\begin{array}{rl} 147 & \tfrac{33}{27} \\ -\ 89 & \tfrac{12}{27} \\ \hline 58 & \tfrac{21}{27} = 58\tfrac{7}{9} \text{ Ans.} \end{array}$$

220. Can a whole number always be expressed in a fractional form? Yes.

EXAMPLE: $8 = \frac{8}{1}$, $5 = \frac{5}{1}$. Denominator is 1.

221. In adding or subtracting two fractions, how can we use cross multiplication to get the same result as with the L.C.D. method?

Ex. (*a*) $\frac{3}{4} + \frac{5}{16}$. Cross-multiply numerators with opposite denominators to get numerator.

$$\frac{(3 \times 16) + (5 \times 4)}{4 \times 16} = \frac{48 + 20}{64} = \frac{68}{64} = 1\frac{4}{64} = 1\frac{1}{16} \text{ Ans.}$$

Multiply denominators to get denominator.

Ex. (*b*) $\frac{7}{9} + \frac{4}{7} = \frac{(7 \times 7) + (4 \times 9)}{9 \times 7} = \frac{49 + 36}{63} = \frac{85}{63} = 1\frac{22}{63}$

Ex. (*c*) $\frac{7}{9} - \frac{4}{7} = \frac{49 - 36}{63} = \frac{13}{63}$

222. What is the procedure in multiplying one proper fraction by another?

Place the product of the numerators over the product of the denominators.

Ex. (*a*) $\frac{3}{4} \times \frac{7}{8} = \frac{3 \times 7}{4 \times 8} = \frac{21}{32}$ Ans.

Ex. (*b*) $\frac{5}{8} \times \frac{18}{35} = \frac{5 \times 18}{8 \times 35} = \frac{90}{280} = \frac{9}{28}$ Ans.

Shorten the work by cancellation when possible.

Ex. (*c*) $\frac{\cancel{8}}{11} \times \frac{\overset{3}{\cancel{9}}}{\underset{3}{\cancel{24}}} = \frac{3}{11}$ Ans.

Ex. (*d*) $\frac{\overset{3}{\cancel{27}}}{\underset{16}{\cancel{32}}} \times \frac{\overset{9}{\cancel{18}}}{\underset{7}{\cancel{63}}} = \frac{3 \times 9}{16 \times 7} = \frac{27}{112}$ Ans.

223. How do we multiply a proper fraction by a whole number?

Either multiply the numerator or divide denominator by the whole number.

Ex. (*a*) $\frac{7}{16} \times 4 = \frac{7 \times 4}{16} = \frac{28}{16} = \frac{7}{4}$ or $\frac{7}{\frac{16}{4}} = \frac{7}{4}$

Ex. (*b*) Multiply $\frac{4}{7}$ by 11.

$$\frac{4}{7} \times \frac{11}{1} = \frac{44}{7} = 6\frac{2}{7}$$

Ex. (*c*) $15 \times \frac{4}{9} = \frac{\overset{5}{\cancel{15}}}{1} \times \frac{4}{\underset{3}{\cancel{9}}} = \frac{20}{3} = 6\frac{2}{3}$

The result is the same when the multiplier and multiplicand are interchanged in position.

224. What is the procedure for multiplying one mixed number by another?

Change the mixed numbers to improper fractions and multiply in the usual way by placing the product of the numerators over the product of the denominators.

Ex. (*a*)

$$2\frac{2}{3} \times 3\frac{2}{3} = \frac{(3 \times 2) + 2}{3} \times \frac{(3 \times 3) + 2}{3} = \frac{8}{3} \times \frac{11}{3} = \frac{88}{9} = 9\frac{7}{9}$$

Ex. (*b*) Multiply $7\frac{3}{4} \times 5\frac{1}{8}$.

$$7\frac{3}{4} = \frac{(4 \times 7) + 3}{4} = \frac{31}{4} \qquad 5\frac{1}{8} = \frac{(8 \times 5) + 1}{8} = \frac{41}{8}$$

$$\frac{31}{4} \times \frac{41}{8} = \frac{1{,}271}{32} = 39\frac{23}{32} \text{ Ans.}$$

225. What is the four-step method of multiplying one mixed number by another?

(*a*) Multiply the fraction in the multiplier by each part of the multiplicand.

(*b*) Then multiply the whole number of the multiplier by each part of the multiplicand.

(*c*) Add the proper columns.

EXAMPLE: Multiply $5\frac{3}{4} \times 7\frac{1}{2}$.

$$\begin{array}{rl} & 5\frac{3}{4} \\ \times & 7\frac{1}{2} \\ \hline & \frac{3}{8} = \frac{1}{2} \times \frac{3}{4} \text{ (step 1)} \\ & 2\frac{1}{2} = \frac{1}{2} \times 5 \text{ (step 2)} \\ & 5\frac{1}{4} = 7 \times \frac{3}{4} \text{ (step 3)} \\ & 35 = 7 \times 5 \text{ (step 4)} \\ \hline & 42\frac{9}{8} = 43\frac{1}{8} \text{ Ans.} \end{array}$$

226. How do we multiply a mixed number by a proper fraction?

(*a*) Change the mixed number to an improper fraction and multiply as usual.

(*b*) Or, multiply the fractions together; then, multiply the whole number by the fraction.

Ex. (*a*)

$$5\frac{1}{3} \times \frac{1}{7} = \frac{(3 \times 5) + 1}{3} \times \frac{1}{7} = \frac{16}{3} \times \frac{1}{7} = \frac{16 \times 1}{3 \times 7} = \frac{16}{21} \text{ Ans.}$$

Ex. (*b*) Multiply $\frac{5}{7}$ by $15\frac{7}{8}$.

$$\begin{array}{rll} & 15\frac{7}{8} & \\ \times & \frac{5}{7} & \\ \hline & \frac{5}{8} & = \frac{5}{\not 7} \times \frac{\not 7}{8} \text{ (multiply fractions)} \\ + & 10\frac{5}{7} & = \frac{5}{7} \times 15 = \frac{75}{7} \text{ (multiply fraction by whole number)} \\ \hline & 10\frac{75}{36} & = 11\frac{19}{56} \text{ Ans.} \end{array}$$

Ex. (*c*) Multiply $16\frac{7}{9}$ by $\frac{3}{5}$.

$$\begin{array}{rll} & 16\frac{7}{9} & \text{(mixed number)} \\ \times & \frac{3}{5} & \text{(fraction)} \\ \hline & \frac{7}{15} & = \frac{3}{5} \times \frac{7}{9} \text{ (multiply fractions)} \\ + & 9\frac{3}{5} & = \frac{3}{5} \times 16 = \frac{48}{5} = 9\frac{3}{5} \text{ (multiply fraction by whole number)} \\ \hline & 9\frac{16}{15} & = 10\frac{1}{15} \text{ Ans.} \end{array}$$

Or, change mixed number to an improper fraction first: $16\frac{7}{9} = \frac{151}{9}$. Then

$$\frac{151}{\underset{3}{\not 9}} \times \frac{\not 3}{5} = \frac{151}{15} = 10\frac{1}{15}$$

227. What word is frequently used instead of the multiplication sign or the word "multiply"?

The word "*of*."

EXAMPLE: $\frac{1}{4}$ of $\frac{1}{7} = \frac{1}{4} \times \frac{1}{7} = \frac{1}{28}$ Ans.

228. What is meant by the reciprocal of a number?

The reciprocal of a number is 1 *divided by the number*.

Ex. (*a*) The reciprocals of 3, 8, 10, and 25 are $\frac{1}{3}$, $\frac{1}{8}$, $\frac{1}{10}$, and $\frac{1}{25}$ respectively.

Since 3, 8, 10, and 25 are equivalent to $\frac{3}{1}$, $\frac{8}{1}$, $\frac{10}{1}$, and $\frac{25}{1}$ respectively in fraction form, we obtain the reciprocal of a fraction by inverting the fraction.

Ex. (*b*) The reciprocals of $\frac{3}{8}$, $\frac{8}{10}$, and $\frac{9}{25}$ are $\frac{8}{3}$, $\frac{10}{8}$, and $\frac{25}{9}$ respectively.

229. When the product of two numbers equals 1, what is each of the two numbers called?

Each is called the *reciprocal* of the other.

Ex. (*a*) $4 \times \frac{1}{4} = 1$. Hence, 4 is the reciprocal of $\frac{1}{4}$ and $\frac{1}{4}$ is the reciprocal of 4.

Ex. (*b*) $\frac{9}{5} \times \frac{5}{9} = \frac{45}{45} = 1$. Hence, $\frac{9}{5}$ is the reciprocal of $\frac{5}{9}$ and $\frac{5}{9}$ is the reciprocal of $\frac{9}{5}$. To get the reciprocal of a fraction we invert the fraction.

230. How can we show that to multiply by the reciprocal of a number is the same as to divide by that number?

We have seen, above, that $\frac{9}{5} \times \frac{5}{9} = 1$. We here *multiply* by $\frac{5}{9}$ to get 1.

It is also true that $\dfrac{\frac{9}{5}}{\frac{9}{5}} = 1$. Here we divide by $\frac{9}{5}$ to get 1.

But $\frac{5}{9}$ is the reciprocal of $\frac{9}{5}$.
Therefore *multiplying* by $\frac{5}{9}$ is the same as *dividing* by $\frac{9}{5}$.

231. How many times are: (*a*) $\frac{1}{4}$, $\frac{2}{4}$, and $\frac{3}{4}$ contained in 1? (*b*) $\frac{1}{4}$, $\frac{2}{4}$, and $\frac{3}{4}$ contained in 2?

(*a*) $\dfrac{1}{\frac{1}{4}} = 1 \times \frac{4}{1} = 4$ times, multiply by the reciprocal ($\frac{4}{1}$).

$\dfrac{1}{\frac{2}{4}} = 1 \times \frac{4}{2} = 2$ times, multiply by the reciprocal ($\frac{4}{2}$).

$\dfrac{1}{\frac{3}{4}} = 1 \times \frac{4}{3} = \frac{4}{3}$ times, multiply by the reciprocal ($\frac{4}{3}$).

(*b*) $\dfrac{2}{\frac{1}{4}} = 2 \times \frac{4}{1} = 8$ times, multiply by the reciprocal ($\frac{4}{1}$).

$\dfrac{2}{\frac{2}{4}} = 2 \times \frac{4}{2} = 4$ times, multiply by the reciprocal ($\frac{4}{2}$).

$\dfrac{2}{\frac{3}{4}} = 2 \times \frac{4}{3} = \frac{8}{3}$ times, multiply by the reciprocal ($\frac{4}{3}$).

232. In each case, what can we do when we want to divide a whole number by a fraction, or a fraction by a whole number, or a fraction by a fraction?

Multiply by its reciprocal.

EXAMPLE: Divide $\frac{4}{5}$ by $\frac{4}{9}$.

$$\frac{\frac{4}{5}}{\frac{4}{9}} = \frac{\cancel{4}}{5} \times \frac{9}{\cancel{4}} = \frac{9}{5} = 1\frac{4}{5}$$

This means that $\frac{4}{5}$ goes into $\frac{4}{9}$ one and four-fifths times.

233. Specifically, how do we divide a proper fraction by a whole number?

Divide its numerator or multiply its denominator by the whole number.

Ex. (*a*) Divide $\frac{8}{9}$ by 2.

$$\frac{\frac{8}{9}}{2} = \frac{\frac{8}{2}}{9} = \frac{4}{9} \text{ Ans.} \qquad \text{or} \qquad \frac{\frac{8}{9}}{2} = \frac{\overset{4}{\cancel{8}}}{9 \times \cancel{2}} = \frac{4}{9} \text{ Ans.}$$

(divide numerator by 2) (multiply denominator by 2)

Multiplying the denominator by the whole number is equivalent to multiplying by the reciprocal of the whole number.

$$\frac{\frac{8}{9}}{2} = \frac{\overset{4}{\cancel{8}}}{9} \times \frac{1}{\cancel{2}} = \frac{4}{9} \text{ Ans.}$$

Ex. (*b*) $\frac{3}{7} \div 4 = \frac{3}{7} \times \frac{1}{4} = \frac{3}{28}$.

234. How do we divide a whole number by a fraction?

Divide the whole number by the numerator and multiply by the denominator.

Ex. (*a*) Divide 24 by $\frac{3}{8}$.

$$\frac{24}{\frac{3}{8}} = \frac{\overset{8}{\cancel{24}}}{\cancel{3}} \times 8 = 64 \text{ Ans.}$$

Ex. (*b*) Divide 17 by $\frac{4}{9}$, or $\frac{17}{\frac{4}{9}}$.

$$\frac{17}{\frac{4}{9}} = \frac{17}{4} \times 9 = \frac{153}{4} = 38\frac{1}{4} \text{ Ans.}$$

In each case the method is equivalent to multiplying by the reciprocal of the fraction.

235. How do we divide one mixed number by another?

Change the mixed numbers to improper fractions, invert the divisor, and multiply. (Inverting the divisor gives the reciprocal of the divisor.)

EXAMPLE: Divide $9\frac{4}{5}$ by $2\frac{7}{8}$.

$$9\tfrac{4}{5} \div 2\tfrac{7}{8} = \tfrac{49}{5} \div \tfrac{23}{8} = \tfrac{49}{5} \times \tfrac{8}{23} = \tfrac{392}{115} = 3\tfrac{47}{115}$$

236. How do we divide a mixed number by a whole number?

Change mixed number to an improper fraction and divide the numerator or multiply denominator by the whole number.

EXAMPLE: Divide $8\frac{5}{8}$ by 3.

$$8\frac{5}{8} \div 3 = \frac{69}{8} \div 3 = \frac{\frac{69}{3}}{8} = \frac{23}{8} = 2\frac{7}{8} \quad \text{or} \quad \frac{\overset{23}{\cancel{69}}}{8 \times \cancel{3}} = \frac{23}{8} = 2\frac{7}{8}$$

Here also the method is equivalent to multiplying by the reciprocal of the whole number.

237. What is another method to use for the above case when the dividend is a large number?

Divide as in whole numbers and simplify the remaining complex fraction.

EXAMPLE: Divide $1{,}184\frac{5}{8}$ by 6.

$$6\overline{)1{,}184\tfrac{5}{8}} \qquad 197\frac{2\frac{5}{8}}{6}$$

$$\frac{2\frac{5}{8}}{6} = \frac{\frac{(2 \times 8) + 5}{8}}{6} = \frac{\overset{7}{\cancel{21}}}{8} \times \frac{1}{\underset{2}{\cancel{6}}} = \frac{7}{16}$$

$$\text{Ans.} = 197\tfrac{7}{16}$$

238. What are some other methods of dividing whole mixed numbers?

Ex. (*a*) Divide 482 by $26\frac{3}{5}$.

$$482 \div 26\frac{3}{5} = \frac{482 \times 5}{26\frac{3}{5} \times 5} = \frac{2{,}410}{133} = 18\frac{16}{133} \text{ Ans.}$$

Multiplying both numerator and denominator by 5 does away with the mixed number in the divisor but does not change the value of the fraction.

Ex. (*b*) Divide $862\frac{2}{3}$ by $29\frac{3}{4}$.

$$862\frac{2}{3} \div 29\frac{3}{4} = \frac{862\frac{2}{3} \times 12}{29\frac{3}{4} \times 12} = \frac{10{,}352}{357} = 28\frac{356}{357} \text{ Ans.}$$

To change to whole numbers multiply numerator and denominator by the common multiple of the denominators of the fractions. L.C.M. here is 12.

239. What is the difference between a fraction applicable to an abstract number and one applicable to a concrete number?

The fraction $\frac{3}{4}$ means that an abstract unit is divided into 4 equal parts and 3 parts are expressed.

The expression "$\frac{3}{4}$ of a dozen" is applicable to 12, because that is the number of units in a dozen, and may be expressed as 9.

The fraction "$\frac{1}{2}$ of a gallon" may be expressed as 2 quarts because there are 4 quarts in a gallon.

240. How do we find what part the second of two numbers is of the first?

Divide the second by the first.

Ex. (*a*) What part of 63 is 9?

$$9 \text{ is } \tfrac{9}{63} \text{ of } 63 \quad \text{or} \quad \tfrac{1}{7} \text{ Ans.}$$

Ex. (*b*) What part of 74 is 18?

$$\tfrac{18}{74} = \tfrac{9}{37} \text{ part of } 74 \quad \text{Ans.}$$

Ex. (*c*) What part of $8\frac{4}{9}$ is $2\frac{2}{9}$?

$$\frac{2\frac{2}{9}}{8\frac{4}{9}} = \frac{\frac{20}{9}}{\frac{76}{9}} = \frac{\overset{5}{\cancel{20}}}{\cancel{9}} \times \frac{\cancel{9}}{\underset{19}{\cancel{76}}} = \frac{5}{19} \text{ part of } 8\tfrac{4}{9} \text{ Ans.}$$

Ex. (*d*) What part of $\frac{7}{8}$ is 7?

$$\frac{7}{\frac{7}{8}} = \cancel{7} \times \frac{8}{\cancel{7}} = 8 \text{ times } \tfrac{7}{8}.$$

241. If you are given a number that is a certain fraction of a whole, how would you find the whole?

Divide the given number by the fraction.

Ex. (*a*) 6 is $\frac{2}{3}$ of what number?

$$\frac{6}{\frac{2}{3}} = \overset{3}{\cancel{6}} \times \frac{3}{\cancel{2}} = 9 \text{ Ans.}$$

Ex. (*b*) 72 is $\frac{6}{7}$ of what number?

$$\frac{72}{\frac{6}{7}} = \overset{12}{\cancel{72}} \times \frac{7}{\cancel{6}} = 84 \text{ Ans.}$$

Ex. (*c*) 99 is $\frac{3}{4}$ of what number?

$$\frac{99}{\frac{3}{4}} = \overset{33}{\cancel{99}} \times \frac{4}{\cancel{3}} = 132 \text{ Ans.}$$

Note that in each case you multiply by the reciprocal of the fraction.

Ex. (*d*) If 78 is $\frac{3}{4}$ of the lot, what is the whole lot?

$$\frac{78}{\frac{3}{4}} = \overset{26}{\cancel{78}} \times \frac{4}{\cancel{3}} = 104 = \text{the whole lot.}$$

Ex. (*e*) Find the number of which 40 is $\frac{4}{5}$.

$$\frac{40}{\frac{4}{5}} = \overset{10}{\cancel{40}} \times \frac{5}{\cancel{4}} = 50 \text{ Ans.}$$

Ex. (*f*) $\frac{7}{8}$ of some radio equipment is worth \$350. What is the value of the entire stock?

$$\frac{\$350}{\frac{7}{8}} = \overset{50}{\cancel{350}} \times \frac{8}{\cancel{7}} = \$400 \text{ Ans.}$$

242. How do we tell which one of two fractions is the greater?

Reduce the fractions to their lowest terms by cancellation.

Get the L.C.D. (lowest common denominator) and change each fraction to have this L.C.D. Compare numerators.

EXAMPLE: Which of the following is greater, $\frac{124}{288}$ or $\frac{385}{665}$?

$$\frac{\overset{31}{\cancel{124}}}{\underset{72}{\cancel{288}}} = \frac{31}{72} \qquad \frac{\overset{\overset{11}{\cancel{77}}}{\cancel{385}}}{\underset{\underset{19}{\cancel{133}}}{\cancel{665}}} = \frac{11}{19}$$

(L.C.D. = 72 × 19 = 1,368)

$$\frac{31}{72} = \frac{31 \times 19}{1{,}368} = \frac{589}{1{,}368} \qquad \frac{11}{19} = \frac{11 \times 72}{1{,}368} = \frac{792}{1{,}368}$$

We see that 792 is greater. Thus, $\frac{385}{665}$ is greater than $\frac{124}{288}$.

243. What is a chain (or a continued) fraction?

One in which the denominator has a fraction, the denominator of which has a fraction, the denominator of which has a fraction, etc.

EXAMPLE:

$$\frac{14}{47} = \cfrac{1}{3 + \cfrac{1}{2 + \cfrac{1}{1 + \cfrac{1}{4}}}}$$

(fraction in denominator)
(fraction in denominator)
(fraction in denominator)

244. What chain fractions are of interest to us?

Only those in which all numerators are 1 or unity—the so-called integer chain fractions.

245. How is a proper fraction converted into a chain fraction?

We know that dividing both numerator and denominator of a fraction by the same quantity does not change the value of the fraction.

Divide both numerator and denominator by the numerator. The numerator becomes 1 and the denominator becomes a whole number and a fraction as a remainder.

Convert the fractional remainder by dividing both its terms by the numerator. Again the numerator becomes 1 and the denominator becomes a whole number and a fraction as a remainder.

Continue this process until the fractional remainder has 1 as a numerator.

EXAMPLE: Convert $\frac{89}{136}$ to a chain fraction.

$$\frac{89}{136} = \frac{\frac{89}{89}}{\frac{136}{89}} = \frac{1}{1+\frac{47}{89}}$$

$$= \frac{1}{1+\frac{\frac{47}{47}}{\frac{89}{47}}} = \cfrac{1}{1+\cfrac{1}{1+\frac{42}{47}}}$$

$$= \cfrac{1}{1+\cfrac{1}{1+\frac{\frac{42}{42}}{\frac{47}{42}}}} = \cfrac{1}{1+\cfrac{1}{1+\cfrac{1}{1+\frac{5}{42}}}}$$

$$= \cfrac{1}{1+\cfrac{1}{1+\cfrac{1}{1+\frac{\frac{5}{5}}{\frac{42}{5}}}}} = \cfrac{1}{1+\cfrac{1}{1+\cfrac{1}{1+\cfrac{1}{8+\frac{2}{5}}}}}$$

$$= \cfrac{1}{1+\cfrac{1}{1+\cfrac{1}{1+\cfrac{1}{8+\frac{\frac{2}{2}}{\frac{5}{2}}}}}} = \cfrac{1}{1+\cfrac{1}{1+\cfrac{1}{1+\cfrac{1}{8+\cfrac{1}{2+\frac{1}{2}}}}}}$$

fractional remainder has 1 in numerator ↙

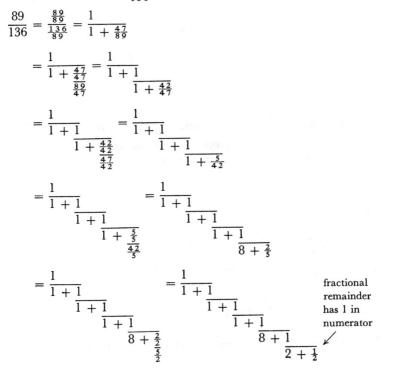

246. How can the above be simplified?

Each time divide the previous divisor by the remainder. The quotients become the denominators of the chain fraction with units for numerators. The denominators 1, 1, 1, 8, 2 are the integral parts of the quotients.

Convert $\frac{89}{136}$:

```
Divisor→ 89)136(1
             89
             47)89(1
                47
                42)47(1
                   42
                    5)42(8
                      40
                       2)5(2
                         4
         Remainder →   1)2(2
```

Remainder becomes new divisor → ; Previous divisor ↓

Hence

$$\frac{89}{136} = \cfrac{1}{1 + \cfrac{1}{1 + \cfrac{1}{1 + \cfrac{1}{8 + \cfrac{1}{2 + \cfrac{1}{2}}}}}}$$

247. How is a chain fraction converted to a proper fraction?

By inverse process, start from the end and go up. In the above start with the last fractional denominator:

$$\frac{1}{2 + \frac{1}{2}} = \frac{1}{\frac{5}{2}} = \frac{2}{5}$$

The next fractional denominator is

$$\frac{1}{8 + \frac{2}{5}} = \frac{1}{\frac{42}{5}} = \frac{5}{42}$$

Next $$\frac{1}{1 + \frac{5}{42}} = \frac{1}{\frac{47}{42}} = \frac{42}{47}$$

Next $$\frac{1}{1 + \frac{42}{47}} = \frac{1}{\frac{89}{47}} = \frac{47}{89}$$

Finally $$\frac{1}{1 + \frac{47}{89}} = \frac{1}{\frac{136}{89}} = \frac{89}{136}$$ (= Original fraction above).

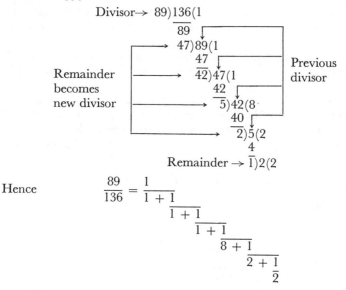

248. Of what practical use are chain fractions?

For one thing they enable us to find another fraction expressed in simpler terms (smaller numbers) and of a value near or very near the one with large numbers.

EXAMPLE: What fraction expressed in smaller numbers is near in value to $\frac{31}{157}$?

Dividing both terms by 31, we get

$$\frac{\frac{31}{31}}{\frac{157}{31}} = \frac{1}{5 + \frac{2}{31}},$$

expressed as a chain fraction.

Now if we reject the $\frac{2}{31}$, the fraction $\frac{1}{5}$ will be larger than $\frac{1}{5 + \frac{2}{31}}$ because the denominator was decreased.

To compare $\frac{31}{157}$ with $\frac{1}{5}$ get the L.C.D. of both or

$$157 \times 5 = 785 = \text{L.C.D.}$$

Then $\frac{31}{157} = \frac{155}{785}$ and $\frac{1}{5} = \frac{157}{785}$; difference $= \frac{2}{785}$.

Thus, $\frac{1}{5}$ is seen to be near the value of $\frac{31}{157}$.

249. What fraction in smaller terms nearly expresses $\frac{3937}{100,000}$?

Divide numerator and denominator by 3,937:

$$\frac{\frac{3937}{3937}}{\frac{100,000}{3937}} = \frac{1}{25 + \frac{1575}{3937}}$$

$\frac{1}{25} = 0.04$ is a little larger than $\frac{3937}{100,000} = 0.03937$ but it gives us a pretty good idea of its value.

250. How can we get a closer approximation?

Continue the chain fraction.

```
3,937)100,000(25
       78 74
       21,260
       19,685
        1,575)3,937(2
              3,150
                787)1,575(2
                    1,574
                        1)787(787
                          787
                            0
```

or

$$\frac{3,937}{100,000} = \cfrac{1}{25 + \cfrac{1}{2 + \cfrac{1}{2 + \cfrac{1}{787}}}}$$

Consider $$\frac{1}{25 + \frac{1}{2}} = \frac{1}{\frac{51}{2}} = \frac{2}{51} = 0.0392+$$

which is smaller than 0.03937.

To get still nearer, take the next part of the chain fraction:

$$\cfrac{1}{2+\cfrac{1}{2+\frac{1}{2}}}$$

Start from the bottom

$$2+\frac{1}{2}=\frac{5}{2} \quad\text{and}\quad \frac{1}{2+\frac{1}{2}}=\frac{1}{\frac{5}{2}}=\frac{2}{5} \quad\text{and}\quad \frac{1}{25+\frac{2}{5}}=\frac{1}{\frac{127}{5}}=\frac{5}{127}$$

$\frac{5}{127} = 0.03937007$. This is the nearest fraction to 0.03937 unless we reduce the entire chain fraction, which would give us 0.03937 itself. $\frac{5}{127}$ is only $\frac{7}{100,000,000}$ larger than $\frac{3937}{100,000}$ which is quite close.

We thus see that a chain fraction can give us a series of successive approximations.

251. What feature of a chain fraction makes it valuable to us?

The approach to the true value is extremely rapid. It gives very rapidly converging approximations.

EXAMPLE: Of above values of $\frac{3937}{100,000} = 0.03937$

			Difference	
$\frac{1}{25}$	$=\frac{1}{25}$	$= 0.04(-0.03937)$	$= +0.00063$	$= +1.6\%$
$\frac{1}{25+\frac{1}{2}}$	$=\frac{2}{51}$	$= 0.039215686\ldots$	$= -0.0001543$	$= -0.39\%$
$\cfrac{1}{2+\cfrac{1}{2+\frac{1}{2}}}$	$=\frac{5}{127}$	$= 0.03937007$	$= +0.00000007$	$= +0.000177\%$
$\cfrac{1}{25+\cfrac{1}{2+\cfrac{1}{2+\frac{1}{787}}}}$		$= 0.03937$		

We see that the second approximation brings us within 0.39 per cent of its true value. Very rapid indeed.

PROBLEMS

1. If there are four weeks in a month, three weeks are equal to what part of three months?

2. If a unit is divided into ten equal parts, what is one part called?

3. Read the following: $\frac{1}{3}, \frac{1}{6}, \frac{1}{16}, \frac{1}{12}, \frac{1}{20}$. What part of these fractions shows the number of parts into which the unit is divided?

4. In $\frac{15}{31}$, what shows how many parts are taken?

5. Which are proper fractions, improper fractions, and mixed numbers in the following?

(*a*) $\frac{1}{4}$, $\frac{1}{7}$, $\frac{5}{6}$, $\frac{7}{9}$ (*b*) $\frac{5}{4}$, $\frac{2}{3}$, $4\frac{1}{8}$, $\frac{18}{16}$

(*c*) $\frac{17}{19}$, $\frac{19}{15}$, $\frac{4}{7}$, $\frac{8}{3}$ (*d*) $8\frac{1}{5}$, $\frac{9}{2}$, $16\frac{1}{2}$, $\frac{3}{5}$

(*e*) $\frac{27}{4}$, $8\frac{2}{7}$, $17\frac{3}{4}$, $\frac{9}{10}$ (*f*) $\frac{27}{39}$, $\frac{18}{5}$, $6\frac{1}{2}$, $\frac{14}{24}$

6. Write as common fractions or mixed numbers:

(*a*) Twenty-nine tenths (*b*) Forty-nine elevenths
(*c*) Eight fifteenths (*d*) Nine one-hundredths
(*e*) Ninety-two and three-fourths
(*f*) One hundred and thirty-five fifty-sixths
(*g*) Eighty-seven and nine tenths
(*h*) Six hundred tenths (*i*) Twenty-three thirty-sevenths
(*j*) Eighteen and six twenty-firsts
(*k*) Thirty-one and seventeen nineteenths
(*l*) One hundred forty-five and one hundred thirty-three one hundred thirty-fifths

7. Change to whole or mixed numbers:

(*a*) $\frac{48}{6}$ (*b*) $\frac{20}{4}$ (*c*) $\frac{18}{3}$ (*d*) $\frac{49}{6}$ (*e*) $\frac{129}{23}$
(*f*) $\frac{63}{7}$ (*g*) $\frac{88}{12}$ (*h*) $\frac{83}{8}$ (*i*) $\frac{318}{12}$ (*j*) $\frac{69}{11}$
(*k*) $\frac{580}{24}$ (*l*) $\frac{693}{480}$ (*m*) $\frac{66}{66}$ (*n*) $\frac{1728}{24}$ (*o*) $\frac{259}{19}$

8. Change to improper fractions:

(*a*) $13\frac{1}{2}$ (*b*) $16\frac{1}{3}$ (*c*) $19\frac{2}{9}$ (*d*) $53\frac{3}{12}$ (*e*) $11\frac{4}{17}$
(*f*) $28\frac{4}{5}$ (*g*) $42\frac{1}{3}$ (*h*) $21\frac{1}{8}$ (*i*) $19\frac{5}{11}$ (*j*) $14\frac{17}{59}$
(*k*) $83\frac{31}{48}$ (*l*) $35\frac{9}{30}$ (*m*) $55\frac{4}{11}$ (*n*) $19\frac{31}{83}$ (*o*) $176\frac{49}{68}$
(*p*) $156\frac{39}{83}$

9. (*a*) How many fourteenths in one unit?
(*b*) How many fourteenths in two units?
(*c*) How many fourteenths in one half unit?
(*d*) Does changing $\frac{7}{14}$ to its lower term, $\frac{1}{2}$, change its value?

10. Reduce the fractions to lowest terms:

(*a*) $\frac{26}{98}$ (*b*) $\frac{136}{248}$ (*c*) $\frac{285}{765}$ (*d*) $\frac{565}{760}$ (*e*) $\frac{685}{1280}$ (*f*) $\frac{51}{102}$ (*g*) $\frac{168}{224}$

11. Change to higher terms:

(*a*) $\frac{3}{4}$ to 20ths (*b*) $\frac{5}{8}$ to 64ths (*c*) $\frac{2}{7}$ to 84ths
(*d*) $\frac{13}{16}$ to 96ths (*e*) $\frac{9}{20}$ to 100ths (*f*) $\frac{5}{6}$ to 24ths

12. Find the missing numerators:

(*a*) $\frac{1}{9} = \frac{\quad}{72}$ (*b*) $\frac{5}{6} = \frac{\quad}{30}$ (*c*) $\frac{5}{9} = \frac{\quad}{27}$

(*d*) $\frac{5}{7} = \frac{\quad}{28}$ (*e*) $\frac{3}{5} = \frac{\quad}{75}$ (*f*) $\frac{23}{34} = \frac{\quad}{374}$

(*g*) $\frac{17}{28} = \frac{\quad}{84}$ (*h*) $\frac{3}{4} = \frac{\quad}{24}$ (*i*) $\frac{7}{10} = \frac{\quad}{50}$

13. Reduce to fractions having an L.C.D.

(*a*) $\frac{3}{4}, \frac{2}{5}, \frac{3}{10}$ (*b*) $\frac{1}{2}, \frac{5}{8}, \frac{7}{12}$ (*c*) $\frac{2}{5}, \frac{3}{7}, \frac{2}{3}$
(*d*) $\frac{5}{8}, \frac{2}{5}, \frac{1}{40}$ (*e*) $\frac{6}{15}, \frac{5}{45}, \frac{6}{30}$ (*f*) $\frac{9}{132}, \frac{12}{144}$
(*g*) $\frac{7}{222}, \frac{15}{430}$ (*h*) $\frac{7}{15}, \frac{13}{45}, \frac{23}{60}$ (*i*) $\frac{7}{11}, \frac{15}{99}, \frac{9}{121}$

14. Change to improper fractions and reduce to L.C.D.:

(*a*) $23\frac{5}{21}, 38\frac{3}{28}$ (*b*) $72\frac{4}{27}, 56\frac{13}{71}$
(*c*) $18\frac{3}{65}, 25\frac{14}{52}$ (*d*) $44\frac{13}{18}, 62\frac{23}{28}$

15. (*a*) Increase the value of $\frac{5}{8}$ three times.
(*b*) Increase the value of $\frac{11}{25}$ two and one-half times.
(*c*) Increase the value of $\frac{7}{8}$ four and one-sixth times.

16. (*a*) Decrease the value of $\frac{11}{13}$ to $\frac{1}{7}$ the value.
(*b*) Decrease the value of $\frac{7}{16}$ to $\frac{1}{10}$ the value.
(*c*) Decrease the value of $\frac{86}{100}$ to $\frac{5}{100}$ the value.

17. Change to a simple fraction:

(*a*) $\frac{5}{8}$ of $\frac{7}{9}$ (*b*) $\frac{5}{6}$ of $\frac{8}{11}$ (*c*) $\frac{11}{27}$ of $\frac{15}{31}$
(*d*) $\frac{13}{24}$ of $\frac{3}{8}$ (*e*) $\frac{5}{9}$ of $\frac{11}{12}$ (*f*) $\frac{9}{23}$ of $\frac{5}{16}$

18. Change to a simple fraction:

(*a*) $\dfrac{5\frac{1}{8}}{13}$ (*b*) $\dfrac{5\frac{5}{6}}{7\frac{3}{4}}$ (*c*) $\dfrac{3\frac{1}{3}}{4\frac{3}{4}}$ (*d*) $\dfrac{6\frac{5}{8}}{9\frac{3}{4}}$ (*e*) $\dfrac{8\frac{2}{5}}{13\frac{4}{15}}$ (*f*) $\dfrac{14\frac{1}{2}}{22\frac{2}{3}}$

19. Add:

(*a*) $\frac{3}{7} + \frac{5}{7} + \frac{6}{7}$ (*b*) $2\frac{3}{7} + 5\frac{4}{7} + 7\frac{5}{7}$ (*c*) $\frac{2}{3} + \frac{3}{4} + \frac{5}{6}$
(*d*) $6\frac{3}{5} + 42\frac{1}{4} + 18\frac{7}{10}$ (*e*) $\frac{7}{9} + \frac{5}{18}$ (*f*) $\frac{5}{7} + \frac{1}{3}$ (*g*) $\frac{3}{5} + \frac{9}{10} + \frac{7}{20}$
(*h*) $\frac{1}{3} + \frac{5}{9} + \frac{3}{14}$ (*i*) $3\frac{2}{3} + 7\frac{1}{6} + 17\frac{11}{12}$ (*j*) $8\frac{7}{9} + 9\frac{13}{30} + 24\frac{3}{4}$
(*k*) $22\frac{4}{7} + 32\frac{9}{21} + 84\frac{1}{6}$ (*l*) $15\frac{7}{8} + 51\frac{7}{10} + 62\frac{3}{10}$

20. Subtract:

(*a*) $\frac{5}{12}$ from $\frac{5}{6}$ (*b*) $3\frac{3}{5}$ from $5\frac{3}{8}$ (*c*) $26\frac{5}{12}$ from $49\frac{7}{8}$
(*d*) $87\frac{5}{9}$ from $142\frac{5}{27}$ (*e*) $9\frac{1}{2}$ from $13\frac{1}{5}$ (*f*) $150\frac{14}{15}$ from $634\frac{3}{5}$
(*g*) $\frac{3}{7}$ from $\frac{8}{9}$

21. Multiply:

(*a*) $\frac{3}{4}$ by $\frac{5}{8}$ (*b*) $\frac{3}{8}$ by $\frac{17}{35}$ (*c*) $\frac{7}{11}$ by $\frac{8}{24}$ (*d*) $\frac{9}{16}$ by 4
(*e*) $\frac{5}{7}$ by 12 (*f*) 17 by $\frac{7}{9}$ (*g*) $3\frac{1}{3}$ by $4\frac{1}{3}$ (*d*) $6\frac{3}{4}$ by $4\frac{1}{8}$
(*i*) $6\frac{3}{8}$ by $9\frac{1}{2}$ (*j*) $6\frac{2}{3}$ by $\frac{1}{5}$ (*k*) $\frac{5}{9}$ by $16\frac{7}{8}$ (*l*) $17\frac{7}{11}$ by $\frac{4}{5}$

22. Express the reciprocals of:

(*a*) 4, 9, 11, 35 (*b*) $\frac{5}{8}, \frac{9}{10}, \frac{8}{25}$

23. How many times are:

(*a*) $\frac{1}{5}, \frac{2}{5}, \frac{3}{5}, \frac{4}{5}$ contained in 1?

(*b*) $\frac{1}{7}, \frac{2}{7}, \frac{3}{7}, \frac{4}{7}, \frac{5}{7}, \frac{6}{7}$ contained in 2?

24. Divide:

(*a*) $\frac{7}{9}$ by 2 (*b*) $\frac{5}{7}$ by 3 (*c*) 27 by $\frac{3}{8}$
(*d*) 19 by $\frac{5}{9}$ (*e*) $8\frac{3}{5}$ by $2\frac{5}{8}$ (*f*) $9\frac{5}{8}$ by 4
(*g*) $1{,}276\frac{5}{8}$ by 7 (*h*) 574 by $24\frac{3}{5}$ (*i*) $786\frac{2}{3}$ by $28\frac{3}{4}$

25. What part of:

(*a*) 72 is 9? (*b*) 86 is 16? (*c*) $7\frac{5}{9}$ is $2\frac{7}{9}$?
(*d*) $\frac{15}{16}$ is 15? (*e*) $5\frac{1}{4}$ is $10\frac{1}{2}$? (*f*) $\frac{1}{3}$ is 72?
(*g*) $6\frac{4}{5}$ is $3\frac{2}{5}$? (*h*) $\frac{4}{9}$ is $\frac{1}{9}$? (*i*) $1\frac{1}{2}$ is $\frac{1}{3}$?
(*j*) $8\frac{4}{9}$ is $2\frac{2}{9}$?

26. (*a*) 8 is $\frac{2}{3}$ of what number?
(*b*) 84 is $\frac{6}{7}$ of what number?
(*c*) 144 is $\frac{3}{4}$ of what number?

27. (*a*) Find the number of which 60 is $\frac{3}{5}$.
(*b*) Five-eighths of a shipment is worth $430, what is the value of the entire shipment?

28. Which fraction has a greater value, $\frac{136}{394}$ or $\frac{376}{432}$?

29. Express $\frac{15}{62}$ as a chain (or continued) fraction.

30. Convert $\frac{73}{142}$ to a chain fraction.

31. Convert

$$\cfrac{1}{1+\cfrac{1}{1+\cfrac{1}{3+\cfrac{1}{2+\cfrac{1}{5}}}}}$$

to a proper fraction.

32. What fraction in smaller numbers is near in value to $\frac{43}{176}$?

33. What fraction in smaller terms nearly expresses $\pi = 3.1416$ or $3 + \frac{1416}{10{,}000}$? (Use chain-fraction method.)

34. The width of a door opening is $\frac{1}{4}$ of its height. What is the width when the height is $9\frac{1}{3}$ ft?

35. I find that I spent $88, which represents $\frac{2}{3}$ of my total allowance. How much do I have left?

36. Three cases of merchandise weighing $343\frac{1}{4}$, $478\frac{3}{4}$, and $506\frac{1}{2}$ lb. were shipped. The cases weighed $17\frac{3}{4}$, $18\frac{1}{2}$, and $19\frac{3}{4}$ lb. What is the total weight of the cases, gross weight, and the net weight of the merchandise?

37. If a lb. of bread had 9 slices, how many ounces are there per slice?

38. How many reams of paper are listed on this invoice: $15\frac{3}{4}$, $24\frac{1}{2}$, $67\frac{2}{3}$, $58\frac{2}{5}$, and $19\frac{4}{5}$ reams?

39. If in a test run a car traveled 26 miles in 30 minutes, how many miles will it travel in $5\frac{1}{2}$ hours at this rate?

40. A crate of apples containing 148 apples was bought at $3\frac{3}{4}$¢ an apple and sold at $\frac{7}{5}$ of the cost. What was the profit?

41. Two partners bought a parcel of land for $3,600, each paying $\frac{1}{2}$. Then each sold $\frac{1}{4}$ of his interest to a third party at cost. What fractional part of the total investment does each party now own and how much is each worth?

42. A man spends $\frac{1}{4}$ of his salary for a suit of clothes, $\frac{3}{8}$ for an overcoat, $\frac{1}{8}$ for shoes, and $\frac{1}{12}$ for a hat. What part has he left?

43. If the above person has $41 left, how much had he to begin with and what does each item cost?

44. The sides of an irregularly shaped yard have the following measurements: $28\frac{1}{5}$ yd., $37\frac{7}{8}$ yd., $43\frac{2}{5}$ yd., $26\frac{6}{7}$ yd. How many yards of fencing will be needed to enclose it?

45. If the mineral matter of the organs of the body is : bones $\frac{11}{50}$, muscles $\frac{3}{200}$, lungs $\frac{11}{1000}$, brain $\frac{1}{100}$, how much more mineral matter is there in bone than in each of the other organs given?

46. If a boy of 10 years needs daily $1\frac{2}{5}$ grams of protein, $\frac{3}{4}$ gram of fat, and $4\frac{1}{2}$ grams of carbohydrates for each pound of weight, how much of each will a boy of 10 weighing 69 lb. require?

47. A lot is $54\frac{1}{4}$ feet wide by $86\frac{2}{3}$ feet deep. How many rods ($16\frac{1}{2}$ ft. to a rod) of wire will be needed to fence the lot?

CHAPTER VII

DECIMAL FRACTIONS

252. What is decimal division?

Division of units into tenths, hundredths, thousandths, etc.

EXAMPLES: $\frac{5}{10}$ $\frac{25}{100}$ $\frac{536}{1000}$

253. What is a decimal fraction?

The part of a unit obtained by decimal division. Decimal fractions are often called decimals. It is a fractional value expressed in tenths, hundredths, thousandths, etc. This means that the denominator is 10 or some multiple of 10.

254. What do we call the decimal point?

The period placed at the left of tenths, hundredths, etc.

EXAMPLES:

$\frac{3}{10} = 0.3$ (three tenths)

$\frac{7}{100} = 0.07$ (seven hundredths)

$\frac{5}{1000} = 0.005$ (five thousandths)

255. How may decimal fractions be expressed?

(*a*) By the position of the decimal point.

(*b*) By a decimal denominator in the form of a common fraction.

Ex. (*a*) 0.2 0.7 0.08 0.24 .017

Ex. (*b*) $\frac{2}{10}$ $\frac{7}{10}$ $\frac{8}{100}$ $\frac{24}{100}$ $\frac{17}{1000}$

256. What are the names of the decimal places and how are decimals written?

Units	Decimal point		Hundredths		Ten thousandths		Millionths
↓	↓		↓		↓		↓
4	.	2	8	7	5	8	3
		↑		↑		↑	
		Tenths		Thousandths		Hundred thousandths	

EXAMPLES:

To express tenths, one place is pointed off, as .2.

To express hundredths, two places are pointed off, as .28.

To express thousandths, three places are pointed off, as .287.

To express ten thousandths, four places are pointed off, as .2875.

Read above: "Four and two hundred eighty-seven thousand five hundred eighty-three millionths."

257. How is a decimal read?

The decimal point is read "and." Read a decimal exactly as if it were a whole number, and then add the fractional name of the lowest place.

EXAMPLE: 5.631,056,923.

Read: "Five and six hundred thirty-one million fifty-six thousand nine hundred twenty-three billionths." The lowest decimal place here is billionths.

258. What is the relation of the number of figures in a decimal to the number of zeros in its denominator when expressed as a common fraction?

They are the same.

Ex. (*a*) 0.345 has three figures; therefore, $\frac{345}{1000}$ has three zeros in the denominator.

Ex. (*b*) 0.01679 has five figures; therefore, $\frac{1679}{100,000}$ has five zeros in the denominator.

259. Is the value of a decimal fraction changed by adding or omitting zeros on the right? No.

EXAMPLE: $.4 = .40 = .400$. Also, $\frac{4}{10} = \frac{40}{100} = \frac{400}{1000}$

Adding zeros to the right does not change the value.

260. What is the effect on decimal fractions of moving the decimal point to the left?

Moving the point one place to the left *divides* the decimal by 10; two places divides it by 100; three places divides it by 1,000, etc.

EXAMPLES: $\frac{4}{10} = .4$ $\quad \frac{4}{100} = .04$ $\quad \frac{4}{1000} = .004$

The decimal point is moved to the left for division by 10's, to make the decimal smaller.

261. What is the effect of moving the decimal point to the right?

Moving the point one place to the right *multiplies* the decimal by 10; two places by 100; three places by 1,000, etc.

EXAMPLES: $\frac{4}{1000} = .004$ $\quad \frac{4}{100} = .04$ $\quad \frac{4}{10} = .4$

The decimal point is moved to the right for multiplication by 10's, to make the decimal larger.

262. What must be done when there is not a sufficient number of figures in the numerator to indicate the denominator of a decimal fraction?

Zeros are placed between the decimal point and the figure or figures in the numerator.

Ex. (*a*) To write nine hundredths as a decimal, place a zero between the 9 and the decimal point; otherwise the fraction would be nine tenths.

$$\tfrac{9}{100} = .09$$

Place sufficient zeros to the right of the decimal point to make up as many figures in the numerator as there are zeros in the denominator when the fractional value is written as a common fraction.

Ex. (*b*) To write $\frac{17}{100,000}$, note that the denominator has five zeros; therefore, the numerator must have five figures to the right of the decimal point. It already has two figures, so add three zeros to the right of the decimal point, or

$$.00017 = \tfrac{17}{100,000}$$

263. How are decimals classified?

(*a*) A *simple* decimal has a whole number to the right of the decimal point, as .0483, .86, .356.

(*b*) A *complex* decimal has a whole number and a common fraction written to the right of the decimal point, as $.07\frac{1}{4}$, $.495\frac{3}{5}$, $.1478\frac{5}{7}$.

264. Do we need a decimal point after every whole number?

No. The decimal point is understood as at the right of the units place.

EXAMPLE: 6 = 6. = 6.0 = 6.00

265. How do we divide any number by a decimal number?

Shift the decimal point one place to the *left* for every zero in the divisor.

EXAMPLES:

(*a*) 132 ÷ 10 = 13.2. One zero in divisor. Move 1 place to left.
(*b*) 132 ÷ 100 = 1.32. Two zeros in divisor. Move 2 places to left.
(*c*) .132 ÷ 10 = .0132. Move 1 place to left.
(*d*) .132 ÷ 100 = .00132. Move 2 places to left.

266. How do we multiply any number by a decimal number?

Shift the decimal point one place to the *right* for every zero in the multiplier.

EXAMPLES:

(*a*) $132 \times 10 = 1{,}320.$ Shift 1 place to right.
(*b*) $132 \times 100 = 13{,}200.$ Shift 2 places to right.
(*c*) $132 \times 1{,}000 = 132{,}000.$ Shift 3 places to right.
(*d*) $.132 \times 10 = 1.32.$ Shift 1 place to right.
(*e*) $.132 \times 100 = 13.2.$ Shift 2 places to right.
(*f*) $.132 \times 1{,}000 = 132.$ Shift 3 places to right.
(*g*) $.132 \times 10{,}000 = 1{,}320.$ Shift 4 places to right.

267. What is a mixed number in decimal form and how do we multiply and divide it by a decimal?

A number that consists of a whole number and a decimal fraction, as 132.465.

The same rules apply as above.

EXAMPLES:

(*a*) $132.465 \times 10 = 1{,}324.65.$ Move 1 place to right.
(*b*) $132.465 \times 100 = 13{,}246.5.$ Move 2 places to right.
(*c*) $132.465 \div 10 = 13.2465.$ Move 1 place to left.
(*d*) $132.465 \div 100 = 1.32465.$ Move 2 places to left.

268. How can we change a common fraction to a decimal?

Annex zeros to the numerator and divide by the denominator.

EXAMPLES:

(*a*) $\frac{1}{5} = \frac{1.0}{5} = .2 \quad \text{or} = 5\overline{)1.0}$ quotient $.2$

(*b*) $\frac{1}{4} = \frac{1.00}{4} = .25 \quad \text{or} = 4\overline{)1.00}$ quotient $.25$

(*c*) $\frac{3}{4} = \frac{3.00}{4} = .75 \quad \text{or} = 4\overline{)3.00}$ quotient $.75$

(*d*) $\frac{3}{8} = \frac{3.000}{8} = .375 \quad \text{or} = 8\overline{)3.000}$ quotient $.375$

(*e*) $\frac{4}{7} = \frac{4.000}{7} = .571\frac{3}{7}$

(*f*) $\frac{5}{9} = \frac{5.0000}{9} = .5555\frac{5}{9}$

When the result is a complex decimal, two places are usually far enough to carry out the decimal. For most purposes three or four places will suffice.

269. How can we extend a complex decimal?

Add zeros to the numerator of the fraction and divide by the denominator. When the division comes out even, the fraction is thereby removed; otherwise the decimal may be extended as many places as are desired.

Ex. (*a*) Extend the complex decimal $.9\frac{5}{8}$.

$$\begin{array}{r} .625 \\ 8\overline{)5.000} \end{array}$$

Add three zeros to the numerator 5 and divide by denominator 8.

Ans. = .9625. The division came out even.

Ex. (*b*) Extend $.39\frac{5}{12}$ to 6 decimal places.

$$\begin{array}{r} 12\underline{)5.0000} \\ .4166 \end{array}$$

Add four zeros to the 5 and divide by 12.

Ans. = .394166 = six decimal places.

270. How can we convert a decimal expression to a common fraction?

Express the decimal as a numerator over a denominator and reduce to lowest terms. The denominator is a multiple of 10 as indicated by the decimal point. The numerator is a whole number.

Ex. (*a*) Change .5 to a common fraction.

The decimal point indicates 10 as the denominator. Thus, $.5 = \frac{5}{10} = \frac{1}{2}$ reduced to lowest terms.

Ex. (*b*) Change .125 to a common fraction.

tenths, hundredths, thousandths

Denominator is 1,000. Thus

$$.125 = \frac{125}{1{,}000} = \frac{\cancel{5} \times \cancel{25}}{\cancel{5} \times 8 \times \cancel{25}} = \frac{1}{8}$$

reduced to lowest terms.

Ex. (*c*) Change .5736 to a common fraction.

There are four places to the right of the decimal point; therefore, there are four zeros in the denominator. Thus,

$$.5736 = \frac{5{,}736}{10{,}000} = \frac{\cancel{8} \times 717}{\cancel{8} \times 1{,}250} = \frac{717}{1{,}250} \text{ Ans.}$$

271. What is the procedure for adding whole numbers and simple decimals?

Place the numbers in columns with the decimal points directly under one another, and add in the usual way. The decimal point of the sum is directly under the points in the column.

EXAMPLE: Add 263.874, 52.0953, 728.3, and 9.35.

$$\begin{array}{r} 263.8740 \\ 52.0953 \\ 728.3000 \\ 9.3500 \\ \hline 1{,}053.6193 \end{array} \text{ Ans.}$$

Adding zeros at the right of the decimal does not affect the value.

272. What is the procedure for adding whole numbers and complex decimals?

Extend the complex decimals the same number of places, and then add in the usual way.

$$\begin{array}{r} 14.4\frac{1}{6} \\ 19.78\frac{1}{3} \\ 495.6\frac{2}{3} \\ 67.06\frac{2}{9} \end{array} \quad \text{becomes} \quad \begin{array}{r} 14.4166\frac{2}{3} \\ 19.7833\frac{1}{3} \\ 495.6666\frac{2}{3} \\ 67.0622\frac{2}{9} \\ \hline 596.9287\frac{17}{9} \end{array} = 596.9288\frac{8}{9}$$

273. What is the procedure for subtracting simple decimals?

Place the decimal point in the subtrahend directly under the decimal point in the minuend and subtract as usual. The decimal point of the remainder is directly under the points above it.

EXAMPLE: Subtract 52.0953 from 728.3.

$$\begin{array}{rrl} \text{From} \rightarrow & 728.3000 & = \text{minuend} \\ \text{Subtract} \rightarrow & 52.0953 & = \text{subtrahend} \\ \hline & 676.2047 & = \text{remainder} \end{array}$$

274. What is the procedure for subtracting a simple decimal and a complex decimal?

Extend the shorter complex decimal until the fraction is removed or there are the same number of places in the minuend and subtrahend, and then subtract in the usual way.

EXAMPLE: From $67.2\frac{1}{16}$ subtract $24.86\frac{4}{15}$

$$\begin{array}{rcll} 67.2\frac{1}{16} & = & 67.2062\frac{1}{2} & \text{Extend decimal to 4 places.} \\ 24.86\frac{4}{15} & = & \underline{24.8626\frac{2}{3}} & \text{Extend decimal to 4 places.} \\ & & 42.3435\frac{5}{6} & \end{array}$$

275. What is the procedure for multiplying simple decimals?

Multiply in the usual way and point off in the product as many places as there are places in both the multiplier and multiplicand.

Ex. (*a*) Multiply .38 by .6.

$$\begin{array}{ll} .38 & 2 \text{ places} \\ \underline{.6} & \underline{1 \text{ place}} \\ .228 \leftarrow & 3 \text{ places} \end{array}$$

$$\text{Proof: } \frac{38}{100} \times \frac{6}{10} = \frac{228}{1000} = .228$$

Ex. (*b*) .043 × .7.

$$\begin{array}{llcl} 3 \text{ places} + & 1 \text{ place} & = & 4 \text{ places} \\ .043 \times & .7 & = & .0301 = 4 \text{ places in product} \end{array}$$

Ex. (*c*) Multiply 93.4216 by 47.897.

$$\begin{array}{rl} 93.4216 & \leftarrow 4 \text{ places in multiplicand} \\ \underline{47.897} & \leftarrow 3 \text{ places in multiplier} \\ 6539512 & \\ 8407944 & \\ 7473728 & \\ 6539512 & \\ \underline{3736864} & \\ 4{,}474.6143752 & \leftarrow 7 \text{ places in product} \end{array}$$

276. What is the procedure for multiplying complex decimals?

Extend the decimal to remove the fraction when it can be done, or change to improper fractions.

EXAMPLE: Multiply $7.21\frac{1}{2}$ by $5.2\frac{1}{4}$.

$$\begin{array}{lcrl} 7.21\frac{1}{2} & = & 7.215 & \leftarrow 3 \text{ places} \\ 5.2\frac{1}{4} & = & \underline{5.225} & \leftarrow 3 \text{ places} \\ & & 36075 & \\ & & 14430 & \\ & & 14430 & \\ & & \underline{36075} & \\ & & 37.698375 & \leftarrow 6 \text{ places} \end{array}$$

Also, $7.21\frac{1}{2} = \frac{14.43}{2}$ as an improper fraction

$5.2\frac{1}{4} = \frac{20.9}{4}$ as an improper fraction

$$\therefore \frac{14.43}{2} \times \frac{20.9}{4} = \frac{301.587}{8} = 37.698\tfrac{3}{8} \quad \text{or} \quad 37.698375$$

277. What is the procedure for dividing one simple decimal by another?

The terms in a division are:

$$\frac{\text{dividend}}{\text{divisor}} = \text{quotient} \quad \text{or} \quad \text{divisor}\overline{)\text{dividend}}^{\;\text{quotient}}$$

(1) The *divisor* must be made a whole number by moving the decimal point to the extreme right (or the end of the number). Count the number of places you moved the point.

(2) Move the decimal point in the dividend an equal number of places. If the dividend is a whole number then add as many zeros instead, and place the point at the end.

(3) Place the decimal point in the quotient just above the point in the dividend.

Remember that a decimal point is understood after every whole number.

Ex. (*a*) Divide 192 by .06.

$$\frac{192}{.06} = .06\overline{)192} = .06.\overline{)192.00.} = 6\overline{)19{,}200}^{\;3{,}200}$$

Six one-hundredths is contained in 192 thirty-two hundred times.

Proof: Multiply 3,200 by .06 (2 places).

$$3{,}200 \times .06 = 192.00 \quad \text{(2 places)}$$

Ex. (*b*) What is the result of dividing .06118 by 14?

$$\frac{.06118}{14} = 14\overline{).06118}^{\;.00437}$$

```
     .00437
14).06118
      56
       51
       42
        98
        98
         0
```

The decimal point in the quotient is always directly above the decimal point in the dividend.

Ex. (*c*) Divide 403.0496 by 4.78.

$$\frac{403.0496}{4.78} = 4.78.\overline{)403.04.96}$$

move two places

```
          84.32 (= quotient)
    478)40304.96
        3824
         2064
         1912
          1529
          1434
           956
           956
             0
```

278. What is the procedure for dividing one complex decimal by another?

Change the complex decimals to simple decimals, if possible, and then divide; otherwise multiply both numbers by the L.C.D. of the denominators of the fractions before you divide.

EXAMPLE: Divide $45.58\frac{1}{2}$ by $2.4\frac{2}{3}$ (L.C.D. = 6).

$45.58\frac{1}{2}$	$2.4\frac{2}{3}$
6	6
273.51	14.8

$$18.48\tfrac{6}{148} = 18.48\tfrac{3}{74} \text{ Ans.}$$

```
14.8.)273.5.1 = 148)2735.10
                    148
                    1255
                    1184
                      711
                      592
                      1190
                      1184
                         6
```

279. How is a decimal number shortened for all practical purposes?

If a rejected or discarded decimal is 5 or over, 1 is added to the next figure to the *left*.

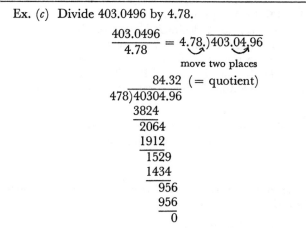

EXAMPLE: 4,474.6143752 can be shortened to 4,474.6144, which is considered to be correct to 4 decimal places (or four significant figures). Since the fifth place, which is 7, is greater than 5, then 1 is added to the number to the left of it, 3, which becomes 4.

Now in 4,474.6144 the fourth place is 4. This is less than 5 and is dropped, leaving 4,474.614, which is said to be correct to three decimal places.

4,474.61 is correct to 2 decimal places.
4,474.6 is correct to 1 decimal place.

280. What other method of decimal approximation has been internationally approved?

That of making the decimal even.

Ex. (*a*) 48.655 is shortened to 48.66.

The last 5 is dropped and 1 is added to the 5 to its left to make the decimal even.

Ex. (*b*) 48.645 is shortened to 48.64.

Since 4 is an even number, you merely drop the 5. It is claimed that a closer average result is obtained when a decimal is made even.

281. What is the least number of significant figures that must be kept when the decimal is purely fractional and contains a number of zeros to the right of the decimal point?

At least one significant figure must be kept.

EXAMPLE: .000072184 may be shortened to .00007.

282. What is the result of 0.3024 × 0.196, correct to 2 significant figures?

```
    0.3024 ← 4 places
    0.196  ← 3 places
    ------
     18144
    27216
    3024
 ---------
 0.0592704 ← 7 places
```

One can tell at once that 0.06 is correct to 2 places (by adding 1 to the 5 to get 6 because 9 is so large).

283. Why is it the rule to work a problem to one more decimal place than we need?

It helps us to determine whether the next figure would be greater or less than 5, and enables us to know whether or not the figure we use is sufficiently accurate.

284. What can we do to simplify things when we want to get an answer correct to two decimal places, in multiplying 4.879 by 3.765?

There is no need to go through the multiplication of the entire numbers.

If we were to multiply 5 × 4 (= 20) we thus drop all decimals, and we guess our answer to be somewhat less than 20. This gives us *no* decimal places.

Now 4.9 × 3.8 = 18.62. If we retain one decimal place in the multiplier and multiplicand, we get an answer with two decimal places, but we are not sure of the .62.

So our rule says to retain one more place than required, and we get 4.88 × 3.77 = 18.3976, or 18.40, approximately correct to 2 places.

The complete multiplication would be:

$$4.879 \times 3.765 = 18.369435.$$

We see that this lengthy multiplication is not justified.

285. What is another way of approximating the desired result involving decimals?

Contracted multiplication. Since the figures to the left of the decimal point are most important,

(1) Multiply all of the multiplicand by the left-hand digit of the multiplier.

(2) Drop right digit of multiplicand and multiply remainder by next figure of multiplier.

(3) Drop two digits of multiplicand and multiply remainder by next figure of multiplier.

(4) Continue successively dropping one more digit of multiplicand each time you multiply by another figure of the multiplier.

EXAMPLE:

```
  4.879                                            4.879 × 3.765
  3.765                                              = 18.369435
 -------                                             = 18.37
 14 637 → (3 × 4.879)
  3 409 → (7 × 4.87)   (drop 9 of multiplicand)
    288 → (6 × 48)     (drop 79 of multiplicand)
     20 → (5 × 4)      (drop 879 of multiplicand)
 -------
 18.354 = 18.35  (a close approximation)
```

286. What is a recurring decimal?

When in some cases of decimal division the calculation can be carried on indefinitely with repeating numbers or sets of numbers, such a decimal is known as a recurring decimal.

EXAMPLES:

$$\begin{aligned}
&(a)\ \tfrac{1}{3} = 1 \div 3 = .333333\cdots = \text{recurring decimal}\\
&(b)\ \tfrac{1}{6} = 1 \div 6 = .166666\cdots = \text{recurring decimal}\\
&(c)\ \tfrac{1}{9} = 1 \div 9 = .111111\cdots = \text{recurring decimal}
\end{aligned}$$

287. How are recurring, circulating, or repeating decimals denoted?

(*a*) By a dot over the recurring figure. Thus $.40\dot{4}$ means $.404444\cdots$ etc., to infinity.

(*b*) By dots placed over the first and last figures of the recurring group. Thus,

$$.\overset{\frown}{\dot{4}\dot{8}} \quad \text{means} \quad .484848, \text{ etc.} \quad (48 \text{ recurs})$$

$$.\overline{\dot{6}47\dot{4}} \quad \text{means} \quad .64746474, \text{ etc.} \quad (6474 \text{ recurs})$$

$$\tfrac{1}{7} = 1 \div 7 = .\overline{\dot{1}4285\dot{7}} \quad \text{means} \quad .142857142857, \text{ etc.}$$

288. How can we convert pure recurring decimals to fractions?

Use nines in the denominator—one 9 for every decimal place in the recurring group.

EXAMPLES:

(*a*) Recurring decimal $.\dot{3} = \frac{3}{9} = \frac{1}{3}$ = fraction.

(*b*) Recurring decimal $.\overline{\dot{1}4285\dot{7}} = \frac{142857}{999999} = \frac{1}{7}$
($142857 \times 7 = 999999$)

Note that a pure recurring decimal is one in which all the digits recur.

289. How can we convert mixed recurring decimals to fractions?

In a mixed recurring decimal the decimal point is followed by some figures which do not recur.

(1) Subtract the nonrecurring figures from all the figures and make the result the numerator.

(2) The denominator consists of as many nines as there are recurring figures, followed by as many zeros as nonrecurring figures.

EXAMPLES:

$$(a)\ .1\dot{6} = \frac{16 - 1}{90} = \frac{15}{90} = \frac{1}{6}$$

$$(b)\ .5\dot{7}\dot{5} = \frac{575 - 5}{990} = \frac{570}{990} = \frac{19}{33}$$

$$(c)\ .24\dot{7}5\dot{6} = \frac{24756 - 24}{99900} = \frac{6183}{24975}$$

$$(d)\ .16\dot{5}3\dot{2} = \frac{16532 - 16}{99900} = \frac{16516}{99900} = \frac{4129}{24975}$$

$$(e)\ .3\dot{9} = \frac{39 - 3}{90} = \frac{36}{90} = \frac{2}{5}$$

290. Why in particular should you know the decimal equivalents of $\frac{1}{2}$, $\frac{1}{4}$, $\frac{1}{8}$, $\frac{1}{16}$, $\frac{1}{32}$, and $\frac{1}{64}$?

$$\tfrac{1}{2} = .5 \qquad \tfrac{1}{4} = .25 \qquad \tfrac{1}{16} = .0625$$
$$\tfrac{1}{32} = .03125 \qquad \tfrac{1}{64} = .015625$$

It is then simple to find other fractional equivalents in this series. Thus,

$$\tfrac{3}{64} = 3 \times .015625 = .046875$$
$$\tfrac{1}{8} = \tfrac{1}{2} \times \tfrac{1}{4} = \tfrac{1}{2} \times .25 = .125$$

291. How can we sometimes produce a decimal equivalent by multiplying both numerator and denominator by a suitable number?

$$\frac{3}{25} = \frac{3 \times 4}{25 \times 4} = \frac{12}{100} = .12$$ (multiply numerator and denominator by 4)

$$\frac{21}{125} = \frac{21 \times 8}{125 \times 8} = \frac{168}{1{,}000} = .168$$ (multiply numerator and denominator by 8)

292. How do we find the whole number when a decimal part of it is given?

Ex. (*a*) 56 is .8 of what number?

$$\frac{56}{.8} = .8\overline{)56} = .8_{\curvearrowright}\overline{)56.0_{\curvearrowright}} = 8\overline{)560}\ \ 70 \text{ Ans.}$$

Ex. (*b*) If .4 of a number is 64, what is the number?

$$\frac{64}{.4} = \frac{64.0}{.4} = \frac{640}{4} = 160 \text{ Ans.}$$

293. How is United States money related to decimal fractions?

The unit is the dollar expressed by the sign \$, as \$15 = fifteen dollars. Dollars may be divided into tenths, hundredths, and thousandths:

$\$\frac{1}{10}$ = 10¢ = \$.10 = one dime (ten-cent piece)
$\$\frac{1}{100}$ = 1¢ = \$.01 = 1 cent = 10 mills
$\$\frac{1}{1000}$ = 1 mill = .001 = 1 mill

$\$\frac{3}{10}$ = 30¢ = \$.30 = 3 dimes (3 ten-cent pieces)
$\$\frac{3}{100}$ = 3¢ = \$.03 = 3 cents = 30 mills
$\$\frac{3}{1000}$ = 3 mills = \$.003 = 3 mills

$\$\frac{10}{10}$ = 100¢ = \$1.0 = ten dimes (10 ten-cent pieces)
$\$\frac{100}{100}$ = 100¢ = \$1.0 = 100 cents = 1,000 mills
$\$\frac{1000}{1000}$ = 1,000 mills = \$1.0 = 1,000 mills

294. If a British pound (£) is worth \$2.80 and there are 20 shillings to the pound and 12 pence to the shilling, how much is (*a*) 1 shilling worth, (*b*) 1 penny worth?

Remember: If you want to get the value of one unit of any element in a problem you should divide by that element.

(*a*) You want to find the value of 1 shilling; then divide by shillings

$$\frac{\$2.80}{20 \text{ shillings}} \quad (= \text{Two dollars and eighty cents})$$

Divide numerator and denominator by 10, or what is the same thing, move the decimal point 1 place to the left in numerator and denominator.

$$\frac{\$.2.80}{2.0} = \frac{\overset{.14}{\$.28\not{0}}}{\underset{1}{\not{2}.0}} = \$.14 = 14¢ \text{ per shilling}$$

(*b*) $$\frac{\$.14}{12 \text{ pence}} = \frac{\$.07}{6} = \$.01\tfrac{1}{6} = 1\tfrac{1}{6}¢ \text{ per penny}$$

295. A manufacturer submitted a bid to the United States government for military insignia in the sum of \$68,399.70

at 31 cents $7\frac{2}{5}$ mills per dozen. How many dozen would be delivered?

31 cents $7\frac{2}{5}$ mills = \$.3174

Ans. = 215,500 dozen

```
             215,500
.3174.)68,399.7000.
       6348
        4919
        3174
        17457
        15870
         15870
         15870
             0
```

PROBLEMS

1. Write in decimal form:

(*a*) Six tenths, four tenths, two and one tenth.

(*b*) Seven and nine thousandths, nine and fifty-three thousandths, three ten-thousandths, eleven millionths

(*c*) One hundred fifty-five thousandths, four hundred ninety-two thousandths, six ten-thousandths, three hundred and four hundredths

(*d*) Six and seven tenths, eight and two tenths, eighty-six hundredths, five hundred and six thousandths

(*e*) Four and three-eighths hundredths, thirty-six and five-sevenths thousandths, eight and two-thirds of a thousandth, eight and four and two-thirds thousandths

2. Write the following fractions as decimal fractions: $\frac{3}{10}$, $\frac{9}{10}$, $\frac{6}{10}$, $\frac{6}{100}$, $\frac{2}{100}$, $\frac{8}{1000}$, $\frac{7}{10000}$.

3. Read 12.584,062,018.

4. Distinguish between 0.400 and 0.00004.

5. What is the denominator of .456, .02763 expressed in fraction form?

6. Express as common fractions 0.25, 0.250, 0.2500.

7. Annexing a cipher to a whole number increases its value how many times?

8. Does annexing a cipher to a decimal affect its value?

9. Select the quantities that have the same value in the following:

(*a*) .08 .8 .008 .80 .800 8.0

(*b*) .046 .460 .0460 .0046 .04600 4.600

(*c*) .738 .7380 .73800 .00738 0.738 .0738

10. Arrange the following in ascending values:

2.60 .260 .026 260 26.0

11. Move the decimal point in .4 so as to make the decimal smaller by $\frac{1}{10}$, by $\frac{1}{100}$.

12. Move the decimal to multiply .004 by 10, by 100, by 1,000.

13. Divide 246 by 10, by 100.

14. Divide .246 by 10, by 100.

15. Multiply 246 by 10, by 100, by 1,000.

16. Multiply .246 by 10, by 100, by 1,000, by 10,000.

17. Multiply 246.576 by 10, by 100.

18. Divide 246.576 by 10, by 100.

19. Change the following to decimals:

(*a*) $\frac{8}{25}$ (*b*) $\frac{5}{8}$ (*c*) $\frac{7}{50}$ (*d*) $\frac{49}{125}$
(*e*) $\frac{15}{80}$ (*f*) $\frac{13}{20}$ (*g*) $\frac{2}{5}$ (*h*) $\frac{7}{40}$
(*i*) $\frac{5}{16}$ (*j*) $\frac{23}{200}$ (*k*) $\frac{15}{32}$ (*l*) $\frac{29}{125}$

20. Extend the complex decimals:

(*a*) $.7\frac{3}{8}$ (*b*) $.56\frac{7}{12}$ (*c*) $.6\frac{1}{16}$ to 5 places (*d*) $.19\frac{5}{12}$ to 6 places

21. Change to common fractions:

(*a*) .6 (*b*) .86 (*c*) .625 (*d*) .1875
(*e*) .0125 (*f*) .750 (*g*) .4765 (*h*) $.09\frac{1}{3}$

22. Add:

(*a*) 7.486, 6.9, 22.536, 24.5, and 62.86
(*b*) 48.6, 65.236, 68.036, 4.3986, and 25.7
(*c*) 374.985, 63.0964, 839.4, and 8.24

23. Add:

(*a*) $4.8\frac{7}{8}$, $25.4\frac{1}{8}$, $34.77\frac{3}{5}$, and $62.39\frac{3}{4}$
(*b*) $9.22\frac{1}{6}$, $84.95\frac{1}{3}$, $78.036\frac{1}{12}$, and $9.04\frac{1}{2}$
(*c*) $48.93\frac{2}{25}$, $54.09\frac{2}{3}$, $23.79\frac{5}{8}$, and $94.53\frac{1}{6}$
(*d*) $25.5\frac{1}{3}$, $29.89\frac{1}{6}$, $586.7\frac{2}{9}$, and $78.07\frac{2}{3}$

24. Subtract:

(*a*) 63.0842 from 839.4 (*b*) 28.84 from 49.836
(*c*) 49.486 from 239.57 (*d*) 81.564 from 128.096
(*e*) 148.9736 from 197.134 (*f*) .3874 from 4

25. (*a*) From $78.3\frac{3}{16}$ subtract $35.97\frac{7}{15}$
(*b*) From $8.8\frac{3}{8}$ subtract $4.34\frac{5}{16}$
(*c*) From $7.44\frac{5}{12}$ subtract $3.6\frac{1}{8}$
(*d*) From $5.89\frac{1}{3}$ subtract $2.35\frac{1}{12}$
(*e*) From $89.7\frac{5}{6}$ subtract $34.45\frac{1}{4}$

26. Multiply:

(*a*) .49 by .7 (*b*) .054 by .8 (*c*) 84.5327 by 58.986
(*d*) 12.32 by 9.8736 (*e*) 18.4236 by 4.9

27. Multiply:

(*a*) $8.32\frac{1}{2}$ by $6.3\frac{1}{4}$ (*b*) $.748\frac{1}{4}$ by $3.45\frac{1}{8}$ (*c*) $8.63\frac{1}{6}$ by $4.5\frac{1}{3}$
(*d*) $4.59\frac{1}{12}$ by $.43\frac{5}{6}$ (*e*) 6.836 by $.09\frac{1}{3}$

28. Divide:

(*a*) 283 by .07 (*b*) .07229 by 16 (*c*) 504.0587 by 5.89
(*d*) 48.735 by 6.486 (*e*) 645.75 by 16.5 (*f*) .9686 by 136

29. Divide:

(*a*) $56.69\frac{1}{2}$ by $3.5\frac{2}{3}$ (*b*) $79.45\frac{1}{3}$ by $2.04\frac{1}{3}$ (*c*) $95.47\frac{1}{4}$ by $5.4\frac{1}{2}$
(*d*) $593.19\frac{2}{3}$ by $5.27\frac{1}{3}$ (*e*) $239\frac{5}{6}$ by $4.3\frac{2}{3}$ (*f*) 99.57 by $4.7\frac{1}{4}$

30. Shorten 5785.7254863 to be correct to:

(*a*) 4 decimal places (*b*) 3 decimal places (*c*) 2 decimal places
(*d*) 1 decimal place

31. Shorten: (*a*) 59.767 (*b*) 59.755

32. Shorten .0000083273 to the least number of significant figures.

33. Find the result of 0.4035×0.287 correct to 2 significant figures.

34. Get the result of 5.987×4.876 correct to 2 decimal places by shortened multiplication.

35. Get the approximate result of 5.987×4.876 by contracted multiplication.

36. Convert the following recurring decimals to fractions:
(*a*) $.\dot{4}$ (*b*) $.\dot{6}$ (*c*) $.\dot{7}$ (*d*) $.\dot{1}5396\dot{8}$ (*e*) $.1\dot{7}$ (*f*) $.6\dot{8}\dot{6}$ (*g*) $35\dot{8}6\dot{7}$
(*h*) $.17\dot{6}4\dot{3}$ (*i*) $.4\dot{9}$

37. What is the decimal equivalent of: (*a*) $\frac{5}{64}$? (*b*) $\frac{5}{32}$? (*c*) $\frac{3}{8}$? (*d*) $\frac{5}{16}$? (*e*) $\frac{9}{32}$? (*f*) $\frac{11}{64}$? (*g*) $\frac{7}{25}$? (*h*) $\frac{23}{125}$?

38. (*a*) 78 is .7 of what number?
(*b*) If .6 of a number is 86, what is the number?
(*c*) 81 is .9 of what number?
(*d*) 99 is .75 of what number?

39. If the British pound (£) is worth $2.80 and there are 20 shillings to the pound, how much are three shillings worth? If there are 12 pence to a shilling, how much is sixpence worth?

40. If the total cost of a shipment is $79,488.65 at 43¢, $8\frac{3}{4}$ mills per dozen items, what is the number of dozens in the shipment?

41. If a family found that at the end of the year it had saved $455, and during the year it had spent .29 of its income for food, .17 for rent, .25 for clothing, and .21 for miscellaneous items, what was the amount of its income?

42. In a college the registration was .33 in pure science courses, .26 in liberal arts, .21 in social science, and the remainder in engineering. The number of students in engineering was 520. What is the total registration of the college? How many students in each category?

43. A man invests .22 of his money in bonds, .32 in common stocks, .36 in real estate, and he has $3,400 in cash left over. How much is his total equity? How much has he in each category?

44. Specifications for phosphor bronze require .86 copper, .065 tin, .0007 iron, .002 lead, .0035 phosphorus, and the remainder zinc. How many lb. of each element are required to make 1,200 lb. of phosphor bronze?

45. A farmer sold 8,460 pounds of apples (each bushel weighing 60 lb.) for \$1.80 a bushel. With the proceeds he bought 9,000 lb. of fertilizer. What is the cost of the fertilizer per 100 lb.?

46. The distance round a wheel is 3.1416 times its height. What is the distance round a wheel 3.85 feet high? Round a 32-inch high wheel?

47. If 100 lb. of milk yield 5.563 lb. of butter, and a gallon of milk weighs 8.7 lb., how much butter will 2 gallons of milk yield?

48. What is the cost of a railroad ticket at \$.045 a mile if the distance you are to travel is 475 miles?

49. If 6,370 pieces of cutlery cost \$753.69 to manufacture, what is the cost of each in cents and mills?

50. If you made \$260 on an investment of \$4,000, what fractional part of the investment did you make?

51. If 2 lb. of coffee cost \$1.65, how many lb. can you buy for \$26.40?

52. If you bought six \$1,000 bonds for $96\frac{3}{4}$ and sold them for $99\frac{5}{8}$; (*a*) what is the total amount paid for the bonds? (*b*) the amount received for them? (*c*) the profit? (*d*) the profit expressed decimally in thousandths? (Note: $96\frac{3}{4}$ means \$$96\frac{3}{4}$ on each \$100 of the bond or \$967.50 for each bond.)

53. Two ball teams, A and B, each having played 46 games, have a respective standing of .826 and .739. If A wins only 4 of the next 10 games and B wins 6 of the next 10 games, how will the clubs stand?

CHAPTER VIII

PERCENTAGE

296. What is meant by (*a*) per cent, (*b*) percentage?

(*a*) Per cent means "by the hundred," the number of hundredths of a number. In Latin *per centum* means "by the hundred."

EXAMPLE: If out of 100 students, 30 failed in the final examination, then 30 per cent failed and 70 per cent passed.

(*b*) Percentage means "by hundredths," and includes the process of computing by hundredths. In dealing with percentage, we are thus working with decimals whose denominator is 100.

EXAMPLE: 30 per cent $= \frac{30}{100} = .30 =$ 30 hundredths.

297. What is the symbol used to represent the denominator 100?

The term per cent is expressed by the sign [%].

EXAMPLES:

(*a*) 30 per cent $= 30\% = \frac{30}{100} = .30$

(*b*) $18\% = \frac{18}{100} = .18$

(*c*) $3\frac{1}{2}\% = \dfrac{3\frac{1}{2}}{100} = .03\frac{1}{2} = .035$

(*d*) $\frac{1}{4}\% = \dfrac{\frac{1}{4}}{100} = .00\frac{1}{4} = .0025$

(*e*) $2.6\% = \dfrac{2.6}{100} = \dfrac{26}{1{,}000} = .026$

(*f*) $100\% = \frac{100}{100} = 1$

(*g*) $130\% = \frac{130}{100} = 1.30$

The per cent sign [%] takes the place of the fraction line and the denominator 100.

298. In what ways may a given per cent, or a given number of hundredths, of a number be expressed?

(*a*) As a whole number, 6%.

(*b*) As a decimal, .06.

(*c*) As a fraction, $\frac{6}{100}$.

299. When do we express quantities as percentages?

When we wish to compare two quantities which are not easily commensurable, it is more convenient to express them as percentages.

EXAMPLE: It is obvious that 4% of a quantity is greater than $3\frac{3}{4}\%$, while it is not so apparent that 268 is a greater proportion of 6,700 than 315 of 8,400.

300. How do we reduce a number written with a per cent sign to a decimal?

Drop the per cent sign and move the decimal point two places to the left. This is equivalent to dividing by 100, which is the meaning of per cent. *Dropping the % means dividing by 100.*

EXAMPLES:

(*a*) 35% = .35 (move 2 places to left to divide by 100)
(*b*) 135% = 1.35 (move 2 places to left to divide by 100)

301. How do we convert to a decimal when the per cent is expressed as a number and a fraction?

Carry out the fraction in order to convert it to a decimal.

EXAMPLES:

(*a*) $37\frac{1}{2}\% = 37\% + \frac{1}{2}\%$

$37\% = .37 \qquad \frac{1}{2}\% = \frac{.5}{100} = .005$ (carry out $\frac{1}{2}\%$)

$\therefore\ 37\frac{1}{2}\% = .37 + .005 = .375$ (dropping % = dividing by 100)

(*b*) $43\frac{7}{8}\% = 43\% + \frac{7}{8}\% = .43 + \frac{\frac{7}{8}}{100}$ (carry out $\frac{7}{8}\%$)

$$= .43 + \frac{.875}{100} = .43 + .00875 = .43875$$

(*c*) $.02\frac{1}{4}\% = .02\% + .00\frac{1}{4}\% = .0002 + \frac{.0025}{100}$

$$= .0002 + .000025$$
$$= .000225$$

Note: You may carry out the fraction directly and add it to the digit numbers.

(*d*) $.02\frac{1}{25}\% = .000204$ (move 2 places to left to divide by 100)

302. How can we convert a whole number, a decimal fraction, a fraction, or a mixed number to a per cent?

In each case *multiply by 100* to annex a % sign.

EXAMPLES:

(*a*) Convert 8 to a per cent $8 \times 100 = 800\%$
(*b*) Convert .4 to a per cent $.4 \times 100 = 40\%$
(*c*) Convert .04 to a per cent $.04 \times 100 = 4\%$
(*d*) Convert $\frac{1}{3}$ to a per cent $\frac{1}{3} \times 100 = 33\frac{1}{3}\%$
(*e*) Convert $\frac{1}{8}$ to a per cent $\frac{1}{8} \times 100 = 12\frac{1}{2}\%$
(*f*) Convert $\frac{1}{4}$ to a per cent $\frac{1}{4} \times 100 = 25\%$
(*g*) Convert $7\frac{1}{9}$ to a per cent $7\frac{1}{9} \times 100 = \frac{64}{9} \times 100 = \frac{6400}{9} = 711\frac{1}{9}\%$

Note: To multiply by 100, move decimal point 2 places to the right whenever that can be done directly.

(*h*) $\frac{6}{25} = \frac{6}{\cancel{25}_1} \times \cancel{100}^4 = 24\%$

(*i*) $\frac{13}{20} = \frac{13}{\cancel{20}_1} \times \cancel{100}^5 = 65\%$

(*j*) $\frac{11}{12} = \frac{11}{\cancel{12}_3} \times \cancel{100}^{25} = \frac{275}{3} = 91\frac{2}{3}\%$

(*k*) $.02\frac{1}{8} = .02\frac{1}{8} \times 100 = 2\frac{1}{8}\% = 2.125\%$

303. What are the per cent equivalents of very common fractions?

Fraction	Per cent equivalent	Fraction	Per cent equivalent
$\frac{1}{100}$	1%	$\frac{1}{8}$	$12\frac{1}{2}\%$
$\frac{1}{10}$	10%	$\frac{1}{4}$	25%
$\frac{1}{5}$	20%	$\frac{3}{8}$	$37\frac{1}{2}\%$
$\frac{3}{10}$	30%	$\frac{1}{2}$	50%
$\frac{2}{5}$	40%	$\frac{5}{8}$	$62\frac{1}{2}\%$
$\frac{3}{5}$	60%	$\frac{3}{4}$	75%
$\frac{7}{10}$	70%	$\frac{7}{8}$	$87\frac{1}{2}\%$
$\frac{4}{5}$	80%	$\frac{1}{6}$	$16\frac{2}{3}\%$
$\frac{9}{10}$	90%	$\frac{1}{3}$	$33\frac{1}{3}\%$
$\frac{1}{1} = 1$	100%	$\frac{2}{3}$	$66\frac{2}{3}\%$
		$\frac{5}{6}$	$83\frac{1}{3}\%$

EXAMPLE: Find $12\frac{1}{2}\%$ of \$3,200.

$\frac{1}{8} \times \$3,200 = \$400$

304. What per cent of the large square is the shaded part?

Large square contains 25 small squares.
Shaded part contains 6 small squares.

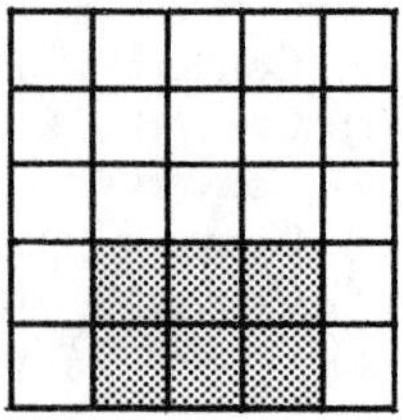

$$\therefore \frac{6}{25} = \frac{6}{\cancel{25}_{\,1}} \times \cancel{100}^{\,4} = 24\%$$

Shaded part is 24% of large square.

305. What is the most common method of finding a given per cent of a number?

Write the per cent as a decimal and multiply.

Ex. (*a*) Find 6% of \$6,700 (6% = .06). Then,

$$.06 \times \$6{,}700 = \$402.00$$

(.06: two places; \$402.00: two places)

Ex. (*b*) Find 14% of \$97.51 (14% = .14).

```
   $97.51  (2 places)
      .14  (2 places)
   ------
    39004
    9751
  -------
 $13.6514  (4 places)
```

Ans. = \$13.65 (to nearest cent)

306. What is another method of finding a given per cent of a number?

Find 1% of the number first, and then multiply by the given per cent.

Ex. (*a*) Find 6% of \$6,700.

$$1\% \text{ of } \$6{,}700 = \tfrac{1}{100} \times \$6{,}700 = \$67$$

(Move 2 places to left to divide by 100.) Then

$$6\% \text{ of } \$6{,}700 = 6 \times \$67 = \$402$$

Ex. (*b*) Find 4% of \$1,860.

$$1\% \text{ of } \$1{,}860 = \$18.60$$

$$\therefore 4\% \text{ is } 4 \times \$18.60 = \$74.40$$

Ex. (*c*) Find $2\frac{1}{2}\%$ of \$7,000.

$$1\% \text{ of } \$7{,}000 = \$70$$

$$\therefore 2\tfrac{1}{2}\% \text{ is } 2\tfrac{1}{2} \times \$70 = \$175$$

307. What is the third method of finding a given per cent of a number?

Write the given per cent as a common fraction and multiply.

This method is useful when the given per cent is the equivalent of a simple common fraction.

Ex. (*a*) Find 25% of \$51.

$$25\% = \tfrac{1}{4}$$

$$\therefore \tfrac{1}{4} \times \$51 = \$\tfrac{51}{4} = \$12\tfrac{3}{4} = \$12.75$$

Ex. (*b*) Find $37\tfrac{1}{2}\%$ of \$84.75.

$$37\tfrac{1}{2}\% = \tfrac{3}{8}$$

$$\therefore \tfrac{3}{8} \times \$84.75 = \frac{3 \times \$84.75}{8} = \frac{\$254.25}{8} = \$31.78\tfrac{1}{8} \quad \text{or} \quad \$31.78$$

308. What terms are commonly used in percentage?

(*a*) The number of which so many hundredths or a certain per cent is to be taken is called the *base* (= B).

(*b*) The per cent to be taken is the *rate* (= R).

(*c*) The result of the rate times the base is the *percentage* (= P).

$$\text{P (percentage)} = \text{R (rate)} \times \text{B (base)} \quad \text{or} \quad P = R \times B$$

Ex. (*a*) Find the percentage when the rate is 4% and the base is \$1,860.

$$P = \underset{\text{(rate)}}{4\%} \times \underset{\text{(base)}}{\$1,860}$$

\$1,860	(= base)
.04	(= rate)
\$74.40	(= percentage)

Ex. (*b*) Find 9% of 50.

50	(= base)
.09	(= rate)
4.5	(= percentage)

309. What is the rule for finding the percentage when the base and rate are given?

Multiply the base by the rate expressed either as a decimal or a common fraction.

Ex. (*a*) In testing a certain ore, 25% of it was found to be iron. How much iron was contained in 552 pounds of ore?

$$\underset{\text{(rate)}}{25\%} \times \underset{\text{(base)}}{552 \text{ lb.}} = \tfrac{1}{4} \times 552 = \underset{\text{(percentage)}}{138 \text{ lb. iron}}$$

Ex. (*b*) Suppose 27% was iron. How much iron was there in 578 pounds of ore?

$$\underset{\text{(rate)}}{27\%} \times \underset{\text{(base)}}{578 \text{ lb.}} = .27 \times 578 = \underset{\text{(percentage)}}{156.06 \text{ lb. iron}}$$

310. What is the rule for finding the rate when the percentage and base are given?

Divide the percentage by the base to get the rate. Since

$$P = B \times R$$

$$\therefore R = \frac{P}{B} = \text{rate} = \frac{\text{percentage}}{\text{base}}$$

Note that rate is a per cent and is a fraction or a division (= a comparison between percentage and base).

Ex. (*a*) $114 is what per cent of $3,800?

Divide the quantity by that with which it is being compared.

$$\text{Rate} = \frac{\$114}{\$3{,}800} = 3{,}800\overline{)114} = \begin{array}{r} .03 \\ 38\overline{)1.14} \\ 114 \\ \hline 0 \end{array}$$

$$\therefore .03 = \tfrac{3}{100} = 3\% = \text{rate}$$

Ex. (*b*) An investor received $382.50 on an investment of $8,500. What rate per cent did the investment pay?

$$\text{Rate} = \frac{\$382.50}{\$8{,}500} = 8{,}500\overline{)382.5} = \begin{array}{r} .045 \\ 85\overline{)3.825} \\ 340 \\ \hline 425 \\ 425 \\ \hline 0 \end{array}$$

$$\therefore .045 = 4\tfrac{1}{2}\% = \text{rate}$$

You are comparing the percentage with the base.

Ex. (*c*) A man earns $9,000 a year. He pays $1,800 a year for rent. What per cent of his salary is his rent? Compare the percentage of $1,800 with the base $9,000.

$$\text{Rate} = \frac{\$1{,}800}{\$9{,}000} = \frac{18}{90} = \frac{9}{45} = \frac{1}{5}$$

$$\therefore \tfrac{1}{5} = 20\% = \text{rate}$$

311. What is the rule for finding the base when the rate and the percentage are given?

Divide the percentage by the rate expressed either as a common fraction or as a decimal. Since

$$P = B \times R$$

$$\therefore\ B = \frac{P}{R} = \frac{\text{percentage}}{\text{rate}} = \text{base} = \text{whole amount}$$

Note: Dividing by the per cent gives you the percentage for 1 per cent (or 1 part in a hundred). Then multiplying by 100 gives you the whole amount.

Ex. (*a*) \$435 is 20% of what amount?

$$\text{Base} = \frac{P}{R} = \frac{\$435}{\frac{1}{5}} = \$435 \times 5 = \$2{,}175$$

or

$$\frac{\$435}{20\%} = \$21.75 = \text{value of 1 per cent.}$$

Remember: If you want to get the value of one unit of any element in the problem, you should divide by that element. Divide by per cent to find value of 1 per cent. Therefore

$$\$21.75 \times 100 = \$2{,}175 = \text{value of } 100\% = \text{base}$$

Ex. (*b*) \$187.20 is 16% of what amount?

$$\frac{\$187.20}{16\%} = \$11.70 = \text{value of 1 per cent}$$

$$\therefore\ \$11.70 \times 100 = \$1{,}170 = \text{value of } 100\% = \text{base}$$

Ex. (*c*) What is the amount of a bill if 2% discount for cash comes to \$2.85?

If \$2.85 is 2%, then

$$\frac{\$2.85}{2\%} = \$1.425 = \text{value of 1 per cent}$$

$$\therefore\ \$1.425 \times 100 = \$142.50 = \text{value of } 100\% = \text{base}$$

or

$$\text{Base} = \frac{P}{R} = \frac{\$2.85}{.02} = \$142.50$$

312. What is meant by (*a*) amount, (*b*) difference, in percentage problems?

(*a*) Amount = base + percentage.
(*b*) Difference = base − percentage.

313. How can we find the base when the rate and amount are given?

$$\text{Base} = \frac{\text{amount}}{1 + \text{rate (as a decimal)}}$$

Ex. (*a*) The rent of an apartment is \$1,848 per year and this is an increase of 10% over the previous year. What was the rent the previous year?

Base = rent previous year   Amount = \$1,848 rent this year
10% = rate increase = .10

$$\therefore \text{Base} = \frac{\$1{,}848}{1 + .10} = \frac{\$1{,}848}{1.10} = \frac{\$18{,}480}{11}$$

$$= \$1{,}680 = \text{Rent previous year}$$

or

$$\begin{aligned} 100\% &= \text{base} = \text{Rent previous year} \\ 10\% &= \text{Advance this year} \\ 110\% &= \$1{,}848\ (= \text{Rent this year}) \\ 1\% &= \$1{,}848 \div 110 = \$16.80 \end{aligned}$$

$$\therefore \text{Base} = 100\% = 100 \times \$16.80 = \$1{,}680$$

Ex. (*b*) A storekeeper sells a TV set for \$270 and makes a profit of $12\frac{1}{2}\%$ on the transaction. What did the TV set cost him?

Base = cost to him   Amount = \$270 = selling price

$$12\tfrac{1}{2}\% = \text{profit} = \text{rate increase} = .125$$

$$\therefore \text{Base} = \frac{\$270}{1 + .125} = \frac{\$270}{1.125} = \frac{270{,}000}{1{,}125} = \$240 = \text{Cost to him}$$

or

$$\begin{aligned} 100\% &= \text{base} = \text{Cost to him} \\ 12\tfrac{1}{2}\% &= \text{Profit} \\ 112\tfrac{1}{2}\% &= \$270 = \text{Selling price} \\ 1\% &= \$270 \div 112.5 = \$2.40 \end{aligned}$$

$$\therefore \text{Base} = 100\% = 100 \times \$2.40 = \$240 = \text{Cost to him}$$

314. How do we find the base when the rate and difference are given?

$$\text{Base} = \frac{\text{difference}}{1 - \text{rate (as a decimal)}}$$

Ex. (*a*) If the waste in mining and handling coal amounts to 4%, how many tons would have to be mined to load 20 cars with 30 tons each?

Base = tons to be mined Difference = 20 × 30 = 600 tons

$$4\% = \text{rate decrease} = \text{waste}$$

$$\therefore \text{Base} = \frac{600}{1 - .04} = \frac{600}{.96} = \frac{60{,}000}{96} = 625 \text{ tons to be mined}$$

or

$$\begin{aligned} 100\% &= \text{base} = \text{Tons to be mined} \\ 4\% &= \text{Loss} \\ 96\% &= 600 \text{ tons} \\ 1\% &= 600 \div 96 = 6.25 \text{ tons} \\ 100\% &= 6.25 \times 100 = 625 \text{ tons to be mined} \end{aligned}$$

Ex. (*b*) A man sells his car for $1,500 and loses 25% on the transaction. What did he pay for it?

$$\text{Base} = \frac{\$1{,}500}{1 - .25} = \frac{\$1{,}500}{.75} = \frac{\$1{,}500}{\frac{3}{4}} = \$1{,}500 \times \frac{4}{3}$$

$$= \$2{,}000 \text{ cost to him}$$

315. On what do we always base the per cent of increase in some quantity?

It is based on the original quantity and not on the increased quantity.

Ex. (*a*) If the price of a newspaper was raised from 5 cents to 10 cents, what was the per cent of increase in price? The original price is 5 cents. The increase is 5 cents.

$$\therefore \frac{\text{increase}}{\text{original}} = \frac{5}{5} = 1 \quad \text{(Written as a per cent, 100\%)}$$

Thus, there was a 100% increase in price.

Ex. (*b*) If at the beginning of the year you had a bank balance of $4,500 and at the end of the year you had $5,400, by what per cent had your balance increased?

$$\text{Original} = \$4{,}500$$

$$\text{Increase} = \$5{,}400 - \$4{,}500 = \$900$$

$$\therefore \frac{\text{increase}}{\text{original}} = \frac{\$900}{\$4{,}500} = \frac{1}{5} = 20\% \text{ Ans.}$$

316. On what do we always base the per cent of decrease in some quantity?

It is based on the original quantity and not on the decreased quantity.

Ex. (*a*) A new automobile which cost \$2,200 was worth \$1,800 a year later. By what per cent has it decreased in value?

$$\text{Original} = \$2{,}200$$
$$\text{Decrease} = \$2{,}200 - \$1{,}800 = \$400$$
$$\therefore \frac{\text{decrease}}{\text{original}} = \frac{\$400}{\$2{,}200} = \frac{2}{11} = 18\tfrac{2}{11}\% \text{ Ans.}$$

Ex. (*b*) If a bank dropped its interest rate from $3\frac{3}{4}$ to $3\frac{1}{4}\%$, what would be the per cent of decrease in the interest rate?

$$\text{Original} = 3\tfrac{3}{4}\%$$
$$\text{Decrease} = 3\tfrac{3}{4}\% - 3\tfrac{1}{4}\% = \tfrac{1}{2}\%$$
$$\therefore \frac{\text{decrease}}{\text{original}} = \frac{\frac{1}{2}\%}{3\frac{3}{4}\%} = \frac{1}{2} \times \frac{4}{15} = \frac{2}{15} = 13\tfrac{1}{3}\% \text{ Ans.}$$

317. How are per cents less than 1 per cent, or fractional parts of 1 per cent, written and used in business and financial matters?

$$\tfrac{3}{100} \text{ of } 1\% = \tfrac{3}{100}\% = .03\% = .0003$$
$$\tfrac{3}{10} \text{ of } 1\% = \tfrac{3}{10}\% = .3\% = .003$$
$$\tfrac{1}{10} \text{ of } 1\% = \tfrac{1}{10}\% = .1\% = .001$$
$$\tfrac{1}{8} \text{ of } 1\% = \tfrac{1}{8}\% = .12\tfrac{1}{2}\% = .00125$$
$$\tfrac{1}{4} \text{ of } 1\% = \tfrac{1}{4}\% = .25\% = .0025$$
$$\tfrac{1}{2} \text{ of } 1\% = \tfrac{1}{2}\% = .5\% = .005$$

EXAMPLE: If the tax on a house is increased by $\frac{1}{4}\%$, what is the amount of increase on a house assessed at \$15,750?

$$\$15{,}750 \times .0025 = \$39.375 = \$39.38 \text{ Ans.}$$

318. How is the expression of "so much per hundred" commonly used in business?

It is used in each of the following examples:

Ex. (*a*) What is the amount of the premium on a \$12,000 fire insurance policy, at 55¢ a hundred dollars?

$$120 \times \$.55 = \$66 \text{ Ans.}$$

Ex. (*b*) A broker charges you \$12.50 per 100 shares. How much will it cost you to buy 500 shares of stock?

$$5 \times \$12.50 = \$62.50 \text{ Ans.}$$

Ex. (*c*) A bankrupt firm pays you 43¢ on the dollar. How much do you get when your claim amounts to \$463.75?

$$\$463.75 \times .43 = \$199.41 \text{ Ans.}$$

319. How is the mill used in tax matters?

$$1 \text{ mill} = \tfrac{1}{10}¢ = \$\tfrac{1}{1000} = .001$$
$$1{,}000 \text{ mills} = \$1$$
$$\therefore\ 2\tfrac{1}{2} \text{ mills} = 2.5 \text{ mills} = \frac{\$2.5}{1{,}000} = \$.0025$$
$$= \tfrac{1}{4}¢ \text{ on every dollar}$$
$$= 25¢ \text{ on every } \$100$$
$$= \$2.50 \text{ on every } \$1{,}000$$

EXAMPLE: A property assessed at $12,500 is taxed at 28.7 mills per dollar. How much is the tax?

$$28.7 \text{ mills} = \frac{\$28.7}{1{,}000} = \$.0287 \text{ per dollar} = \$2.87 \text{ per } \$100$$
$$\therefore\ 125 \times \$2.87 = \$358.75 \text{ Ans.}$$

320. How are per cents added, subtracted, multiplied, or divided?

They are converted to decimals first and carried out in the same manner as similar operations involving decimals.

321. If a number is increased by a certain per cent to get an amount, what per cent must be subtracted from this amount to get the original number again?

To get back to the original number, a different per cent must be subtracted from the amount.

EXAMPLE: If 6% of 85 is added to it, we get,

$$.06 \times 85 + 85 = 5.1 + 85 = 90.1 = \text{Amount}$$

Now, what per cent of 90.1 must be subtracted from it to get 85 again?

$$\frac{5.1}{90.1} = 90.1.\overline{)5.1.0000}\ \ .0566$$

```
                    .0566
5.1    =  90.1.)5.1.0000
90.1             4505
(move 1 place    5950
 to right in     5406
 each case)      5440
```

$$= .0566 = 5.66\% \text{ Ans.}$$

We see that 5.1 is only 5.66% of 90.1, whereas 5.1 is 6% of 85, the original number.

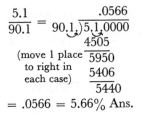

322. If Boston has a population of 2,000,000, and Philadelphia is 50% larger, how much smaller is Boston than Philadelphia?

Boston = 2,000,000
$50\% = \frac{1}{2}$
$\frac{1}{2} \times 2{,}000{,}000 = 1{,}000{,}000$ = Difference between the two cities
$2{,}000{,}000 + 1{,}000{,}000 = 3{,}000{,}000$ = Population of Philadelphia

$$\frac{1{,}000{,}000}{3{,}000{,}000} = \frac{1}{3} = 33\tfrac{1}{3}\%$$

∴ Boston is $33\frac{1}{3}\%$ smaller than Philadelphia
(Also Philadelphia is 50% larger than Boston.)

This again emphasizes the rule that the per cent of increase or decrease of some quantity is always based on the original quantity. For Boston the original quantity is 2,000,000 and for Philadelphia it is 3,000,000.

323. If a man spends 30% of his income for rent and 10% of the remainder for clothes, what is his salary if the landlord gets $1,150 more than the clothier?

30% of income = Rent
10% of remainder $(100\% - 30\%) = .1 \times .7 = .07 = 7\%$ = Clothes
30% of income = 7% of income + $1,150 or
23% of income = $1,150

∴ $1,150 ÷ .23 = $5,000 = Income

324. A man sells his car to his friend and takes a loss of 20%. His friend sells the car later to a third party for $1,500, losing 25%. How much did the original owner pay for the car?

$1,500 represents 75% of his cost

$$\therefore \frac{\$1{,}500}{.75} = \frac{150{,}000}{75} = \$2{,}000 \text{ (cost to friend)}$$

$2,000 represents 80% of original owner's cost

$$\therefore \frac{\$2{,}000}{.8} = \frac{20{,}000}{8} = \$2{,}500 = \text{Cost to original owner}$$

PROBLEMS

1. What does .27 mean in terms of percentage?
2. What per cent of 4,000 is 1,800?
3. 1,400 is what % of 3,600?

4. Reduce to a decimal:

(*a*) 5% (*b*) $4\frac{1}{2}$% (*c*) $9\frac{1}{3}$% (*d*) 6% (*e*) 75% (*f*) $5\frac{3}{4}$%
(*g*) $83\frac{1}{3}$% (*h*) 115% (*i*) $63\frac{1}{2}$% (*j*) 9.26% (*k*) .003%
(*l*) $135\frac{1}{2}$% (*m*) 2.25% (*n*) .6% (*o*) 250% (*p*) 7.3% (*q*) .03%
(*r*) $\frac{2}{3}$% (*s*) $1\frac{1}{4}$% (*t*) $\frac{1}{16}$% (*u*) .60% (*v*) $\frac{1}{4}$% (*w*) $\frac{1}{10}$%
(*x*) $.03\frac{1}{8}$% (*y*) $34\frac{7}{8}$%

5. Express $\frac{1}{4}$%, $\frac{1}{5}$%, $\frac{2}{10}$% as decimals of a per cent, and as decimals.

6. Express as common fractions in lowest terms:

(*a*) 12%, .12%	(*b*) 25%, .25%	(*c*) 36%, .36%
(*d*) 75%, .75%	(*e*) $16\frac{2}{3}$%, $\frac{1}{6}$%	(*f*) $12\frac{1}{2}$%, $\frac{1}{8}$%
(*g*) 150%, .15%	(*h*) 375%, 3.75%	(*i*) $\frac{3}{4}$%, 1.4%
(*j*) $\frac{1}{2}$%, .05%	(*k*) $33\frac{1}{3}$%, $\frac{1}{3}$%	(*l*) $62\frac{1}{2}$%, $\frac{5}{8}$%

7. Change to a per cent: (*a*) 9 (*b*) .6 (*c*) $\frac{1}{4}$ (*d*) $\frac{1}{16}$ (*e*) $8\frac{1}{7}$ (*f*) $\frac{7}{25}$ (*g*) $\frac{17}{20}$ (*h*) $\frac{7}{12}$ (*i*) $.03\frac{1}{8}$ (*j*) $\frac{16}{100}$ (*k*) $\frac{4}{1000}$ (*l*) $\frac{7}{8}$ (*m*) .84 (*n*) $\frac{1}{3}$ (*o*) .65 (*p*) $\frac{3}{5}$ (*q*) .8 (*r*) $.87\frac{1}{2}$ (*s*) .07 (*t*) .0425 (*u*) $.16\frac{2}{3}$ (*v*) $\frac{1}{10}$

8. What per cent of the circle is the shaded part?

9. What per cent of the large square is the shaded part?

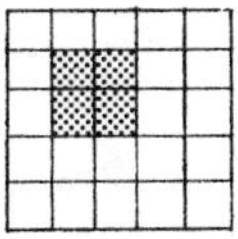

10. Find:

(*a*) 4% of $4,800	(*b*) 16% of $86.42	(*c*) 6% of $8,500
(*d*) 7% of $1,940	(*e*) $3\frac{1}{4}$% of $6,000	(*f*) 25% of $62
(*g*) $33\frac{1}{3}$% of $76.25	(*h*) $87\frac{1}{2}$% of $1,600	(*i*) $66\frac{2}{3}$% of 1,500
(*j*) 150% of 500	(*k*) $\frac{5}{6}$% of 7,254	(*l*) $38\frac{1}{2}$% of 6,542

11. Find the result by first finding 1% of the given number in the following:

(*a*) $\frac{1}{4}$% of 10,000	(*b*) 4% of 1,600	(*c*) $2\frac{1}{2}$% of 4,000
(*d*) $3\frac{1}{4}$% of 10,000	(*e*) $\frac{1}{2}$% of 6,000	(*f*) 6% of 7,000

12. A man owned 960 acres of land. He sold $37\frac{1}{2}$% of it. How many acres did he sell?

13. A man had $24,000 in cash. He invested $12\frac{1}{2}$% of it in bonds and 46% in stocks. How much did he invest in each and how much money had he left?

14. In testing a certain ore, 27% of it was found to be iron. How much iron was contained in 645 lb. of ore?

15. There are 2,760 voters in a certain town. If 69% of the voters go to the polls, how many votes will be cast?

16. An investor received $460.50 on an investment of $9,200. What rate per cent did the investment pay?

17. A man earns $8,000 a year. He pays $1,600 a year for rent. What per cent of his salary is his rent?

18. $565 is 20% of what amount?

19. $238.30 is 18% of what amount?

20. What is the amount of a bill if 2% discount for cash comes to $3.45?

21. What per cent of:

(*a*) 138 is 56? (*b*) 495 is 65? (*c*) 9,860 is 1,260?
(*d*) .125 is .05? (*e*) .03 is .0085? (*f*) $15\frac{1}{2}$ is $3\frac{3}{5}$?
(*g*) 47,830 is 64.58? (*h*) 273.6 is 59.85? (*i*) 93 is 154.6?
(*j*) 66 is 24? (*k*) .107 is 765? (*l*) .1235 is .05486?
(*m*) 289 is 148.5?

22. Find the number of which:

(*a*) 360 is 15% (*b*) 459 is 40% (*c*) 56 is $33\frac{1}{3}$%
(*d*) 420 is 125% (*e*) 52 is $133\frac{1}{3}$% (*f*) 112 is $6\frac{1}{4}$%
(*g*) 306 is $67\frac{1}{7}$% (*h*) 132 is $44\frac{4}{9}$% (*i*) .89653 is 6%

23. What is 4% of $\frac{1}{8}$ of an acre of land?

24. If a merchant's scales weigh 14 oz. for a pound, what per cent does the purchaser lose?

25. What per cent is $33\frac{1}{3}$% of 6%?

26. 6 is 5% of what number? 10% of what number?

27. 8 is 2% of what number? 25% of what number?

28. $2.50 is $16\frac{2}{3}$% of what? $33\frac{1}{3}$% of what?

29. 53.2 is 105% of what number? 90% of what number?

30. 80 is 125% of what number? 75% of what number?

31. 95 is .05% of what number? 176% of what number?

32. The rent of an apartment is $1,656 and this is an increase of 12% over the previous year. What was the rent the previous year?

33. A man sells a refrigerator for $340 and makes a profit of $16\frac{2}{3}$% on the transaction. What did the refrigerator cost him?

34. If the waste in mining and handling coal amounts to $4\frac{1}{2}$%, how many tons would have to be mined to load 40 cars with 30 tons each?

35. A man sells his house for $12,000 and loses 12% on the transaction. What did the house cost him?

36. If the price of a magazine was raised from 15¢ to 25¢, what was the per cent increase in price?

37. If at the beginning of the year your bank balance was $3,800, and at the end of the year $4,600, by what per cent had your balance increased?

38. A new car which cost $3,100 was worth $2,700 a year later. By what per cent had it decreased in value?

39. If a bank dropped its interest rate from $3\frac{1}{2}$% to 3%, what would be the per cent decrease in the interest rate?

40. Express in fractions of a per cent and in decimals:

(*a*) $\frac{7}{100}$ of 1% (*b*) $\frac{7}{10}$ of 1% (*c*) $\frac{1}{10}$ of 1%

(*d*) $\frac{1}{16}$ of 1% (*e*) $\frac{3}{8}$ of 1% (*f*) $\frac{1}{5}$ of 1%

41. If the tax on a house is increased by $\frac{1}{8}$%, what is the amount of increase on a house worth $14,700?

42. What is the premium on an $18,000 fire insurance policy at 64¢ per hundred dollars?

43. If you are charged $12.50 per 100 shares to buy stocks, how much will it cost you to buy 1,200 shares of stock?

44. A bankrupt firm pays you 67¢ on the dollar. How much do you receive when your claim amounts to $585.45?

45. A property assessed at $14,500 is taxed at 24.3 mills per dollar. How much is the tax?

46. If 8% is added to $96 to get $103.08, what per cent of $103.08 must be subtracted from it to get back to $96?

47. If university *A* has an enrollment of 12,000 students, and university *B* is 35% larger, how much smaller is university *A* than *B*?

48. If a man spends 25% of his income for food, and 12% of the remainder for education, what is his salary if the landlord gets $960 more than the school?

49. A man sells his house and takes a loss of 15%. The purchaser later sells the house to a third party for $14,000, losing 20%. How much did the original owner pay for the house?

50. The price of eggs dropped from 63¢ a dozen to 56¢ a dozen. What was the per cent decrease in price?

51. An article that cost $12 was sold for $16. What per cent of the cost was the difference between the selling price and the cost? What per cent of the selling price was the difference between the selling price and the cost?

52. A college had an enrollment of 2,600 in 1950, 22% more than in 1940. At the same rate of increase, how many students were enrolled in 1960? What was the enrollment in 1940?

53. What is:

(*a*) 64 increased by 12½% of itself?
(*b*) 45 increased by 33⅓% of itself?
(*c*) .054 increased by 24% of itself?

54. What number increased by:

(*a*) 10% of itself is 462?
(*b*) 62½% of itself is 299?
(*c*) 8% of itself is 3.024?

55. What number decreased by:

(*a*) 12½% of itself is 266?
(*b*) 37½% of itself is 450?
(*c*) 7% of itself is 21.39?

CHAPTER IX

INTEREST

325. What is meant by interest?

Interest is the amount paid for the use of borrowed money, or the amount received for the use of money loaned or invested. In bookkeeping these go under the items of *interest cost* and *interest earned.*

326. What are the three factors to consider in calculating interest?

(*a*) Principal = the sum loaned or the capital invested.

(*b*) Time = duration of the period. One year is the customary unit of time. For a part of a year, the subdivision used is the month or the day.

(*c*) Rate = rate per cent = number of units paid upon each hundred units of borrowed sum. The units are expressed in the money of the country concerned, as dollars, pounds sterling, francs, marks, kroner, florins, or pesos.

EXAMPLE: If \$6 are paid as interest for every hundred dollars loaned at the end of each year, then the rate = 6 per 100 or 6 per cent or 6%.

Thus the rate = the ratio of the interest to the principal for each unit of time.

327. How do we express a rate of interest?

(*a*) As an integral or a mixed number with a per cent sign after it.

EXAMPLE: 5% = five per cent = an integral with a % sign

$6\frac{3}{4}\%$ = six and three-quarters per cent = a mixed number with a % sign

(*b*) As a decimal, the correct way to write it.

EXAMPLE: 0.05 = five per cent = $\frac{5}{100}$

0.0675 = six and three-quarters per cent = $\frac{6\frac{3}{4}}{100}$

328. What is meant by simple interest?

Interest calculated on the original principal for the time the principal is used. Simple interest is nothing more than percentage with a time element involved. The original principal remains constant and the quantity of interest for each unit time interval remains unchanged.

EXAMPLE: 6% interest on \$100 for 1 year = \$6 = simple interest

$$.06 \text{ of } \$100 = \$6$$
$$6\% \text{ of } \$100 = \$6$$

Thus simple interest = a percentage with a time element.

329. What is meant by compound interest?

It is interest calculated upon both the principal and the interest which has already accrued. The interest is compounded quarterly, semi-annually, or annually, according to agreement. You merely compute simple interest on the new principal at the various periods agreed upon.

EXAMPLE: Find the interest for 3 years at 6% on \$200, with interest compounded annually.

$$\text{For first year, interest} = 6\% \text{ of } \$200 = .06 \times \$200 = \$12$$
$$\text{New principal} = \$200 + \$12 = \$212$$
$$\text{For second year, interest} = 6\% \text{ of } \$212 = .06 \times \$212 = \$12.72$$
$$\text{New principal} = \$212 + \$12.72 = \$224.72$$
$$\text{For third year, interest} = 6\% \text{ of } \$224.72 = \$13.48$$
$$\text{New principal} = \$224.72 + \$13.48 = \$238.20$$
$$\text{Original principal} = \$200.00$$
$$\text{Compound interest for 3 years} = \$38.20$$

Note that the simple interest for the 3 years would be

$$\$200 \times .06 \times 3 = \$36.00$$

330. What is the formula for figuring simple interest?

$$\text{Interest} = \text{principal} \times \text{rate} \times \text{time}$$
$$\text{I} = \text{P} \times r \times t = \text{P}rt$$

EXAMPLE: What is the interest on \$2,000 at 6% per year for a half year?

$$\text{I} = \text{P}rt = \$2{,}000 \times .06 \times \tfrac{1}{2} = \$60$$

331. What is meant by the "amount" and what is its symbol?

The sum obtained by adding the interest to the principal = amount = S

or $\text{S} = \text{Principal} + \text{Interest} = \text{P} + \text{I}$
or $\text{S} = \text{P} + \text{P}rt$, since $\text{I} = \text{P}rt$
or $\text{S} = \text{P}(\text{I} + rt)$, since P is a common factor of P and Prt

EXAMPLE: If you borrowed \$500 at simple interest for 3 years at 5%, how much will the creditor receive in all?

$$S = \text{amount} = P(1 + rt) = \$500(1 + .05 \times 3)$$
$$= \$500\ (1.15) = \$575 \text{ Ans.}$$

Creditor will receive \$575, of which \$500 is the principal and \$75 is the interest.

332. In figuring simple interest for less than a year what is the rule for establishing (*a*) the terminal days, (*b*) the due date?

(*a*) In the United States we exclude the first day and include the last day.

EXAMPLE: For a bank loan made January 4 and falling due January 27 interest would be charged for 23 days.

(*b*) Date of maturity of a loan is determined by the wording of the agreement: if time is stated in months, payment is due on the same date of due month; if time is stated in days then the exact number of days is counted to get due date.

EXAMPLE: If in a transaction on July 5 a debtor agrees to repay a loan in *five months*, the money is due December 5. If the agreement is to run 150 days, the due date would be December 2.

Note: Generally, in the United States loans falling due on Saturday, Sunday, or a holiday are payable on the next business day, and this extra time is counted.

333. How are the methods for figuring simple interest commonly referred to?

(*a*) The ordinary method.
(*b*) The exact or accurate method.
(*c*) The bankers' method.

The difference in these methods is in the way the time is figured.

334. How do we find the time by the ordinary method?

In the ordinary method, a year is considered to have 12 months of 30 days each, or 360 days.

The time is found easily by compound subtraction.

EXAMPLE: Find the time between February 8, 1959, and May 15, 1957.

Year	*Month*	*Day*
1959	2	8
1957	5	15
1	8	23

Borrow 1 month = 30 days and add it to the 8 days to make 38 days.
Subtract 15 days from 38 days to get 23 days.
Borrow 1 year = 12 months and add it to the 1 month to make 13 months.
Subtract 5 months from 13 months to get 8 months.
Now, 1957 from 1958 leaves 1 year.
The result is 1 year, 8 months, and 23 days.

335. How do we find the time by the exact method?

(*a*) The actual number of days in each month is counted.

EXAMPLE: Find the exact time from May 8, 1958, to January 12, 1959.

May	23 days
June	30 days
July	31 days
August	31 days
September	30 days
October	31 days
November	30 days
December	31 days
January	12 days
	249 days

(*b*) Use Table 1 in Appendix B. Each day of the year is indicated as the total number of days from January 1 to the day in question, inclusive. Find the number opposite the last date and from this subtract the number opposite the first date to get the number of days between the dates.

EXAMPLE: Use above dates: May 8 is the 128th day; December 31 is the 365th day. Then, 365 − 128 = 237 days in 1958. Now add 12 days in January, 1959, to 237 days to get 249 days in all.

336. How do we figure time by the bankers' method?

Time is expressed in months and days, or in exact days only. This method is used to find the time for short periods.

EXAMPLE: What is the time from June 4 to October 21?

From June 4 to October 4 is 4 months.
From October 4 to October 21 is 17 days.
Ans. = 4 months, 17 days.

Or (from Table 1 in Appendix B):

October 21 = 294 June 4 = 155

Ans. = 294 − 155 = 139 days

The 360-day year is used with exact days.

337. Find the interest on $3,000 at 6% from November 18, 1958, to April 6, 1959: (*a*) by the ordinary method, (*b*) by the exact method, (*c*) by the bankers' method.

(*a*)

Year	*Month*	*Day*	
1959	4	6	
1958	11	18	
	4	18	= 138 days

A year = 12 months, 30 days each, or 360 days = ordinary method.

$\$3{,}000 \times .06 \times \frac{138}{360} = \$69$ interest = Ordinary method

(*b*) Table 1, Appendix B: November 18 is 322nd day of the year.
365 − 322 = 43 days in 1958.
April 6 is the 96th day of the year.
Then, 43 + 96 = 139 days (exact).

$\therefore\ \$3{,}000 \times .06 \times \frac{139}{365} = \$68.55$ interest = Exact method

(*c*) $\$3{,}000 \times .06 \times \frac{139}{360} = \$69.50$ interest = Bankers' method
(Exact days and 360-day year are used.)

Note: Exact method produces the least interest of the three and the bankers' method produces the most (because the denominator is smaller while the number of days is exact).

338. What is the constant relationship of exact interest to ordinary or bankers' interest, based on exact number of days?

Let N = exact number of days.

Then $\frac{N}{365}$ = exact interest

and $\frac{N}{360}$ = ordinary or bankers' interest based on exact days.

$$\therefore \frac{\text{exact interest}}{\text{ordinary interest}} = \frac{\frac{N}{365}}{\frac{N}{360}} = \frac{\cancel{N}}{365} \times \frac{360}{\cancel{N}} = \frac{360}{365} = \frac{72}{73}$$

Then

$$\textit{Exact} = \tfrac{72}{73} \times \text{ordinary} = \text{ordinary} - \tfrac{1}{73}\,\text{ordinary}$$

and

$$\textit{Ordinary} = \tfrac{73}{72} \times \text{exact} = \text{exact} + \tfrac{1}{72}\,\text{exact}$$

We can remember this by noting that exact is always less than ordinary interest, so

$$\frac{\text{exact}}{\text{ordinary}} = \frac{72\ (\text{less})}{73\ (\text{greater})} = \text{Ratio}$$

Therefore, to get exact we subtract $\frac{1}{73}$ of ordinary from ordinary. To get ordinary we add $\frac{1}{72}$ of exact to exact.

339. What is the 60-day, 6 per cent method of calculating interest?

60 days are $\frac{60}{360} = \frac{1}{6}$ of a year.

Then if the interest rate is 6 per cent a year, the interest rate for 60 days is $\frac{1}{6} \times 6\% = 1\% = .01$.

Therefore, to find the interest for 60 days at 6 per cent on any principal, point off two places to the left.

Ex. (*a*) The interest on \$1,360 for 60 days at 6% is \$13.60.

Now 6 days are $\frac{1}{10} \times 60$ days.

Then interest for 6 days at $6\% = \frac{1}{10} \times 1\% = .001$.

Therefore, to find the interest for 6 days at 6% on any principal, point off three places to the left.

Ex. (*b*) The interest on \$1,360 for 6 days at 6% is \$1.36.

Ex. (*c*) Find the interest on \$570 for 66 days at 6%.

\$5.70	= interest for 60 days	(point off 2 places to left)
.57	= interest for 6 days	(point off 3 places to left)
\$6.27	= interest for 66 days	

340. A businessman borrowed \$850 for 75 days at 6%. How much interest did he pay?

\$8.50	= interest for 60 days	(point off 2 places)
2.13	= interest for 15 days	(= $\frac{1}{4}$ of 60 days)
\$10.63	= interest for 75 days	

341. How are the aliquot parts of 60 used when the time is greater or less than 60 days in finding interest by the 60-day 6% method?

EXAMPLE: What are the aliquot parts of 60 days contained in: (*a*) 49 days? (*b*) 58 days? (*c*) 77 days?

(*a*)	(*b*)	(*c*)
30 days	30 days	60 days
15 days	20 days	15 days
4 days	6 days	2 days
49 days	2 days	77 days
	58 days	

342. What is the interest on $953.70 for 124 days at 6%?

$9.537	= interest for 60 days	(point off 2 places)
9.537	= interest for 60 days	(point off 2 places)
.6358	= interest for 4 days	(point off 3 places for 6 days and take $\frac{2}{3}$ of this for 4 days)
$19.7098		

∴ $19.71 Ans. (to nearest whole cent)

(60 days + 60 days + 4 days = 124 days)

343. What is the interest on $598.60 for 48 days at 6%?

To get interest for 30 days, first get $5.986 interest for 60 days, and divide by 2.

$2.993	= interest for 30 days	
1.4965	= interest for 15 days	(= $\frac{1}{2}$ of 30 days)
.2993	= interest for 3 days	(= $\frac{1}{10}$ of 30 days)
$4.7888	= interest for 48 days	

∴ $4.79 Ans. (to nearest whole cent)

(30 + 15 + 3 = 48 days)

344. How can we sometimes simplify the 60-day 6% process?

(*a*) By exchanging the amount of the principal and the number of days.

EXAMPLE: Find the interest on $120 for 176 days at 6%. Make it $176 for 120 days by exchanging one for the other.

$1.76	= interest for 60 days	(point off 2 places)
1.76	= interest for 60 days	
$3.52	= interest for 120 days	

Ans. = $3.52 interest on $120 for 176 days

(*b*) By deducting from the interest for 60 days, the interest for the difference in time between the time given and 60 days.

EXAMPLE: Find the interest on \$170 for 50 days at 6%.

	\$1.70	= interest for 60 days	(point off 2 places)
Deduct	.2833	= interest for 10 days	(= $\frac{1}{6}$ of interest for 60 days)
	\$1.4167		

Ans. = \$1.42 interest on \$170 for 50 days

345. How do we find the interest at a rate other than 6%?

First find the interest at 6%; then to get:

(*a*) 3%, take $\frac{1}{2}$ of the interest at 6% $\left(\frac{3\%}{6\%} = \frac{1}{2}\right)$

(*b*) 4%, subtract $\frac{1}{3}$ of the interest at 6% $\left(\frac{4\%}{6\%} = \frac{2}{3}\right)$

(*c*) $4\frac{1}{2}$%, subtract $\frac{1}{4}$ of the interest at 6% $\left(\frac{4\frac{1}{2}\%}{6\%} = \frac{3}{4}\right)$

(*d*) 5%, subtract $\frac{1}{6}$ of the interest at 6% $\left(\frac{5\%}{6\%} = \frac{5}{6}\right)$

(*e*) 7%, add $\frac{1}{6}$ of the interest at 6% $\left(\frac{7\%}{6\%} = \frac{7}{6}\right)$

(*f*) $7\frac{1}{2}$%, add $\frac{1}{4}$ of the interest at 6% $\left(\frac{7\frac{1}{2}\%}{6\%} = \frac{5}{4}\right)$

(*g*) 8%, add $\frac{1}{3}$ of the interest at 6% $\left(\frac{8\%}{6\%} = \frac{4}{3}\right)$

(*h*) 9%, add $\frac{1}{2}$ of the interest at 6% $\left(\frac{9\%}{6\%} = \frac{3}{2}\right)$

EXAMPLE: Find the interest on \$790 for 145 days at $4\frac{1}{2}$% and at $7\frac{1}{2}$%.

\$7.90	= interest for 60 days	
7.90	= interest for 60 days	
2.6333	= interest for 20 days (= $\frac{1}{3}$ of 60 days)	
.6583	= interest for 5 days (= $\frac{1}{4}$ of 20 days)	
4) \$19.0916	= interest for 145 days at 6%	4) \$19.0916
−4.7729	deduct $\frac{1}{4}$ of \$19.0916	+4.7729
\$14.3187	= interest for 146 days at $4\frac{1}{2}$%	\$23.8645 at $7\frac{1}{2}$%
(or \$14.32)		(or \$23.86)

346. How can we make use of the interest formula in finding one of the four factors—interest, principal, rate, and time—when the other three are given?

We have seen that interest is merely a percentage problem with a time factor, or

$$\text{I (interest)} = \text{P}rt \text{ (principal} \times \text{rate} \times \text{time)}$$

Ex. (*a*) On a 360-day-per-year basis, at what rate per cent would you have to invest \$970 for 72 days to earn \$9.70 interest?

$$I = P \times r \times t$$
$$\$9.70 = \$970r \times \tfrac{1}{5} \text{ year} \qquad \left(\frac{72 \text{ days}}{360 \text{ days}} = \frac{1}{5} \text{ year}\right)$$
$$\$9.70 = 194r$$
$$\therefore r = \frac{9.70}{194} = 194\overline{)9.70}^{\;.05} = 5\% \text{ Ans.}$$

Ex. (*b*) How much money would you need to invest at 5% for 96 days to earn \$11.60 interest?

$$I = P \times r \times t$$
$$\$11.60 = P \times \overset{.01}{\cancel{.05}} \times \frac{4}{\underset{3}{\cancel{15}}} \text{ year} \qquad \left(\frac{96 \text{ days}}{360 \text{ days}} = \frac{4}{15} \text{ year}\right)$$
$$\$11.60 = \frac{.04P}{3}$$
$$\$11.60 \times 3 = .04P$$
$$\$34.80 = .04P$$
$$P = \frac{\$34.80}{.04} = \$870 \text{ Ans.}$$

Ex. (*c*) How long will it take to earn \$15.30 interest on an investment of \$1,080 at 6%?

$$I = P \times r \times t$$
$$\$15.30 = \$1{,}080 \times .06t$$
$$\$15.30 = 64.80t$$
$$\therefore t = \frac{1530}{6480} = \frac{\overset{17}{\cancel{153}}}{\underset{72}{\cancel{648}}} = \frac{17}{72} \text{ year} = \frac{17}{\underset{1}{\cancel{72}}} \times \overset{5}{\cancel{360}} = 85 \text{ days Ans.}$$

347. What is the 6-day 6% method of finding interest and what is its principal value?

The interest for 6 days at 6% can be found by moving the decimal point three places to the left because 6 is $\frac{1}{10}$ of 60.

(*a*) Move decimal point 3 places to the left for 6-day interest.

(*b*) Divide the number of days by 6 to get the number of 6-day units.

(*c*) Multiply the results of the above.

This method can be used to check the result of the 60-day method.

Ex. (*a*) Find the interest on \$300 for 27 days at 6%.

\$.30 = interest for 6 days at 6% (move 3 places to left)

$$\frac{27 \text{ days}}{6 \text{ days}} = 4\tfrac{1}{2}$$

∴ \$.30 × $4\frac{1}{2}$ = \$1.35 = interest for 27 days at 6%

Ex. (*b*) What is the interest on \$529.36 for 78 days at 6%?

\$.52936 = interest for 6 days at 6% (move 3 places to left)

$$\frac{78 \text{ days}}{6 \text{ days}} = 13$$

∴ \$.52936 × 13 = \$6.88168 = \$6.88 = interest for 78 days

348. What is the significance of compound interest?

In simple interest, the principal remains constant during the term of a loan.

In compound interest, the principal is increased by the addition of interest at the end of each interest period during the term of a loan.

Whenever the interest is added to the principal at the end of a period, it is said to be *converted* or *compounded.* The principal then becomes larger at the beginning of the second period than it was at the beginning of the first period. In turn, the interest due at the end of the second period is larger than that due at the end of the first period. This condition continues for each successive period during the indebtedness.

349. What is meant by: (*a*) compound amount, (*b*) compound interest, (*c*) conversion period, (*d*) frequency of conversion?

(*a*) Compound amount = principal + compound interest.

(*b*) Compound interest = compound amount − original principal.

(*c*) Conversion period = interval of time at the end of which interest is compounded.

(*d*) Frequency of conversion = number of times a year that the interest is converted into principal.

Most New York savings banks convert interest four times a year. Thus the conversion period is 3 months and the frequency is 4.

350. What will \$450 amount to in three years at 4% if interest is compounded annually?

\$450.00 = principal at beginning of first year

\$450.00 × .04 = \$18 = first year's interest

$\$450.00 + \$18 = \$468$ = principal at beginning of second year
$\$468.00 \times .04 = \$18.72$ = second year's interest

$\$468.00 + \$18.72 = \$486.72$ = principal at beginning of third year
$\$486.72 \times .04 = \$19.47$ = third year's interest

$\$486.72 + \$19.47 = \$506.19$ = principal or compound amount at *end* of third year

351. What is a shorter method of figuring the compound amount?

The amount at the beginning of the second year was seen to be equal to the principal at the beginning of the first year, plus one year's interest upon it (see Question 350).

$\$468 = \$450 + \$450 \times .04$
or $\$468 = \$450(1 + .04)$ ($450 is a common factor)
and $\$468 = \$450 \times 1.04$ = amount at beginning of second year

Thus, to get the amount for one year, multiply the principal by (1 + the annual interest rate):

$\$450.00$ = principal at beginning of first year
$\times\ 1.04$
$\$468.00$ = principal at beginning of second year
$\times\ 1.04$
$\$486.72$ = principal at beginning of third year
$\times\ 1.04$
$\$506.19$ = principal at end of third year

The above multiplications are expressed in one line as

$$\$506.19 = \$450 \times 1.04 \times 1.04 \times 1.04$$

or

$$\$506.19 = \$450 \times (1.04)^3 = \text{amount at compound interest}$$

The small figure 3 at the upper right-hand side of the parenthesis is called an *exponent* and means that the quantity in the parenthesis is to be used as a factor in multiplication that number of times. In this case 3 corresponds to the number of years for which interest was computed and means that (1.04) is to be multiplied 3 times. Similarly $(1.035)^4$ means an interest rate of $3\frac{1}{2}\%$ for 4 years.

352. What is the formula for the amount at compound interest?

S = amount = $506.19 (in Question 351)
P = principal = $450 (in Question 351)
r = interest rate per year = .04 (in Question 351)
t = number of years = 3 (in Question 351)

Therefore

$$S = P(1 + r)^t$$
$$S = \$450(1 + .04)^3 = \$450 \times (1.04)^3 = 450 \times 1.124864$$
$$S = \$506.19$$

Thus the compound amount of \$450 in 3 years with interest at 4% compounded annually is

\$506.19 Ans.

353. In order to have \$6,000 at the end of 3 years, how much must you invest now at 5% compounded annually?

$$S = P(1 + r)^t$$
$$\$6{,}000 = P(1 + .05)^3 = P(1.05)^3$$
$$P = \frac{\$6{,}000}{(1.05)^3} = \frac{\$6{,}000}{1.157625} = \$5{,}183.26$$

You must invest \$5,183.26 now to have \$6,000 at the end of 3 years when the interest is compounded annually at 5%.

354. What is used in actual business and financial practice to save a great deal of time, labor, and computation in figuring compound interest?

A table which has been computed giving the amount of 1 (unity) at compound interest for varying periods of time, and at different rates of interest. This table is called the "Compound Amount of 1" (see Table 2, Appendix B).

$$S = (1 + r)^t = \text{Formula for the compound amount of 1}$$

when interest is compounded annually. Here P = 1. $(1 + r)^t$ is known as the *accumulation factor*, since the compound amount indicates the accumulation of interest.

Accumulation factor × any principal = compound amount to which that principal accumulates at compound interest during a specified time.

You find in the table the compound amount of 1 for the proper time (number of periods) and rate, and then multiply this figure by the principal. The symbol for the time or number of periods is usually given as n. The table can be used for any denomination of currency, such as pounds sterling, francs, marks, lira, pesos, etc., or for any required unit.

Ex. (*a*) To find what \$1 will amount to in one year at 5%, enter column head n at 1, and run horizontally until the column headed 5% is reached, where you will find 1.05.

Ex. (*b*) To find the compound amount on \$1 for 4 years at 5%, enter column n at 4 and go horizontally until you reach the column headed 5%, where you will find \$1.21551.

Ex. (*c*) Which is greater: (1) a sum of money accumulating for 10 years at 2% compound interest, or (2) the same sum accumulating for 5 years at 4% compound interest?

$$\begin{aligned} \text{10 years at } 2\% &\rightarrow \$1.21899 = \text{compound amount of 1} \\ \text{5 years at } 4\% &\rightarrow \$1.21665 = \text{compound amount of 1} \end{aligned}$$

$\therefore$ 10 years at 2% gives a larger compound amount.

355. What would \$12,000 amount to if invested for 7 years at 4% compounded annually?

$$S = \$12{,}000 \times 1.31593 = \$15{,}791.16 \text{ Ans.}$$

(Compound amount of \$1 for $n = 7$ years and 4% = 1.31593 from table.)

356. What amount of money invested at 5% for nine years would amount to \$5,895.05?

$$S = P(1 + r)^t$$

$$\$5{,}895.05 = P(1 + .05)^9 = P \times (1.05)^9 = 1.55133P$$

$$\therefore P = \frac{\$5{,}895.05}{1{,}55133} = \$3{,}800 \text{ Ans.}$$

(According to Table 2, Appendix B, compound amount of \$1 for 9 years and 5% = \$1.55133.)

357. If you deposited \$1,800 in a bank which pays 4% per annum, how long will it take for this deposit to grow to \$2,277.58 if interest is compounded annually?

$$S = P(1 + r)^t$$

$$\$2{,}277.58 = \$1{,}800(1 + .04)^t = \$1{,}800 \times 1.04^t$$

$$1.04^t = \frac{\$2{,}277.58}{\$1{,}800} = 1.26532$$

Refer to Table 2 and go down under column headed 4% and you find 1.26532 is in a horizontal line running out to where

$$n = 6 = t = 6 \text{ years Ans.}$$

If the result had been more or less than 1.26532, then the time would not have been a whole year and the time would have to be interpolated between two integral years.

358. What is meant by the nominal rate of interest?

When interest is compounded or converted more than once a year, the stated rate of interest per year is called the *nominal rate.*

EXAMPLE: If a savings bank pays $3\frac{1}{2}\%$ on deposits compounded every quarter year, the nominal rate which you receive is $3\frac{1}{2}\%$. Actually you get a little more than $3\frac{1}{2}\%$ because each balance is increased at each 3-month interval by the interest added to it.

359. What is meant by the effective annual rate of interest?

Rate of interest actually earned in a year.

EXAMPLE: How much will \$700 amount to in one year if interest is compounded quarterly?

$$\text{Nominal interest} = 4\%$$
$$\text{Now, } .04 \text{ of principal} = \text{a year's interest}$$
$$\therefore \tfrac{1}{4} \times .04 = .01 \text{ of principal} = \text{a quarter's interest}$$

Date	*Principal*	*Amount at end of 3 months*	*Interest*
January 1	\$100		
April 1	\$100 × 1.01	\$101	\$1.00
July 1	\$101 × 1.01	\$102.01	\$1.01
October 1	\$102.01 × 1.01	\$103.03	\$1.02
January 1	\$103.03 × 1.01	\$104.06	\$1.03
			\$4.06

Thus a rate of 4% compounded quarterly for 1 year will produce the same result as a rate of 1% compounded annually for 4 years.

We see that the original \$100 earned \$4.06 in one year. This means

$$\frac{\$4.06}{\$100} = .0406 = 4.06\%$$

actually earned during the year.

4.06% is known as the actual or *effective* annual rate.

Thus a nominal rate of 4% compounded quarterly is equivalent to an effective rate of 4.06% compounded annually because the same amount of money is produced at the end of a year.

360. When are nominal and effective rates equivalent?

When they produce the same amount of money at the end of a year. In above:

$$\$100 \times (1 + .0406) = \$100 \times \left(1 + \frac{.04}{4}\right)\left(1 + \frac{.04}{4}\right)\left(1 + \frac{.04}{4}\right)\left(1 + \frac{.04}{4}\right)$$

or

$$\$100(1 + .0406) = \$100\left(1 + \frac{.04}{4}\right)^4$$

Divide both sides by \$100 to get the compound amount for \$1

$$(1 + .0406) = \left(1 + \frac{.04}{4}\right)^4$$

We see that the effective rate, .0406, is equivalent to the nominal rate, .04, compounded 4 times a year.

361. What is the formula showing the relationship between an effective rate i and an equivalent nominal rate r_p compounded p times a year?

$$1 + i = \left(1 + \frac{r_p}{p}\right)^p$$

In above:

$$(1 + .0406) = \left(1 + \frac{.04}{4}\right)^4$$

362. What is the formula for the compound amount of 1 at a rate r_p compounded p times per annum for t years?

The formula for the compound of 1 was shown in Question 354 to be $S = (1 + r)^t$ when interest is compounded annually.

To obtain a formula for the compound amount of 1 at a rate r_p compounded p times per year,

$$\left(1 + \frac{r_p}{p}\right)^p$$

is merely substituted for $(1 + r)$ in above because i and r_p are taken as equivalent rates. Thus

$$S = \left(1 + \frac{r_p}{p}\right)^{pt}$$

The exponent pt = the total number of conversion periods during the indebtedness.

EXAMPLE: If \$800 is left on deposit for 1 year at a nominal rate of $3\frac{1}{2}\%$ compounded semiannually, what will be the amount at the end of the year?

$$S = P\left(1 + \frac{r_p}{p}\right)^{pt} = \$800\left(1 + \frac{.035}{2}\right)^{2\times 1} = \$800(1 + .0175)^2$$

$$= \$800 \times 1.03530625 = \$828.245024$$

$$\therefore \$828.25 = \text{Amount}$$

363. What is the rule for use of compound-amount-of-1 tables where interest is compounded at a nominal rate more than once a year?

(*a*) Find value of pt = total number of conversion periods during time of indebtedness = n in tables.

(*b*) Find r_p/p = rate per period = per cent interest in tables.

(*c*) Look in the calculated per cent tables for the per cent for a quantity in line horizontally with the n column ($= pt$).

EXAMPLE: What is the amount of 1 at 6% compounded quarterly for 4 years?

$$S = \left(1 + \frac{r_p}{p}\right)^{pt} = \left(1 + \frac{.06}{4}\right)^{4\times 4} = (1 + .015)^{16}$$

Look at $1\frac{1}{2}\%$ = .015, go horizontally across from $n = 16$, and get 1.26898555 Ans.

364. A man invests \$8,000 for 12 years at 5% compounded quarterly. What amount will he get after 12 years?

$$S = P\left(1 + \frac{r_p}{p}\right)^{pt} = \$8{,}000\left(1 + \frac{.05}{4}\right)^{4\times 12} = \$8{,}000(1 + .0125)^{48}$$

Look at $1\frac{1}{4}\%$ interest for $n = 48$ horizontally

$$S = \$8{,}000 \times 1.81535485 = 14{,}522.8388$$

Therefore, he will receive

\$14,522.84 Ans.

PROBLEMS

1. (*a*) What per cent of 100 is 6?
   (*b*) What per cent of \$1 is 6¢?
   (*c*) If \$6 is charged for the use of \$100, what per cent of the sum loaned is the sum charged?
2. Find the interest on:

(*a*) \$5 for 1 year at 4%, at 5%, at 6%
(*b*) \$300 for 2 years at 2%, at 7%, at 9%
(*c*) \$400 for 3 years at 6%, for 2 years 3 months at 7%
(*d*) \$1,200 for 1 year at 3%, for 3 years at 7%, for 6 months at 8%

3. If you borrowed \$800 at simple interest for 4 years at 4%, how much will the creditor receive at the termination of the contract? How much would the interest amount to?

4. For a bank loan made on March 6 and falling due on March 28, interest would be charged for how many days?

5. (*a*) If in a transaction on September 4, a debtor agrees to repay in six months, when is the money due?

(*b*) If the agreement was to run 180 days, when would the due date be?

6. Find the time by compound subtraction between:

(*a*) June 14, 1958, and August 28, 1958
(*b*) September 12, 1957, and July 18, 1958
(*c*) December 14, 1955, and May 12, 1958
(*d*) October 18, 1954, and February 6, 1959
(*e*) July 29, 1955, and May 14, 1959

7. Find the exact time between the following dates, using Table 1, Appendix B:

(*a*) May 10, 1958, and January 14, 1959
(*b*) October 18, and January 10
(*c*) July 16, and November 11
(*d*) March 5, and November 8
(*e*) February 16, 1960, and July 7, 1960 (remember that a leap year has 366 days)

8. Find the interest on \$2,500 at 5% from October 17, 1959, to May 7, 1960: (*a*) by the ordinary method, (*b*) by the exact method, and (*c*) by the banker's method. Which produces the least interest, which the most?

9. Find the exact interest on \$1,000 from January 12 to April 18 at 3%.

10. Find the ordinary interest on \$6,200 from April 6 to July 12 at 3%.

11. Obtain the interest at 4% on \$12,000 for six months from April 15.

12. How much will \$5,000 be worth 120 days after April 21, 1960, if invested at 6% ordinary interest, and what is the due date?

13. Find the exact interest on \$3,800 for 135 days at $3\frac{1}{2}$%.

14. How would you find the exact interest, given the ordinary interest? How would you find the ordinary interest when given the exact interest?

15. Find the exact interest when the ordinary interest is:

(*a*) \$47.83	(*b*) \$386.40	(*c*) \$2.95
(*d*) \$12.02	(*e*) \$290.00	(*f*) \$3.75
(*g*) \$34.79	(*h*) \$3.68	(*i*) \$49.80

16. Find the ordinary interest when the exact interest is:

(*a*) \$3.28	(*b*) \$54.90	(*c*) \$658.60
(*d*) \$81.36	(*e*) \$6.22	(*f*) \$9.04
(*g*) \$227.90	(*h*) \$4,469.00	(*i*) \$64.38

17. What is the principal which at 5% for 146 days will yield an exact interest \$1.20 less than the ordinary interest?

18. Find the ordinary and exact interest on \$6,950 from May 10 to August 23 at 5%.

19. Find the interest for 60 days at 6% on \$1,438.

20. A businessman borrowed \$840 for 75 days at 6%. How much interest did he pay?

21. What is the interest on \$2,470 for 6 days at 6%?

22. Find the interest on \$680 for 66 days at 6%.

23. What are the aliquot parts of 60 in the following?

(*a*) 27 days	(*b*) 75 days	(*c*) 39 days	(*d*) 96 days
(*e*) 40 days	(*f*) 87 days	(*g*) 129 days	(*h*) 105 days
(*i*) 145 days	(*j*) 21 days	(*k*) 126 days	(*l*) 99 days

24. Find the interest on \$953.70 for 124 days at 6%.

25. Find the interest on \$598.90 for 47 days at 6%.

26. Find the interest on \$140 for 191 days at 6% (by interchanging the days and principal).

27. Find the interest on \$180 for 50 days at 6% (by deducting from the interest for 60 days).

28. By proper division of days, find the interest by the 60-day 6% method of:

(*a*) \$6,970.00 for 156 days
(*b*) \$386 for 84 days
(*c*) \$617.75 for 48 days
(*d*) \$5,900 for 222 days
(*e*) \$87.49 for 23 days

29. Find the interest from April 1 to July 9 by the 60-day 6% method on \$5,850.

30. By proper division of days, find the interest by the appropriate method on:

(*a*) \$487 for 142 days at 4.5%
(*b*) \$653 for 180 days at 2½%
(*c*) \$98.25 for 192 days at 1½%
(*d*) \$37.60 for 164 days at 8%
(*e*) \$2,179.75 for 105 days at 5%
(*f*) \$470 for 85 days at 7½%
(*g*) \$2,130 for 120 days at 4%
(*h*) \$423 for 129 days at 9%
(*i*) \$3,570 for 75 days at 3%

31. On a 360-day-per-year basis, at what rate per cent would you have to invest \$860 for 78 days to earn \$8.40 interest?

32. How much money would you need to invest at 4% for 82 days to earn \$12.90 interest?

33. How long will it take to earn \$16.45 interest on an investment of \$1,160 at 6%?

34. What principal will produce:

(*a*) \$18.70 interest at 6% for 72 days?
(*b*) \$8.35 interest at 6% for 126 days?
(*c*) \$14 interest at 6% for 96 days?
(*d*) \$15.74 interest at 6% for 75 days?

35. In what time will:
(*a*) \$700 produce \$14 at 6% (*b*) \$960 produce \$22.35 at 6%?
(*c*) \$1,400 produce \$11 at 6%? (*d*) \$2,200 produce \$84 at 6%?

36. At what rate will:
(*a*) \$1,400 produce \$28.30 in 126 days?
(*b*) \$760 produce \$11.60 in 96 days?
(*c*) \$1,680 produce \$21 in 75 days?
(*d*) \$3,200 produce \$18.20 in 36 days?

37. Find the interest by the 6-day 6% method on:
(*a*) \$300 for 24 days (*b*) \$150 for 27 days
(*c*) \$638.42 for 78 days (*d*) \$400 for 36 days
(*e*) \$25 for 66 days (*f*) \$500 for 51 days

38. What will \$550 amount to in 3 years at 4%, if interest is compounded annually?

39. In order to have \$5,000 at the end of 3 years, how much must you invest now at 4% compounded annually?

40. Find the compound amount on \$1 for 5 years at 4% using Table 2, Appendix B.

41. Which is greater: (1) a sum of money accumulating for 8 years at 2% compound interest, or (2) the same sum accumulating for 4 years at 4% compound interest? (use table).

42. What would \$10,000 amount to if invested for 6 years at $4\frac{1}{2}$% compounded annually?

43. What amount of money invested at 5% for 8 years would amount to \$3,841.40?

44. If you deposited \$2,100 in a bank which pays 5% per annum, how long will it take for this deposit to grow to \$2,954, if interest is compounded annually?

45. If \$1,000 is left on deposit for 1 year at a nominal rate of 4% compounded semiannually, what amount will there be by the end of the year?

46. What is the amount of \$1 at 6% compounded quarterly for 6 years (use table).

47. If a man invests \$10,000 for 10 years at 5% compounded quarterly, what amount will he get after 10 years?

48. Find the compound interest on \$2,000 for 8 years at 5% compounded (*a*) annually, (*b*) semiannually, and (*c*) quarterly.

49. Find the amount of \$5 placed annually for 10 years at 5% compound interest (use table).

50. If interest at 5% is compounded semiannually for 3 years, it amounts to the same as interest at $2\frac{1}{2}$% compounded annually for how many years?

51. A trust fund of \$20,000 earns interest at 3% a year compounded semiannually. What will the fund amount to in 10 years? How much will the interest be in that time?

CHAPTER X

RATIO — PROPORTION — VARIATION

365. What are the two ways of comparing like quantities?

(*a*) Subtracting the smaller from the larger—the *difference* method.

EXAMPLE: If you are 35 years old and your son is 5 years old, you are 30 years older than your son ($35 - 5 = 30$).

(*b*) Dividing one by the other—the *ratio* method.

EXAMPLE: You are 7 times as old as your son ($\frac{35}{5} = 7$).

366. What is meant by a ratio?

A comparison of two like quantities by dividing one by the other. As a ratio is a relationship of two quantities, we must be specific and indicate the *order* of their relationship.

Ex. (*a*) If machine A produces 300 units per hour, while machine B produces 450 units per hour, it is incorrect to say that the production ratio of these machines is $\frac{300}{450}$. We must say the production ratio of machine A to that of machine B is $\frac{300}{450}$.

Ex. (*b*) In Question 365 you must say that the ratio of your age to your son's age is $\frac{35}{5} = 7$, and not that the ratio of the ages is $\frac{35}{7}$. You may also say that the ratio of you son's age to yours is $\frac{5}{35} = \frac{1}{7}$.

367. What two terms are given in all ratio calculations?

The first term given is the numerator and is called the *antecedent*. The second term given is the denominator and is called the *consequent*.

Ex. (*a*) What is the relation between 4 and 12?

Here 4 is the first term = antecedent, and 12 is the second term = consequent

$$\therefore \text{Ratio} = \frac{4}{12} = \frac{1}{3} \qquad \text{or} \qquad \text{Ratio} = 1 \text{ to } 3$$

Ex. (*b*) If one house costs \$54,000 and another costs \$18,000, the ratio between the first and second house is

$$\frac{\$54{,}000}{\$18{,}000} = \frac{3}{1}$$

or, ratio is 3 to 1. One costs three times the other.

368. What symbol is used to indicate ratio?

Colon [:] = "to."

EXAMPLES: \$54,000: \$18,000 = 3 : 1 (to) (to)   4 : 12 = 1 : 3 (to) (to)

The colon is actually an abbreviation for [÷] with the horizontal line omitted.

369. How may ratios be expressed?

(*a*) By a single whole number.
The ratio of 35 years to 5 years is 7 (35 ÷ 5 = 7).

(*b*) As a fractional number.
The ratio of 1 ounce to a pound is $\frac{1}{16}$.

(*c*) As a decimal fraction.
The ratio of one side of a triangle 4 inches long to a second side 5 inches long is $\frac{4}{5}$ or 0.8.

(*d*) In fractional form and treated like a fraction.
$\frac{4}{5}$ may be read as the ratio of 4 to 5.

(*e*) With two dots separating the terms.
4:5 means the ratio of 4 to 5.

Note that when a ratio is expressed by a single integral fractional, or decimal number, the number 1 is the second term of the ratio, but is not written down. The ratio of 35 to 5 is the ratio of 7 to 1 or simply 7.

370. Can there be a ratio of unlike things?

No. The terms must be of like things. There can be no ratio between dollars and beans or between houses and yachts. Unless things can be changed to something that makes them alike there can be no ratio. There can be a ratio between the *cost* of a house and the *cost* of a yacht as expressed in dollars. Also the comparison must not only be between quantities of the same kind but between quantities expressed in the same units. We cannot compare pounds and inches for they are not quantities of the same kind, and we cannot compare a length in inches with a length in yards without first making the units alike, that is, we must either reduce yards to inches or convert inches to yards.

371. Is a ratio dependent upon the units of measure?

No. The ratio itself is always abstract, and the terms may be written as abstract numbers.

EXAMPLE: If two boards are 10 feet and 12 feet long respectively, the ratio of the first to the second board is 5:6, whether we express their lengths as inches, feet, or yards. The units cancel out and the ratio is 5:6.

372. Does multiplying or dividing both terms of a ratio by the same number change its value? No.

Ex. (*a*) $4{:}6 = \frac{4}{6} = \frac{2 \times 4}{2 \times 6} = \frac{8}{12}$ $\quad \therefore\ 4{:}6 = 8{:}12$

Ex. (*b*) $\frac{4}{6} = \frac{\frac{4}{2}}{\frac{6}{2}} = \frac{2}{3}$ $\quad \therefore\ 4{:}6 = 2{:}3$

373. What is the relation between ratio and per cent?

Since a ratio is always a fraction, we may think of a per cent as a ratio. Ratios are frequently expressed as per cents.

EXAMPLE: When we say \$100 is 20% of \$500, we mean that the ratio of \$100 to \$500 is

$$\frac{\$100}{\$500} = \frac{1}{5} = 20\%$$

Problems involving per cent can, however, be solved directly without referring to ratio.

374. How is a ratio simplified?

A ratio is always reduced to its simplest form. Perform the indicated division and reduce the resulting fraction to its lowest terms. Express the fraction as a ratio.

Ex. (*a*) Ratio $18{:}54 = \frac{18}{54} = \frac{1}{3} = 1{:}3$ simplified.

Ex. (*b*) Simplify the ratio $2\frac{2}{3}{:}3\frac{1}{5}$.

$$2\frac{2}{3} : 3\frac{1}{5} = 2\frac{2}{3} \div 3\frac{1}{5} = \frac{8}{3} \div \frac{16}{5} = \frac{\not{8}}{3} \times \frac{5}{\underset{2}{\not{16}}} = \frac{5}{6} \quad \text{or} \quad 5{:}6$$

375. What can be done in order to compare readily two or more ratios?

Reduce the ratios to such forms that the first terms of the ratios to be compared shall be the same, usually 1.

Ex. (*a*) Reduce 9:27 to a ratio having 1 for its first term.

Divide both terms by 9, getting 1:3.

Ex. (*b*) Reduce 16:39 to a ratio having 1 for its first term.

$$16:39 = \frac{16}{39}$$

Divide both terms by 16, getting $\frac{16}{16} = 1$ and $16\overline{)39}$ (quotient 2; 32; remainder 7) $= 2\frac{7}{16}$

$$\therefore\ 16:39 = 1:2\tfrac{7}{16} \text{ Ans.}$$

Ex. (*c*) Reduce .784:9 to a ratio having 1 for its first term.

$$.784:9 = \frac{.784}{9} \quad \text{(Divide both terms by .784)}$$

$.784 \div .784 = 1$ and $\frac{9}{.784} = .784\overline{)9.000}$ (quotient 11; 784; 1160; 784; 376) $= 11\frac{376}{784} = 11\frac{47}{98}$

$$\therefore\ .784:9 = 1:11\tfrac{47}{48} \text{ Ans.}$$

Ex. (*d*) Reduce $\frac{17}{129}:\frac{19}{127}$ to a ratio having 1 for its first term.

Divide both terms by $\frac{17}{129}$.

$$\frac{17}{129} \div \frac{17}{129} = 1 \quad \text{and} \quad \frac{19}{127} \div \frac{17}{129} = \frac{19}{127} \times \frac{129}{17} = \frac{2451}{2150} = 1.14$$

$$\therefore\ \frac{17}{129}:\frac{19}{127} = 1:1.14 \text{ Ans.}$$

376. What would you do when required to work out a complicated ratio containing fractions, per cents, or decimals?

Simplify the ratio first.

(*a*) If the denominators of both fractions are alike, they are in the ratio of their numerators.

EXAMPLE: Find the ratio between $\frac{5}{13}$ and $\frac{9}{13}$.

$$\frac{5}{13}:\frac{9}{13} = 5:9 = \text{Ratio of numerators}$$

(*b*) If the denominators are not alike, make them alike or divide the first fraction by the second fraction.

EXAMPLES:

(1) Find the ratio between $\frac{3}{5}$ and $\frac{11}{15}$. $\left(\frac{3}{5} = \frac{9}{15}\right)$

$$\therefore\ \frac{3}{5}:\frac{11}{15} = 9:11 = \text{Ratio of numerators}$$

(2) Find the ratio between $\frac{8}{25}$ and $\frac{64}{15}$.

$$\frac{8}{25}:\frac{64}{15} = \frac{8}{25} \div \frac{64}{15} = \frac{\overset{1}{\cancel{8}}}{\underset{5}{\cancel{25}}} \times \frac{\overset{3}{\cancel{15}}}{\underset{8}{\cancel{64}}} = \frac{3}{40} = 3:40 \text{ Ans.}$$

377. How do we divide some given number in a given ratio?

(*a*) Add the terms of the ratio and make it the denominator with the given number as the numerator.

(*b*) Multiply the quotient by each term of the ratio.

Ex. (1) Given 65. Divide 65 in the ratio 2:3.

$2 + 3 = 5$ = denominator
$\frac{65}{5} = 13$ = quotient = groups of 5
Now $13 \times 2 = 26$ = quotient × first term of ratio
$13 \times 3 = 39$ = quotient × second term of ratio
$26:39 = 2:3$
As $65 = 26 + 39$, therefore, 65 is divided into two terms, 26 and 39, in the ratio of 2:3.

Ex. (2) A shipment of 1,200 TV sets is to contain color sets in the ratio of 3:5. How many of each kind are there?

$3 + 5 = 8$ = denominator
$\frac{1200}{8} = 150$ = quotient = groups of 8
$150 \times 3 = 450$ = quotient × first term of ratio
$150 \times 5 = 750$ = quotient × second term of ratio
$\frac{450}{750} = \frac{3}{5}$ $(1{,}200 = 450 + 750)$

$\therefore$ 3:5 = 450 color sets : 750 black and white sets. Ans.

Ex. (3) 1,600 books are to be allotted to three classes in the ratio of 4:7:9. How many books will each class receive?

$4 + 7 + 9 = 20$ = denominator
$\frac{1600}{20} = 80$ = quotient = groups of 20
$80 \times 4 = 320$ = quotient × first term of ratio = books to class 1
$80 \times 7 = 560$ = quotient × second term of ratio = books to class 2
$80 \times 9 = 720$ = quotient × third term of ratio = books to class 3
Total = 1,600 books

$\therefore$ 320:560:720 = 4:7:9 Ans.

378. How can we divide 65 in the ratio $\frac{1}{2}:\frac{1}{3}$?

Reduce fractions to a common denominator.
First term $= \frac{1}{2} = \frac{3}{6}$ and $\frac{1}{3} = \frac{2}{6}$ = second term.
Add the numerators of these: $3 + 2 = 5$.
Divide 65 by 5 and use 3 and 2 as numerators.
First term $= \frac{65}{5} \times 3 = 39$ and $\frac{65}{5} \times 2 = 26$ = second term

$\therefore \frac{1}{2}:\frac{1}{3} = \frac{39}{26}$ Ans.

379. How do we solve a ratio problem in which the ratio is not given?

First, we assign a ratio value of 1 to the given basic quantity. We then compute the ratio values of all the other quantities, basing our calculations on the given facts, thus arriving at a ratio.

Then we proceed as in Question 377 as when ratio is given.

EXAMPLE: A company bought 3 trucks. The first cost $1\frac{1}{2}$ times as much as the second. The third cost $2\frac{1}{2}$ times as much as the second. The company paid \$30,000 for the 3 trucks. How much did it pay for each?

$$\begin{aligned}&\text{Basic truck (second truck)} &&= 1\\ &\text{First truck} = 1\tfrac{1}{2} \times 1 &&= \tfrac{3}{2}\\ &\text{Third truck} = 2\tfrac{1}{2} \times 1 &&= \tfrac{5}{2}\end{aligned}$$

$$\therefore \text{ First:second:third} = \tfrac{3}{2}:1:\tfrac{5}{2}$$

Add the terms of the ratio $\frac{3}{2} + 1 + \frac{5}{2} = 5$ (= denominator = 5 parts, one part of which is the basic truck)

$$\frac{\$30{,}000}{5} = \$6{,}000 = \text{quotient} = \text{cost of 1 part} = \text{basic truck}$$

$$\therefore \ \$6{,}000 \times \tfrac{3}{2} = \$9{,}000 = \text{cost of truck No. 1}$$

$$\$6{,}000 \times \tfrac{5}{2} = \$15{,}000 = \text{cost of truck No. 3}$$

$$(\$9{,}000 + \$6{,}000 + \$15{,}000 = \$30{,}000)$$

$$\therefore \ \tfrac{3}{2}:1:\tfrac{5}{2} = \$9{,}000:\$6{,}000:\$15{,}000 \text{ Ans.}$$

380. If the wing span of a plane is 76 ft. 6 in., what will the wing span of a model have to be when the ratio of the length of any part of the model to the length of the corresponding part of the actual plane is 1:72?

The length of model is thus $\frac{1}{72}$ of the corresponding length of the actual plane.

$$\frac{1}{72} \times 76\frac{1}{2} = \frac{1}{\underset{8}{\cancel{72}}} \times \frac{\overset{17}{\cancel{153}}}{2} = \frac{17}{16} = 1\frac{1}{16} \text{ ft.}$$

or

$$\frac{17}{\underset{4}{\cancel{16}}} \times \frac{\overset{3}{\cancel{12}}}{1} = \frac{51}{4} = 12\frac{3}{4} \text{ in.}$$

381. If a bankrupt firm can pay 60¢ on the dollar, and if its assets amount to $28,000, what are its liabilities?

Paying 60¢ on the dollar means that its ratio of assets to liabilities = .60.

$$\$28{,}000 = 60\% \text{ of what it owes}$$

$$\therefore \frac{\$28{,}000}{.60} = \$46{,}666.66 = \text{Liabilities. Ans.}$$

382. What selling price should be placed on a TV set if the cost is $250 and the dealer operates on a margin of 30% of cost?

A margin of 30% of cost = ratio of margin on set to its cost = $\frac{3}{10}$.
Thus margin here = $\frac{3}{10}$ of cost
Or margin = $\frac{3}{10} \times \$250 = \$75$

$\therefore$ Selling price = $250 + $75 = $325 Ans.

383. If you allow 12% of your income for clothing and 21% for rent, (*a*) what is the ratio of the cost of rent to the cost of clothing, (*b*) how much do you spend for rent per month when your income is $8,400 per year?

(*a*) $\frac{21\%}{12\%} = \frac{7}{4}$ or 7:4

(*b*) $\frac{\$,8400}{12 \text{ months}} \times .21 = \$147$ for rent

384. If a town estimates that it has to raise $300,000 in taxes and the assessed valuation of its real property is $9,000,000, what is the tax rate?

Tax rate = ratio of amount to be raised to assessed valuation.

$$\therefore \text{Tax rate} = \frac{\$300{,}000}{\$9{,}000{,}000} = \frac{1}{30} \times 100\% = \frac{10}{3}\% = 3\tfrac{1}{3}\% \text{ Ans.}$$

385. A certain concrete mixture is to be made up of 1 part cement, 3 parts sand, and 5 parts stone. What is (*a*) ratio of sand to stone, (*b*) the ratio of cement to sand, (*c*) per cent of sand in the concrete mixture?

(*a*) Sand:stone = 3:5
(*b*) Cement:sand = 1:3
(*c*) 1 + 3 + 5 = 9 parts in the entire mixture

$\therefore$ Sand:mixture = 3:9 = $\frac{1}{3}$ $\frac{1}{3} \times 100\% = 33\frac{1}{3}\%$ Ans.

386. If the bedroom of a house is shown on the print to be $3\frac{1}{2}$ in. × $4\frac{1}{4}$ in., and if the scale of the blueprint is $\frac{1}{4}$ in. = 1 ft., what are the actual dimensions of the room?

$3\frac{1}{2}$ in. = $\frac{7}{2}$ in. and if $\frac{1}{4}$ in. = 1 ft.,

$$\text{then } \frac{7}{2} \div \frac{1}{4} = \frac{7}{2} \times \frac{4}{1} = 14 \text{ ft.} = \text{Width}$$

$4\frac{1}{4}$ in. = $\frac{17}{4}$ in.,

$$\text{then } \frac{17}{4} \div \frac{1}{4} = \frac{17}{4} \times \frac{4}{1} = 17 \text{ ft.} = \text{Length}$$

387. What is meant by an "inverse ratio"?

It is merely a ratio with reversed terms.

EXAMPLE: What is the inverse ratio of 40:8?

Reverse the ratio, getting $8:40 = \frac{8}{40} = \frac{1}{5}$ = "inverse ratio" of 40:8.

388. What would be your share in an automobile that cost you and your brother \$880, if $\frac{1}{6}$ of your share is equal to $\frac{1}{5}$ of your brother's?

The ratios in this case would be $\frac{1}{6}:\frac{1}{5}$ or $\frac{1}{6} \div \frac{1}{5} = \frac{5}{6}$. Since in division of fractions one fraction is reversed, our answer $\frac{5}{6}$ is the reverse of the true ratio. Therefore the true ratio is the reverse of this, or $\frac{6}{5}$. The ratio $\frac{5}{6}$ is known as an "inverse" ratio.

Now add the terms of the ratio 6 + 5 = 1. Then

$$\frac{\$880}{11} = \$80$$

$$\therefore\ \$80 \times 6 = \$480 = \text{your share}$$

$$\text{and } \$80 \times 5 = \$400 = \text{your brother's share}$$

$$\text{Check: } \frac{\$480}{6} = \$80 \qquad \frac{\$400}{5} = \$80$$

389. What are some general rules for ratio calculation?

(*a*) To get a ratio, divide the first term by the second term.

EXAMPLE: What is the ratio of 1 yard to 1 inch?

$$1 \text{ yard} = 36 \text{ inches} \qquad \therefore \frac{1 \text{ yard}}{1 \text{ inch}} = \frac{36}{1} = 36:1 \text{ Ans.}$$

(*b*) To get the first term, multiply the given second term by the given ratio.

EXAMPLE: ?:3 = 9

$\therefore$ 3 × 9 = 27 = First term. Ans. (Check: 27:3 = 9)

(*c*) To get the second term, divide the first term by the ratio.

EXAMPLE: 36: ? = 12

$\therefore \frac{36}{12} = 3$ = Second term. Ans. (Check: 36 ÷ 3 = 12)

390. How do we compound ratios?

Change the expressions to fraction form. Then treat the two fractions as a problem in multiplication.

EXAMPLE: What is the compound ratio of 8:4 and 24:36?

$$8{:}4 = \tfrac{8}{4} \quad \text{and} \quad 24{:}36 = \tfrac{24}{36}$$

$$\therefore \frac{\overset{2}{\cancel{8}}}{\underset{1}{\cancel{4}}} \times \frac{\overset{2}{\cancel{24}}}{\underset{3}{\cancel{36}}} = \frac{4}{3} = \text{Compound ratio of } 8{:}4 \text{ and } 24{:}36$$

The product of two or more simple ratios is a compound ratio.

391. How do we solve, in a manner similar to that of a ratio problem, a problem in which the same number of articles are bought, each at a different price?

Add the various prices and divide this sum into the total price.

EXAMPLE: If you buy the same number of oranges at 6¢, 8¢, and 10¢, and you spend $2.88, how many at each price did you buy?

To buy one of each would cost 6 + 8 + 10 = 24¢

$$\therefore \frac{\$2.88}{24¢} = 12 \text{ of each were bought. Ans.}$$

392. How do we solve, in a manner similar to that of a ratio problem, a problem in which a different number of articles are bought at different prices?

Proceed in the same manner as in a ratio problem when the ratio is not given.

(*a*) Find the basic quantity with which all the others are compared.

(*b*) Assign value 1 to it and compute value of other quantities according to given facts or relations.

(*c*) Multiply the prices by their respective values.

(*d*) Add these products and divide this sum into the total cost to get the basic quantity. Multiply this basic quantity by the ratio value to get the other quantities.

EXAMPLE: If your firm buys 4 times as many truck tires at $37 each as passenger-car tires at $18, and twice as many station-wagon tires at $24 each, how many of each did it buy if it spent $2,354?

The basic quantity is "passenger-car tires." Assign a value 1 to this base.

Value = 4 for truck tires, as there are 4 times as many truck tires.
Value = 2 for station-wagon tires, as there are twice the number of these as compared with the base 1.

Since we cannot compare a ratio of unlike things, the ratio cannot be expressed in tires, but in cost of tires.

Thus, passenger-car tires cost = \$18 each = base 1
Truck tires cost \$37 each × 4 value = \$148
Station-wagon tires cost \$24 each × 2 value = \$48
Therefore, the ratio is \$148: \$18: \$48
\$148 + \$18 + \$48 = \$214 (= cost per group of 4 + 1 + 2 = 7 tires)

$$\frac{\$2,354}{\$214} = 11 \text{ passenger tires } (= \text{basic})$$

$$4 \times 11 = 44 \text{ truck tires}$$
$$2 \times 11 = 22 \text{ station-wagon tires}$$

For each type the firm spent

44 × \$37 =	\$1,628	(truck tires)
11 × \$18 =	\$198	(passenger-car tires)
22 × \$24 =	\$528	(station-wagon tires)
	\$2,354	= Total cost

393. What is meant by a proportion?

A statement that two ratios are equal.

EXAMPLE: $\frac{4}{8} = \frac{1}{2}$ 4:8 = 1:2
ratio = ratio   ratio = ratio

394. How are proportions written?

[::] = "as."

6:8::3:4   6 is to 8 as 3 is to 4
or 6:8 = 3:4   6 is to 8 equals 3 is to 4
or $\frac{6}{8} = \frac{3}{4}$ = fractional form

395. What are the terms of a proportion?

"Extremes" = first and last terms.
"Means" = the two middle terms.

means
6 : 12 = 8 : 16
extremes

extremes
$\frac{6}{12} = \frac{8}{16}$
means

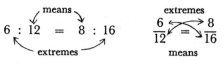

396. What is the test as to whether the terms are in proportion?

The product of the extremes = the product of the means.

means (4, 9); extremes (3, 12)

$$3 : 4 :: 9 : 12 \qquad \text{also} \qquad \frac{3}{4} = \frac{9}{12}$$

extremes (3, 12) means (4, 9)

(also called cross multiplication)

$$3 \times 12 = 4 \times 9 = 36 = \text{Test for a proportion}$$

(extremes) (means)

397. From the above, how do we find either mean that is not given?

Multiply the extremes and divide by the given mean.

$6:? = 8:16 \qquad 6:12 = ?:16$

$$\frac{6 \times \overset{2}{\cancel{16}}}{\underset{1}{\cancel{8}}} = 12 \qquad \frac{\cancel{6} \times \overset{8}{\cancel{16}}}{\underset{2}{\cancel{12}}} = 8$$

398. From the above, how do we find a missing extreme?

Multiply the means and divide by the given extreme.

$?:12 = 8:16 \qquad 6:12 = 8:?$

$$\frac{\overset{6}{\cancel{12}} \times \cancel{8}}{\underset{2}{\cancel{16}}} = 6 \qquad \frac{\overset{2}{\cancel{12}} \times 8}{\underset{1}{\cancel{6}}} = 16$$

399. You buy 8 tons of coal for $208. What will 12 tons cost?

8 tons: $208::12 tons:?

$$\frac{\overset{26}{\cancel{208}} \times 12}{\underset{1}{\cancel{8}}} = \$312 \text{ Ans.}$$

400. A 9-foot-high tree casts a shadow of $16\frac{1}{2}$ feet. What is the height of a radio tower that casts a shadow of 203 feet?

9-ft. tree:16.5-ft. shadow:: ? height of tower:203 ft. shadow

$$\frac{9 \times 203}{16.5} = \frac{1827}{16.5} = 110.73 \text{ ft. high Ans.}$$

401. When are quantities said to be in direct proportion?

When the first is to the second as the second is to the third.

EXAMPLE: 3:6:12 = a direct proportion.

402. What is meant by a mean proportional?

When the second term is equal to the third each is a *mean proportional* to the other two.

Ex. (*a*) 3:6::6:12

6 is a mean proportional to 3 and 12

Ex. (*b*) $2:5::5:12\frac{1}{2}$

5 is a mean proportional between 2 and $12\frac{1}{2}$

Ex. (*c*) $3:x::x:12$

x is a mean proportional between 3 and 12 or $x^2 = 36$

Product of means = product of extremes

$\therefore x = 6 =$ the mean proportional between 3 and 12

This is also known as the geometric mean.

403. How does stating a problem as a simple proportion simplify the finding of an unknown term in a problem?

EXAMPLE: If 36 gallons of gasoline cost \$8.64, how much will 60 gallons cost?

$$\frac{\overset{3}{\cancel{36}}\text{ gal.}}{\underset{5}{\cancel{60}}\text{ gal.}} = \frac{\$8.64}{x}$$

$$3x = 5 \times \$8.64$$

Product of means = Product of extremes

$$\therefore x = \frac{5 \times \$8.64}{3} = \frac{\$43.20}{3} = \$14.40 \text{ Ans.}$$

By elementary arithmetic we can find the cost of one gallon.

```
               $.24
1 gal. = 36) $8.64        ∴ 60 gal. = 60 × $.24 = $14.40
              72
              144
```

This method can be lengthy.

404. An alloy consists of 4 parts of tin and 6 parts of copper. How many pounds of copper would be needed with 120 pounds of tin to maintain the given ratio?

$$\frac{4 \text{ parts}}{6 \text{ parts}} = \frac{120 \text{ lb.}}{x \text{ lb.}}$$

$$4x = 6 \times 120 = 720$$

Product of means = Product of extremes

$$\therefore\ x = \frac{720}{4} = 180 \text{ lb. copper, Ans.}$$

405. What is meant by an inverse proportion?

Quantities are said to vary inversely when one quantity increases as the other decreases. Most of such problems deal with "speed-and-time" or "work-and-time."

Ex. (*a*) As speed increases, time taken decreases.

Ex. (*b*) The greater the number of men employed on a job, the less time it takes for completion.

Ex. (*c*) The distance between two airfields is 1,000 miles. If the average speed of a plane is 100 mph, the trip will take 10 hours. If the average speed is 200 mph, it will take 5 hours.

$$\text{Ratio of "speeds"} = \frac{100 \text{ mph}}{200 \text{ mph}} = \frac{1}{2}$$

$$\text{Corresponding ratio of "times"} = \frac{10 \text{ hr.}}{5 \text{ hr.}} = \frac{2}{1}$$

Inverse ratios

One is the inverse of the other.

406. Driving to your office at 45 mph, you make it in 55 minutes. At what speed would you have to travel to get there in 50 minutes?

$$\text{Ratio of "speeds"} = \frac{45 \text{ mph}}{x \text{ mph}}$$

$$\text{Inverse ratio of "time"} = \frac{50 \text{ min.}}{55 \text{ min.}}$$

$$\frac{45}{x} = \frac{50}{55} \qquad 50x = 45 \times 55$$

$$\therefore\ x = \frac{\overset{9}{\cancel{45}} \times 55}{\underset{10}{\cancel{50}}} = 49.5 \text{ mph Ans.}$$

Note that the speed 45 mph, and the time 55 minutes must be so set up to provide for cross multiplication in the fractional form to give "speed-time" (45 × 55).

407. How is an inverse proportion set up?

EXAMPLE: If 24 men do a job in 15 days, how many men will be required to do it in 5 days?

Set up proportion in fractional form to utilize cross multiplication, so that 24 men and 15 days are multiplied to give "man-days." This will give the setup for an inverse ratio.

Any one of the following will do that.

$$\frac{24}{x} = \frac{5}{15}, \quad \frac{24}{5} = \frac{x}{15}, \quad \frac{15}{5} = \frac{x}{24}, \quad \frac{15}{x} = \frac{5}{24}$$

$$5x = 24 \times 15 \quad \text{or} \quad x = \frac{24 \times \overset{3}{\cancel{15}}}{\underset{1}{\cancel{5}}} = 72 \text{ men Ans.}$$

Further simplification can be obtained by reducing the fraction in which 5 occurs, getting

$$\frac{24}{x} = \frac{5}{15} = \frac{1}{3} \quad \text{or} \quad x = 3 \times 24 = 72 \text{ directly}$$

408. If 130 yards of a copper wire offer 1.8 ohm resistance, what will be the resistance of 260 yards of copper wire of $2\frac{1}{2}$ times the cross-sectional area?

The greater the cross-sectional area of a wire the less the resistance. First, the increased length will increase the resistance.

$$\frac{130 \text{ yd.}}{1.8 \text{ ohm}} = \frac{260 \text{ yd.}}{x \text{ yd.}} \qquad 130x = 260 \times 1.8$$

$$\therefore x = \frac{260}{130} \times 1.8 = 3.6 \text{ ohm Ans.}$$

Second, the larger area will decrease the resistance in the ratio

$$\frac{1}{2\frac{1}{2}} = \frac{1}{\frac{5}{2}} = \frac{2}{5} = \text{inverse ratio}$$

$$\therefore 3.6 \text{ ohm} \times \tfrac{2}{5} = 1.44 \text{ ohm Ans.}$$

409. What is a compound proportion?

One in which either or both ratios are compound.

We sometimes have to deal with units that have to be multiplied.

EXAMPLE: A private nursing home took care of 16 city welfare patients for 5 months and another group of 20 patients for 7 months.

(*a*) What is the ratio of the maintenance charge for the two groups?

(*b*) If the charge for the smaller group was \$16,000, what would the charge for the larger group be?

(*c*) If the charge for the larger group was \$35,000, what would the smaller be?

(*a*) The ratio between the groups would be

$$\frac{\overset{4}{\cancel{16}} \times \cancel{5}}{\underset{5}{\cancel{20}} \times 7} = \frac{4}{7} = \text{Compound ratio}$$

(*b*) Charge for smaller group is thus $\frac{4}{7}$ of the larger, and the charge for larger group is $\frac{7}{4}$ of the smaller.

If smaller charge is \$16,000

$$\frac{7}{\underset{1}{\cancel{4}}} \times \overset{\$4,000}{\$\cancel{16,000}} = \$28,000 = \text{Charge for larger group}$$

(*c*) If charge for larger group is \$35,000

$$\frac{4}{\underset{1}{\cancel{7}}} \times \overset{\$5,000}{\$\cancel{35,000}} = \$20,000 = \text{Charge for smaller group}$$

410. What is the rule for solving a compound proportion?

(*a*) Place the unknown quantity as the fourth term of the proportion.

(*b*) Place as the third term the given quantity expressing the same kind of thing as the unknown quantity.

(*c*) Arrange each of the other ratios according to its relation to the ratio already stated.

(*d*) Get the product of all the means and divide it by the product of all the extremes, except the unknown one, to find the answer.

411. If 20 men working 6 hours per day can dig a trench 80 feet long in 30 days, how many men working 10 hours a day can dig a trench 120 feet long in 12 days?

(*a*) Place x = unknown quantity as fourth term (= men).

(*b*) Place 20 = men as third term. Then $20:x = \frac{20 \text{ men}}{x \text{ men}}$ (which is the third to fourth term ratio).

(*c*) Next ratio: $\frac{12 \text{ days}}{30 \text{ days}}$ is an inverse ratio, and must be set up so that 30 days and 20 men can be cross-multiplied to give "man-days."

Next ratio: $\frac{80 \text{ ft.}}{120 \text{ ft.}}$ is a direct ratio.

Next ratio: $\frac{10 \text{ hours per day}}{6 \text{ hours per day}}$ is an inverse ratio and is so set up that 6 hours per day × 20 men gives "man-hours per day."

Thus

$$\text{Compound ratio} \rightarrow \left\{\begin{matrix} (\frac{12}{30}) \\ (\frac{80}{120}) \\ (\frac{10}{6}) \end{matrix}\right\} = \frac{20}{x} \quad \text{or} \quad \frac{\overset{2}{\cancel{12}}}{\cancel{30}} \times \frac{\overset{2}{\cancel{80}}}{\underset{3}{\cancel{120}}} \times \frac{\cancel{10}}{\underset{1}{\cancel{6}}} = \frac{4}{9}$$

$$\frac{4}{9} = \frac{20}{x}$$

$$\therefore\ 4x = 180 \qquad x = \tfrac{180}{4} = 45 \text{ men Ans.}$$

412. Why is it possible to set up the second member of the proportion as a single ratio?

(*a*) In the above, 20 men dig a trench in 30 days. Then, in 12 days

$$\frac{12 \text{ days}}{30 \text{ days}} = \frac{20 \text{ men}}{? \text{ men}} \qquad \frac{30 \times 20}{12} = \text{(50) men}$$

$$\rightarrow 12:30 = 20:50$$

(*b*) Now, if 50 men dig an 80-ft. trench in 12 days, then, for a 120-ft. trench

$$\frac{80 \text{ ft.}}{120 \text{ ft.}} = \frac{50 \text{ men}}{? \text{ men}} \qquad \frac{\overset{15}{\cancel{120}} \times \cancel{50}}{\underset{1}{\cancel{80}}} = \text{(75) men}$$

$$\rightarrow 80:120 = 50:75$$

(*c*) If 75 men dig a 120-ft. trench in 12 days working 6 hours per day, then, working 10 hours per day

$$\frac{10}{6} = \frac{75 \text{ men}}{? \text{ men}} \qquad \frac{6 \times 75}{10} = \textcircled{45} \text{ men}$$

$$\rightarrow 10:6 = 75:45$$

This method of procedure may be shortened by multiplying the completed proportions (*a*), (*b*), and (*c*) together term by term, to get a new proportion which is expressed as a ratio.

$$\frac{12 \times 80 \times 10}{30 \times 120 \times 6} = \frac{20 \times \cancel{50} \times \cancel{75}}{\cancel{50} \times \cancel{75} \times 45} = \frac{20}{45}$$

We see that the answers obtained from the first two proportions cancel, leaving the second member a simple ratio. The ratio may now be expressed as a proportion

$$\left\{\begin{matrix} 12:30 \\ 80:120 \\ 10:6 \end{matrix}\right\} = 20:? \text{ men}$$

and solved, as follows

$$\frac{30 \times \overset{1}{\cancel{120}} \times \overset{3}{\cancel{6}} \times \overset{1}{\cancel{20}}}{\underset{1}{\cancel{12}} \times \underset{\underset{2}{\cancel{4}}}{\cancel{80}} \times \cancel{10}} = 45 \text{ men Ans.}$$

As the first two answers cancel, it was unnecessary to obtain them to arrive at the final answer.

413. If 2 men cut 8 cords of wood in 4 days, how long will it take 12 men to cut 36 cords?

$$\text{Set up} \rightarrow \frac{4 \text{ days}}{x \text{ days}} = \frac{\text{third term}}{\text{fourth term}}$$

$$\text{Compound ratio} \left\{\begin{matrix} \dfrac{8 \text{ cords}}{36 \text{ cords}} = \text{direct ratio.} & \left\{\begin{matrix}\text{The more “cords,”} \\ \text{The more “time,”}\end{matrix}\right. \\ \dfrac{12 \text{ men}}{2 \text{ men}} = \text{inverse ratio.} & \left\{\begin{matrix}\text{The more “men,”} \\ \text{The less “time.”}\end{matrix}\right. \end{matrix}\right.$$

$$\therefore \frac{\overset{4}{\cancel{8}}}{\underset{\underset{3}{\cancel{6}}}{\cancel{36}}} \times \frac{\overset{\overset{1}{\cancel{6}}}{\cancel{12}}}{\underset{1}{\cancel{2}}} = \frac{4}{x} \qquad \frac{4}{3} = \frac{4}{x} \qquad 4x = 3 \times 4$$

$$\therefore x = 3 \text{ days Ans.}$$

414. If the eggs laid by 30 hens in 15 weeks are worth $108, what will be the value of the eggs laid by 60 hens in 10 weeks?

$$\text{Set up} \rightarrow \frac{\$108}{\$x} = \frac{\text{third term}}{\text{fourth term}}$$

$$\text{Compound ratio} \begin{cases} \dfrac{30 \text{ hens}}{60 \text{ hens}} = \text{direct ratio.} & \begin{cases}\text{The more "hens,"} \\ \text{The more "money."}\end{cases} \\ \dfrac{15 \text{ weeks}}{10 \text{ weeks}} = \text{direct ratio.} & \begin{cases}\text{The less "weeks,"} \\ \text{The less "money."}\end{cases} \end{cases}$$

$$\frac{\overset{1}{\cancel{30}}}{\underset{2}{\cancel{60}}} \times \frac{\overset{3}{\cancel{15}}}{\underset{2}{\cancel{10}}} = \frac{108}{x} \qquad \frac{3}{4} = \frac{108}{x}$$

$$\therefore\ x = \frac{\overset{36}{\cancel{108}} \times 4}{\underset{1}{\cancel{3}}} = \$144 \text{ Ans.}$$

415. What are some of the properties of proportion that can be obtained by elementary algebraic changes in the form of the equation which expresses the proportion?

(*a*) If $\frac{a}{b} = \frac{c}{d}$, where *a*, *b*, *c*, and *d* are numbers in proportion, the product of the means = the product of extremes.

$ad = bc$ by multiplying diagonally "corner to corner."

EXAMPLE: If $\frac{3}{4} = \frac{6}{8}$, 3, 4, 6, and 8 are in proportion, and

$$3 \times 8 = 4 \times 6$$

(*b*) If $\frac{a}{b} = \frac{c}{d}$, then $\frac{b}{a} = \frac{d}{c}$.

The numbers are in proportion by inversion. You merely invert both sides of the proportion.

EXAMPLE: If $\frac{3}{4} = \frac{6}{8}$, then $\frac{4}{3} = \frac{8}{6}$

(*c*) If $\frac{a}{b} = \frac{c}{d}$, then $\frac{a}{c} = \frac{b}{d}$.

The numbers are in proportion by alternation. The first is to the third as the second is to the fourth.

EXAMPLE: If $\frac{3}{4} = \frac{6}{8}$, then $\frac{3}{6} = \frac{4}{8}$

(*d*) If $\frac{a}{b} = \frac{c}{d}$, then $\frac{a+b}{b} = \frac{c+d}{d}$.

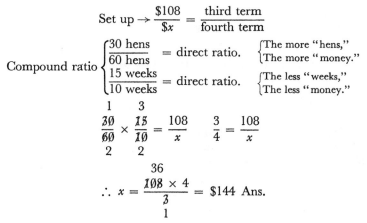

The terms are in proportion by composition. You add the second to the first and the fourth to the third.

EXAMPLE: If $\frac{3}{4} = \frac{6}{8}$, then $\frac{3+4}{4} = \frac{6+8}{8}$ or $\frac{7}{4} = \frac{14}{8}$

(*e*) If $\frac{a}{b} = \frac{c}{d}$, then $\frac{a-b}{b} = \frac{c-d}{d}$.

The terms are in proportion by division. You subtract the second from the first and the fourth from the third.

EXAMPLE: If $\frac{8}{6} = \frac{16}{12}$, then $\frac{8-6}{6} = \frac{16-12}{12}$ or $\frac{2}{6} = \frac{4}{12}$

(*f*) If $\frac{a}{b} = \frac{c}{d}$, then $\frac{a+b}{a-b} = \frac{c+d}{a-b}$.

The terms are in proportion by composition and division.

EXAMPLE: If $\frac{8}{6} = \frac{16}{12}$, then $\frac{8+6}{8-6} \quad \frac{16+12}{16-12}$ or $\frac{14}{2} = \frac{28}{4}$

416. What proportions of 3% milk and 5% milk must be mixed to get $4\frac{1}{2}$% milk?

If you have a unit volume of 5% butterfat milk, you can reduce its concentration by adding x parts of a unit of 3% milk.

The sum of the concentrations over the combined volume = the required concentration. Then

$$\frac{5\% + 3\% x}{1 + x} = 4\tfrac{1}{2}\%$$

or

$$5 + 3x = 4\tfrac{1}{2}(1 + x) = 4\tfrac{1}{2} + 4\tfrac{1}{2}x$$

$$\tfrac{1}{2} = 1\tfrac{1}{2}x$$

$$x = \frac{\frac{1}{2}}{1\frac{1}{2}} = \frac{1}{2} \times \frac{2}{3} = \frac{1}{3} \text{ part of a unit of 3\% milk added}$$

$$5\% \text{ milk} : 3\% \text{ milk} :: 3 : 1 \quad \text{Ans.}$$

This means that for every unit volume of 3% milk you must have 3 unit volumes of 5% milk.

$$\frac{3 \text{ volumes of } 5\% + 1 \text{ volume of } 3\%}{1 + 3} = 4\tfrac{1}{2}\%$$

$$\frac{3 \times 5 + 3}{4} = \frac{18\%}{4} = 4\tfrac{1}{2}\% \text{ (check)}$$

417. How is proportion applied to the principle of the lever?

The lever is a rigid structure, often a straight bar, which turns freely on a fixed point or fulcrum, and which is used to transmit pressure or motion from a source of power (or force) to a weight (or resistance).

When the lever is in equilibrium, the power and the weight (or resistance) are in inverse ratio to their respective distances from the fulcrum.

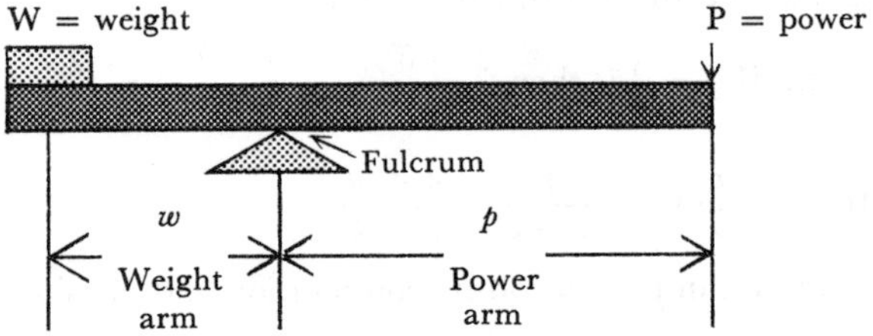

$$\frac{W}{P} = \frac{w}{p} = \text{inverse ratio or proportion} \quad \text{or} \quad Pp = Ww$$

When the setup is such that there is cross multiplication between the corresponding factors, you have an inverse ratio or proportion.

EXAMPLE: Using a 14-foot plank, where must you put the support under the planks so that two children weighing 45 and 55 pounds respectively can play seesaw?

$$W = 45 \text{ lb.} \qquad w = x \text{ ft.}$$
$$P = 55 \text{ lb.} \qquad p = (14 - x) \text{ ft.}$$
$$Ww = Pp \quad \text{or} \quad 45x = 55(14 - x) = 770 - 55x$$
$$100x = 770 \qquad \therefore\ x = 7.7 \text{ ft. Ans.}$$

Support to be placed 7.7 ft. from smaller child.

418. What is the relation between ratio and proportion, and the language of variation?

Ratio and proportion may at times be conveniently stated in the language of variation.

EXAMPLE: If you divide the circumference C of any circle by its diameter d, you will get:

(a) $\frac{C}{d} = \pi = \text{constant} = \text{ratio.}$ This is a statement of a ratio $\frac{C}{d}$.

(b) This ratio, however, may be written as a variation, in the form:

$$C = \pi d = \text{variation form.}$$

This means that circumference C varies as diameter d.
If d is halved, then C is halved (π is constant).
If d is doubled, then C is doubled (π is constant).

419. What may be said about each of the statements of ratio and proportion?

Each implies an equation involving a *constant.*

Ex. (*a*) Hooke's law states that the elongation E of a spring balance varies directly as the weight W is applied.

$$E = kW \quad \text{or} \quad \text{ratio} = \frac{E}{W} = k = \text{constant}$$

Ex. (*b*) Boyle's law states that the volume v of a gas at a constant temperature varies inversely as the pressure p ("inversely as" means "reciprocal of").

$$v = \frac{k}{p} \quad \text{or} \quad vp = k = \text{constant}$$

(supplied to take care of different gases and various temperatures).

A single experiment will determine k. If for a certain gas at a certain temperature a volume of 250 cc. results from a pressure of 20 lb. per sq. in., then

$$250 = \frac{k}{20} \quad \text{or} \quad k = 5{,}000$$

and Boyle's law would for this case be $v = 5{,}000/p$.

420. What is implied in a direct variation and how is a direct variation expressed?

The statement "y varies directly as x" (or abbreviated as "y varies as x" or "$y \propto x$") ($\propto$ means "varies as") implies that there is a constant k such that

$\frac{y}{x} = k$ is true (symbol $\propto$ is replaced by [=] and a constant k)

The direct variation is expressed as $y = kx$.

k in applied work is found numerically by an experiment, and is inserted to get a particular equation for later use.

Example: We know that the surface S of a sphere varies directly as the square of its radius r.

$$S \propto r^2$$

This implies the equation

$$\frac{S}{r^2} = k$$

and the direct variation is expressed as $3 = kr^2$. By theory and measurement we can determine that $k = 4\pi$ and the equation becomes

$$S = 4\pi r^2$$

which is the usual formula for the surface of a sphere.

421. What is implied in an inverse variation and how is an inverse variation expressed?

The statement "y varies inversely as x" or $y \propto 1/x$ implies that there is a constant k such that $y = k/x$ is true (symbol $\propto$ is replaced by [=] and a constant k).

The inverse variation is expressed as $yx = k$.

EXAMPLE: In Question 419, what is the volume of the gas for a pressure of 25 lb. per sq in.?

$$\text{Boyle's law} \quad v = \frac{5{,}000}{p} = \frac{5{,}000}{25} = 200 \text{ cc. Ans.}$$

422. What is meant by a joint variation and how is it expressed?

A joint variation may be any combination of one or more of each of the direct and inverse types.

If z varies as x and inversely as y, or $z \propto x/y$, then we may write

$$z = \frac{kx}{y}$$

by replacing the symbol $\propto$ with [=] and a constant k, and this implies that there is a constant k such that $zy/x = k$ is true. This is an expression of a joint variation.

EXAMPLE: Thus, if we know that when $z = 6$, $x = 4$, $y = 2$, we can find the value of z when $x = 5$ and $y = 3$.

From

$$k = \frac{zy}{x} = \frac{6 \times 2}{4} = 3$$

then

$$z = \frac{kv}{y} = \frac{3 \times 5}{3} = 5$$

423. What is the electrical resistance of 1,000 feet of copper wire $\frac{1}{10}$ inch in diameter, using $k = 10.3$?

The resistance of any round conductor varies jointly as the length and inversely as the square of the diameter.

$$R \propto \frac{L}{d^2} \quad \text{or} \quad R = \frac{kL}{d^2}$$

where

R = resistance in ohms
L = length in feet
d = diameter in mils
k = constant determined by substituting L = 1, d = 1, and getting k = R

Thus k = resistance of 1 ft. of wire which is 1 mil dia.
Hence k = circular mil-ft. constant or mil-ft. resistance.

$$1 \text{ mil} = .001 \text{ in.} = \tfrac{1}{1000} \text{ in.} \quad \text{and} \quad \tfrac{1}{10} \text{ in.} = 100 \text{ mil}$$

$$\therefore\ R = \frac{10.3 \times 1{,}000}{(100)^2} = 1.03 \text{ ohm Ans.}$$

PROBLEMS

1. Express the following common fractions in the form of ratios:

(*a*) $\frac{1}{3}$ (*b*) $\frac{3}{1}$ (*c*) $\frac{1}{7}$ (*d*) $\frac{4}{3}$ (*e*) $\frac{5}{6}$ (*f*) $\frac{6}{5}$ (*g*) $\frac{\frac{1}{2}}{\frac{2}{2}}$ (*h*) $\frac{\frac{3}{4}}{\frac{4}{4}}$ (*i*) $\frac{\frac{5}{6}}{1}$

2. Express the following ratios as fractions:

(*a*) 7:10 (*b*) 10:70 (*c*) 5:9 (*d*) 13:12 (*e*) 1:12 (*f*) 12:1

3. If machine *A* produces 350 units per hour, while machine *B* produces 630 units per hour, what is the production ratio of machine *A* to that of machine *B*?

4. If you are 40 years old, and your son is 8 years old, what is the ratio of your son's age to yours?

5. If one house costs $12,000 and another costs $22,000, what is the ratio between the second and the first house?

6. Write the ratio of:

(*a*) 1 foot to 1 inch (*b*) 1 inch to 1 foot
(*c*) 1 cent to 1 dollar (*d*) 1 dollar to 1 cent

7. If the length of a rectangle is 110 ft. and its width is 80 ft., what is the ratio of its length to its width, and the ratio of its width to its length?

8. If two boards are 8 ft. and 10 ft. long respectively, what is the ratio of the first to the second?

9. If one side of a triangle is 3 ft. and another 5 ft., what is the ratio of the first to the second expressed as a decimal fraction?

10. When we say $200 is 25% of $800, what does that mean in ratio terms?

11. Simplify each of the following ratios:

(*a*) 15:25 (*b*) 24:15 (*c*) 8:24 (*d*) 27:24

12. Simplify:

(*a*) 6:10 (*b*) 36:24 (*c*) $2\frac{1}{2}:6\frac{1}{4}$ (*d*) $\frac{1}{3}:4$
(*e*) $1:\frac{1}{4}$ (*f*) 7:28 (*g*) $3\frac{1}{5}:2\frac{1}{2}$ (*h*) 1.8:4.6

13. Reduce each of the following to a ratio having 1 for its first term:

(*a*) 3:9 (*b*) 6:12 (*c*) 7:21 (*d*) 6:60
(*e*) 19:72 (*f*) 9:81 (*g*) 11:23 (*h*) 96:600
(*i*) .1:4 (*j*) .7:4 (*k*) .695:8 (*l*) .54:12
(*m*) $\frac{1}{2}$:2 (*n*) $\frac{1}{5}$:13 (*o*) $\frac{6}{15}$:24 (*p*) $\frac{9}{127}$:.9642
(*q*) $\frac{18}{137}$:$\frac{20}{133}$

14. What is the ratio between $1\frac{1}{4}$ hours and 45 minutes?

15. What is the ratio of \$6.50 to \$4?

16. If 6 bushels of wheat cost \$9 and 8 bushels of corn cost \$8, find the ratio of the value of 10 bushels of wheat to the value of 10 bushels of corn.

17. If a photograph is 12 in. by 8 in., and it is enlarged so that the larger side becomes 24 in., in what ratio is the area increased?

18. Find the ratios between:

(*a*) $\frac{4}{13}$ and $\frac{7}{13}$ (*b*) $\frac{4}{5}$ and $\frac{7}{15}$ (*c*) $\frac{9}{25}$ and $\frac{72}{15}$ (*d*) $\frac{3}{4}$ and $\frac{7}{8}$
(*e*) $\frac{7}{8}$ and $\frac{9}{64}$

19. Divide 35 into two parts whose ratio is 2:3.

20. 560 children arrive at a camp and are divided between two lodgings in the ratio 3:5. How many are assigned to each lodging?

21. A shipment of 200 radios, TV sets, and record players is received in the ratio 5:7:8 respectively. How many of each are there?

22. 1,200 books are to be allotted to three classes in the ratio 6:9:10. How many books will each class receive?

23. Divide 85 in the ratio $\frac{1}{2}$:$\frac{1}{3}$.

24. A city department bought three business machines. The first cost twice as much as the second. The third cost three times as much as the second. It paid \$4,800 for the three machines. How much did it pay for each?

25. What is the inverse ratio of 25:5?

26. John and Bill bought \$105 worth of merchandise. If $\frac{1}{8}$ of John's share is equal to $\frac{1}{7}$ of Bill's, what was the cost of the merchandise each bought?

27. If the wing span of a plane is 85 ft. 6 in., what is the wing span of a model if the ratio of the length of any part of the model to the length of the corresponding part of the actual plane is 1:72?

28. If a bankrupt firm can pay 55¢ on the dollar, and if its assets amount to \$24,000, what are its liabilities?

29. What selling price should be placed on a refrigerator, if the cost is \$325, and the dealer operates on a margin of 35% of cost?

30. If you allow 22% of your income for food and 18% for rent, (*a*) what is the ratio of the cost of food to the rent, (*b*) how much do you spend for food per month when your income is \$7,200 per year?

31. If a town estimates that it has to raise \$406,250 in taxes, and the assessed valuation of its real property is \$12,500,000, what is its tax rate?

32. A certain concrete mixture is to be made of 1 part cement, $2\frac{1}{2}$ parts sand, and $5\frac{1}{2}$ parts stone. What is: (*a*) the ratio of sand to stone, (*b*) the ratio of cement to sand, and (*c*) per cent of sand in the concrete mixture?

33. If the living room of a house is shown on the blueprint to be $5\frac{1}{2}$ in. by $4\frac{5}{8}$ in., and the scale of the print is $\frac{1}{4}$ in. = 1 ft., what are the actual dimensions of the room?

34. What is the inverse ratio of 7:56?

35. What is the ratio of 1 yard to 1 foot?

36. (*a*) $?:4 = 16$ (*b*) $?:5 = 4$ (*c*) $?:3\frac{1}{2} = 4\frac{1}{2}$

37. (*a*) $24:? = 6$ (*b*) $49:? = 7$ (*c*) $3\frac{3}{4}:? = 1\frac{2}{3}$

38. What is the compound ratio of 12:8 and 4:15?

39. If you buy the same number of cigars at 10¢, 15¢, and 20¢, and you spend $3.60, how many at each price did you buy?

40. If you bought *five* times as many grade A articles at $28 as grade B articles at $16, and three times as many grade C articles at $22 as grade B articles, how many of each grade did you buy if you spent $3,552?

41. Complete the following proportions:

(*a*) 2:4::3:? (*b*) ?:4::4:8 (*c*) 6:9::12:?
(*d*) 1.2:?::2:4 (*e*) 5:15::6:? (*f*) 9:24::?:8
(*g*) 6:?::10:20 (*h*) 3.6:7.2::?:6.4

42. Find the missing terms in the following proportions:

(*a*) $\frac{3}{4} = \frac{156}{x}$ (*b*) $\frac{5}{2} = \frac{x}{600}$ (*c*) $\frac{5}{9} = \frac{35}{x}$

(*d*) $\frac{x}{136} = \frac{84}{336}$ (*e*) $\frac{12}{144} = \frac{x}{1{,}728}$ (*f*) $\frac{17}{51} = \frac{102}{x}$

(*g*) $\frac{17.4}{191.1} = \frac{11.5}{x}$ (*h*) $\frac{18.2}{x} = \frac{7.3}{9.1}$ (*i*) $\frac{.60}{.75} = \frac{.48}{x}$

43. If 10 bushels of apples cost $25, what will 15 bushels cost?

44. If 25 lb. of sugar cost $3.50, what will 75 lb. cost?

45. If at a certain moment a post 32 ft. high casts a shadow 48 ft. long, how long is the shadow of a tree which is 48 ft. high?

46. Measure the height of a post and the length of its shadow. Also at the same time, measure the length of the shadow of any tall object and calculate the height of the tall object.

47. A certain brand of white paint contains 21 parts of titanium dioxide and 37 parts of white lead by weight. If you have 600 lb. of the oxide, how many pounds of white lead would you need to make a batch of paint?

48. Equal sums of money are invested at $3\frac{1}{2}\%$ and $4\frac{1}{2}\%$. If the income at $4\frac{1}{2}\%$ is $819, what is the income at $3\frac{1}{2}\%$?

49. Calculate the mean proportional:

(*a*) $\frac{4}{x} = \frac{x}{25}$ (*b*) $\frac{9}{x} = \frac{x}{25}$ (*c*) $\frac{\frac{1}{8}}{x} = \frac{x}{\frac{1}{2}}$

50. If 42 gal. of gasoline cost $12.60, how much will 85 gal. cost?

51. An alloy consists of $4\frac{1}{2}$ parts tin and $6\frac{1}{2}$ parts copper. How many pounds of copper would be needed with 150 pounds of tin to maintain the given ratio?

52. If it takes you 45 minutes to drive to work at 40 mph, at what speed would you have to travel to make it in 38 minutes?

53. If 28 men do a job in 18 days, how many men will be required to do it in 12 days?

54. If 110 yards of copperwire offer 1.2 ohm resistance, what will be the resistance of 600 yards of copper wire of $2\frac{1}{4}$ times the cross-sectional area?

55. A hotel puts up 8 guests for 12 days, and another group of 12 guests for 21 days. What is the ratio of the two hotel bills? If the smaller bill was $1,344, what would the larger bill be? If the larger bill was $4,032 what would the smaller be?

56. If 16 men working 6 hours per day dig a canal 120 feet long in 40 days, how many men working 8 hours a day can dig a canal 160 feet long in 10 days?

57. If 4 men cut 16 cords of wood in 9 days, how long will it take 10 men to cut 30 cords of wood?

58. If the eggs laid by 24 hens in 12 weeks are worth $80.64, what will be the value of the eggs laid by 48 hens in 8 weeks?

59. What proportion of $3\frac{1}{2}\%$ milk and $4\frac{3}{4}\%$ milk must be mixed to get 4% milk?

60. Using a 12-foot plank, where would you put the support under the plank, so that two children weighing 40 and 50 pounds respectively, can play seesaw?

61. What is the electrical resistance of 800 ft. of copper wire $\frac{1}{10}$ in. dia., using $k = 10.25$?

62. The weight of a body above the surface of the earth varies inversely as the square of its distance from the center of the earth. If a man weighs 160 pounds at sea level, what will he weigh at the top of a mountain 3 miles high? Assume 4,000 miles = radius of the earth.

63. The distance that a body falls from rest varies as the square of the time. If a body falls 16 ft. the first second, how far will it fall in the first 5 seconds?

64. Write the following as equations:

(*a*) x varies as y^3

(*b*) z varies inversely as x^2 and directly as y

(*c*) x varies inversely as y^2

65. If y varies inversely as x and $y = 6$ when $x = 3$, find x when $y = 3$.

66. The velocity, V, of a freely falling body from a resting position is proportional to the time, *t*. If it has a velocity of 32.2 ft./sec. at the end of the first second, what is the velocity at the end of the fifth second?

67. The pressure of a confined gas at constant temperature varies inversely as the volume. If a gas has a pressure of 60 pounds per square inch when confined in a volume of 120 cu. in., what is the pressure when the volume is reduced to 80 cu. in.?

68. If it takes 2 cu. yd. of concrete to make 40 posts 6 in. × $4\frac{1}{2}$ in. × 7 ft., how many cubic yards will it take to make 700 posts 4 in. × 4 in. × 5 ft.?

69. If it takes 17 men working 7 hr. a day to build a bridge in 22 days, how many men working 10 hours a day will it take to build the bridge in 4 days?

70. A map is drawn to a scale of 1:500. What is the distance between two places that are $6\frac{1}{2}$ inches apart on this map?

CHAPTER XI

AVERAGES

424. What is meant by an average in statistics?

An average is a significant representative value for an entire mass of data. It stands for the essential meaning of the detailed facts.

Individual measurements usually have meaning only when they are related to other individual measurements, usually to some typical value which represents a number of such measurements—for example, average cost of living, average wage, average weight for age and height, and average birth rate.

425. What are the uses of averages in statistics?

(*a*) They give us a concise idea of a large group.

EXAMPLE: We do not get a clear mental image when we are given the height of every tree in a forest, but the average height of the trees is something definite and understandable.

(*b*) They give us a basis for comparison of different groups by simple representative facts.

EXAMPLE: Two forests can more readily be compared by means of averages of some kind.

(*c*) They give us an idea of a complete group by using only simple data.

EXAMPLE: It is not necessary to measure the height of each person of a race to get the typical height of that race. An average obtained from a limited number, say a few thousand samples, would generally be sufficient to give a figure close to the exact average.

(*d*) They provide us with a numerical concept of the relationship between different groups.

EXAMPLE: We may say that the people of one race are taller than those of another, but to get any definite ratio of heights we need averages.

426. Why may an average be a more reliable figure to represent a group than a sample figure selected from the group?

It represents many individual measurements. It levels out all differences by disregarding the variations among the items of the series, giving significance to the entire series. Scientists frequently perform a fine measurement a number of times, and then average the result, because by so doing they hope errors will cancel out. If some measurements are too large and others too small, mistakes each way will about balance. Thus, the average describes the series of varying individual values and is presumed to be the best possible representation of the series.

427. Can averages be compared when they are derived from data representing widely different conditions and groups?

No. The data must be homogeneous.

The arithmetical average of a series of wage data, where wages of both men and women are included, is not typical of either men's or women's wages. A useful average must be typical of actual conditions, not merely a result of a mathematical calculation.

428. What is meant by a deviation from the average?

Once a value representative of an entire group is established, the single item can be compared with it. The difference is called the deviation from the average.

429. What is the significance of a small total amount of deviations?

The smaller the total amount of the deviations, the greater is the homogeneity of the data; the closer the grouping about the average, the smaller the variability among the individual items. This can serve to decide whether or not the average is typical.

430. What are the two classes of averages in general?

(*a*) Averages of ordinary numbers representing time, money, and general things.

(*b*) Averages of ratios representing speed and other ratios.

431. How do we find the arithmetic average or mean value of a number of similar quantities?

Add the quantities and divide this sum by the number of the quantities.

Ex. (*a*) If 10 men earn \$80, \$96, \$102, \$78, \$92, \$65, \$59, \$110, \$150, and \$87 respectively per week, what are the average earnings of the 10 men?

$$\frac{\text{Sum of earnings}}{\text{Sum of men}} = \frac{80+96+102+78+92+65+59+110+150+87}{10}$$

$$\therefore \text{ Average} = \frac{\$919}{10} = \$91.90 \text{ Ans.}$$

Ex. (*b*) If a car travels 180 miles in 4 hours, what is its average speed?

$$\frac{180 \text{ mi.}}{4 \text{ hr.}} = 45 \text{ mph Ans.}$$

Ex. (*c*) What is the average of .42, .86, .53, .79, 2.03, 5.93?

$$\frac{\text{Sum}}{\text{Sum of terms}} = \frac{.42 + .86 + .53 + .79 + 2.03 + 5.93}{6}$$

$$\therefore \text{ Average} = \frac{10.56}{6} = 1.76 \text{ Ans.}$$

432. When is an average an excellent way of showing the middle or most typical figure?

When the figures are fairly close together.

Ex. (*a*) What is the average mark of a group of 5 students when their respective grades are 75, 78, 80, 81, and 77?

$$\text{Average} = \frac{75 + 78 + 80 + 81 + 77}{5} = \frac{390}{5} = 78\% \text{ Ans.}$$

Ex. (*b*) What is the average mark when the grades are 75, 88, 100, 50, and 77?

$$\text{Average} = \frac{75 + 88 + 100 + 50 + 77}{5} = \frac{390}{5} = 78\% \text{ Ans.}$$

Example (*a*) describes fairly well the performance of the students. Example (*b*) does not really describe the performance of the group even though the average is the same 78%.

433. If a train takes the following times between stops—48 minutes, 55 minutes, 1 hour 8 minutes, and 42 minutes—what is the average time between stops?

$$48 + 55 + 68 + 42 = 213 \text{ minutes}$$

$$\therefore \text{ Average} = \frac{213}{4 \text{ stops}} = 53\tfrac{1}{4} \text{ minutes Ans.}$$

434. A car travels 10 miles up a steep grade at 30 mph, and then 90 miles on a level road at 50 mph. What is its average speed?

Speed is a ratio of two things, distance and time. 30 mph and 50 mph are ratios and we cannot get the average of the two ratios by dividing their sum by 2. The average speed is *not*

$$\frac{30 + 50}{2} = 40 \text{ mph}$$

To average ratios we must divide the *sum* of one kind of thing by the sum of the other kind of thing.

Here, the sum of the miles traveled is $10 + 90 = 100$ miles.

$$\text{Time up the grade} = \frac{10 \text{ miles}}{30 \text{ mph}} = \frac{1}{3} \text{ hour}$$

$$\text{Time on level road} = \frac{90 \text{ miles}}{50 \text{ mph}} = \frac{9}{5} \text{ hour}$$

$$\text{Sum of the ``times''} = \tfrac{1}{3} + \tfrac{9}{5} = \tfrac{5}{15} + \tfrac{27}{15} = \tfrac{32}{15} \text{ hours}$$

$$\therefore \text{ Average speed} = \frac{\text{total distance}}{\text{total time}} = \frac{100}{\frac{35}{15}} = \overset{25}{\cancel{100}} \times \frac{15}{\underset{8}{\cancel{32}}} = \frac{375}{8}$$

$$= 46\tfrac{7}{8} \text{ mph Ans.}$$

435. Two planes leave at the same time from Seattle, Washington, for El Paso, Texas,—a distance of 1,381 miles. One plane, *A*, flies at 400 mph and returns at 400 mph. The other plane, *B*, flies at 600 mph from Seattle and returns at 200 mph, because of defective engines. If each plane remains 12 hours in El Paso, which comes back first?

We cannot say that they both get back together. While it is true that the average of the numbers 400 and 400 is the same as of 600 and 200, the speeds themselves are ratios and we must in each case divide the total distance by the total time.

$$\text{Total distance} = 1{,}381 \times 2 = 2{,}762 \text{ miles}$$

Plane A takes $\frac{1381}{400} = 3.45$ hr. going and 3.45 hr. returning

$$\therefore \text{ Its average speed} = \frac{2{,}762 \text{ miles}}{6.9 \text{ hr.}} = 400 \text{ mph}$$

Plane B takes $\frac{1381}{600} = 2.3$ hr. going and $\frac{1381}{200} = 6.9$ hr. returning

$$\therefore \text{ Its average speed} = \frac{2{,}762 \text{ miles}}{9.2 \text{ hr.}} = 300 \text{ mph}$$

Thus, Plane A flying at 400 mph returns first.

436. If you paid an income tax of 22% on $3,400 one year and 28% on $4,600 the following year, how much did you pay altogether?

Since 22% and 28% are ratios, you must not figure that

$$\frac{22 + 28}{2} = 25\%$$

is the average on the total income of $8,000. Instead you figure:

$$\begin{aligned} 22\% \times \$3,400 &= \quad \$748 \\ 28\% \times \$4,600 &= \underline{\$1,288} \\ & \quad \$2,036 = \text{the amount you paid} \end{aligned}$$

437. How would you find the total, given the average with ordinary numbers (not ratios)?

Multiply the average by the number of items involved.

EXAMPLE: If the average weight of a person is assumed to be 150 pounds, what would the carrying capacity of a passenger elevator be when only 12 people are permitted to ride?

$$150 \text{ lb.} \times 12 = 1,800 \text{ lb.} = \text{Capacity.}$$

438. An appliance dealer sells 15 TV sets that cost $180 per set at an average profit of 30%, and 20 other TV sets that cost him $260 per set, at an average profit of 35%. What is the total profit, assuming the percentages are based on the cost price?

$$\begin{aligned} \text{Average profit on 15 TV sets} &= 30\% \times \$180 = \$54 \\ \text{Total profit on 15 TV sets} &= 15 \times \$54 = \$810 \\ \text{Average profit on 20 TV sets} &= 35\% \times \$260 = \$91 \\ \text{Total profit on 20 TV sets} &= 20 \times \$91 = \$1,820 \end{aligned}$$

$$\therefore \text{ Total profit on 35 sets} = \$810 + \$1,820 = \$2,630 \text{ Ans.}$$

439. What is meant by a weighted average?

One obtained by first multiplying each item by its appropriate factor before adding and then dividing by the number of items.

EXAMPLE: In a Civil Service examination the weights for the measurements are: Oral 1, Arithmetic 2, Practical 4, Citizenship 1, English 2. What is the average mark of a candidate whose marks are: Oral 85%, Arithmetic 92%, Practical 79%, Citizenship 80%, English 76%?

The ratio of weights = 1:2:4:1:2 which adds up to 10.

The weighted marks are:

$$\begin{array}{rcrcl} 1 \times 85 &=& 85 &=& \text{Oral} \\ 2 \times 92 &=& 184 &=& \text{Arithmetic} \\ 4 \times 79 &=& 316 &=& \text{Practical} \\ 1 \times 80 &=& 80 &=& \text{Citizenship} \\ 2 \times 76 &=& 152 &=& \text{English} \\ \hline 10 & & 817 & & \end{array}$$

$$\therefore \tfrac{817}{10} = 81.7\% = \text{Average mark Ans.}$$

440. How can we find the value of one quantity that is not given, when the weights and the final average are known?

EXAMPLE: In the above, if we are given a minimum passing average of 70%, what must a candidate get for the Practical mark in order to pass?

$$1:2:4:1:2 = 10 = \text{sum of weights}$$

$$10 \times 70\% = 700 = \text{total weighted mark in order to pass}$$

$$\begin{array}{rcrcl} 1 \times 85 &=& 85 &=& \text{Oral} \\ 2 \times 92 &=& 184 &=& \text{Arithmetic} \\ 1 \times 80 &=& 80 &=& \text{Citizenship} \\ 2 \times 76 &=& 152 &=& \text{English} \\ \hline & & 501 &=& \text{points already scored} \end{array}$$

The average must be 700 points in order to pass.
He has already scored 501 points.
Remainder = 199.
But the Practical has a weight of 4.

$$\therefore \tfrac{199}{4} = 49.75 = 49.8\%$$

must be scored on the Practical to get a minimum 70% average.

Usually a minimum is set for each part of the test.

441. There are 8 manufacturing plants having 453, 699, 341, 621, 383, 562, 741, and 214 employees respectively. If the employees in plants 1, 2, and 3 worked 38 hours per week, in plants 4, 5, and 6, 40 hours per week, and in plants 7 and 8, 42 hours per week, how could we (*a*) get a true comparison of their productivity expressed in man-hours, (*b*) determine the average number of hours each man worked in the given week?

(*a*) Multiply the number of employees in each plant by the number of hours each is required to work. Divide by the number of plants to get the average number of man-hours worked per week in each plant.

(*b*) To get the average number of hours each employee worked in the given week, divide the total number of man-hours by the total number of employees.

Plant No. 1	453 × 38 hours =	17,214 man-hours
Plant No. 2	699 × 38 hours =	26,562 man-hours
Plant No. 3	341 × 38 hours =	12,958 man-hours
Plant No. 4	621 × 40 hours =	24,840 man-hours
Plant No. 5	383 × 40 hours =	13,320 man-hours
Plant No. 6	562 × 40 hours =	22,480 man-hours
Plant No. 7	741 × 42 hours =	31,122 man-hours
Plant No. 8	214 × 42 hours =	8,988 man-hours
Total	4,014 employees	157,484 man-hours

$$8\overline{)157{,}484}$$

Average = 19,685½ man-hours per week per plant

There are 4,014 employees who worked 157,484 man-hours.

$$\frac{157{,}484}{4{,}014} = 39.23 \text{ Ans.} \quad \begin{cases}\text{Average number of hours worked}\\ \text{by each employee per week.}\end{cases}$$

442. How can we simplify the process of getting an average of several numbers that differ from one another by a comparatively small amount?

(*a*) Determine mentally the approximate average.
(*b*) Get each deviation above or below this figure.
(*c*) Subtract the sum of the deviations below this amount from the sum of the deviations above the amount.
(*d*) Find the average deviation and add it to the original approximate value.

EXAMPLE: What is the average daily sales figure if the daily sales record is:

		+ Deviation (above)	− Deviation (below)
	$324.78	$24.78	
	$296.53		$3.47
	$285.27		$14.73
	$318.06	$18.06	
	$325.68	$25.68	
	$294.46		$5.54
Totals	$1,844.78	$68.52	$23.74

We see at once that the average is approximately \$300 a day.
Deviations, + \$68.52 (above) − \$23.74 (below) = \$44.78

$$\text{Average deviation} = \frac{\$44.78}{6} = \$7.46$$

$$\therefore\ \$300 + \$7.46 = \$307.46 = \text{Average}$$

$$\text{As a check}\quad 6\overline{)\$1{,}844.78}\qquad \$307.46 = \text{Average}$$

443. For scattered data, what two other ways are there of finding the "middle" that stand for more than an average?

The *median* and the *mode* are two ways of sometimes getting a more representative picture of the "middle."

444. What is meant by the median?

The median is the middle score in a series of scores after they have been arranged in order from lowest to highest. The median score is such that there are as many scores above it as there are scores below it.

445. How is the median located?

When there is an odd number of scores, the median value is that of the middle case. When there is an even number of scores, the median value is located between the two middle items. If the two middle values are identical, then either may be chosen as the median value.

Ex. (*a*) What is the median of 8, 15, 12, 3, 18, 2, 23, 13, and 9?

Arrange these in the order of their magnitude, getting 2, 3, 8, 9, 12, 13, 15, 18, and 23 (9 values = odd number). The median is 12 because it is the fifth or middle value. There are four numbers in this series higher than the median and there are four numbers lower than the median.

The *mean* average is

$$\frac{2 + 3 + 8 + 9 + 12 + 13 + 15 + 18 + 23}{9} = \frac{103}{9} = 11.44$$

Ex. (*b*) What is the median of 12, 38, 49, 18, 5, 23, 8, 11, and 30?

Arrange these in order of magnitude, getting 4, 5, 8, 9, 11, 12, 18, 23, 30, and 38 (10 values = an even number). The two middle numbers are 11 and 12. The median is half way between them at 11.5.

The mean average is

$$\frac{4 + 5 + 8 + 9 + 11 + 12 + 18 + 23 + 30 + 38}{10} = \frac{158}{10} = 15.8$$

446. If 25 salesmen in an organization report their average weekly incomes as \$260, \$200, \$95, \$200, \$220, \$160, \$160, \$800, \$240, \$240, \$235, \$350, \$150, \$260, \$200, \$275, \$450, \$275, \$175, \$200, \$500, \$225, \$250, \$650, and \$200, what is the average weekly income of the group, and is this average representative of the group?

$$\text{Average} = \frac{\begin{array}{l}260 + 260 + 95 + 200 + 220 + 160 + 160 + 800 \\ \quad + 240 + 240 + 235 + 350 + 150 + 260 + 200 \\ \quad + 275 + 450 + 275 + 175 + 200 + 500 + 225 \\ \quad + 250 + 650 + 200\end{array}}{25}$$

$$= \frac{6{,}970}{25} = \$278.80 \text{ Ans.}$$

This average does not give a true picture of what the salesmen get, because the \$800 and the \$650 incomes throw it off.

447. What is the median of the above and does this median give a reasonable idea of the group income?

Arrange the incomes in order of magnitude: 95, 150, 160, 160, 175, 200, 200, 200, 200, 200, 220, 225, 235, 240, 240, 250, 260, 260, 275, 275, 350, 450, 500, 650, and 800.

The median value is the thirteenth value, or \$235. As many salesman have incomes more than \$235 as have less than \$235. This gives us a reasonable idea of how much this group earns as compared with a group whose median is \$500 a week. Actually, however, only one person earns \$235 and therefore this cannot be considered as the most typical figure.

448. What is meant by the mode?

It is the most frequent size of item, the position of greatest density. When we speak of the average man, the average income, we usually mean the modal man, or the modal income. We might say the modal tip at a restaurant is 15%, the modal workingman's house has five rooms—in each instance, that is the most usual occurrence, the common thing. The figure having the highest frequency is the mode. The mode means the single most typical figure.

449. What is the mode of the weekly incomes of Question 446?

Make a frequency table showing how many salesmen receive each weekly amount:

$800	(1)	$260	(2)	$200	(5)
$650	(1)	$250	(1)	$175	(1)
$500	(1)	$240	(2)	$160	(2)
$450	(1)	$235	(1)	$150	(1)
$350	(1)	$225	(1)	$95	(1)
$275	(2)	$220	(1)		

More salesman have incomes of $200 a week than any other amount. This figure having the highest frequency is the mode for this table.

450. How can we widen the concept that the mode is the most typical figure and get a better measure of the group?

Group the frequencies of Question 449:

$700	and over		(1)
$400	to	$699	(3)
$300	to	$399	(1)
$250	to	$299	(5)
$200	to	$249	(10)
$150	to	$199	(4)
	less than	$150	(1)

The largest group receives from $200 to $249 and that is the mode for this table.

451. What are the best measures of typical earnings of the group of salesmen?

We have seen that the mean or average is $278.80.

The median is $235.

The mode is $200 for the frequency table.

The mode is $200–249 for the grouped frequency table.

Thus, here the median and mode are the best measures of what typically this group receives per week. They give us a better idea of individual incomes than does the average.

452. What are the advantages of the arithmetic mean or average?

(*a*) It is located by a simple process of addition and division.

(*b*) Extreme deviations are given weight, which is desirable in certain cases.

(*c*) It is affected by every item in the group.

453. What are the disadvantages of the arithmetic mean or average?

(*a*) Average is affected by the exceptional and the unusual. One or two large contributions in a church collection conceal the usual or typical contribution. A few very large incomes produce an average income far above a representative of the majority.

(*b*) The average emphasizes the extreme variations, which in most cases is undesirable.

(*c*) It may fall where no data actually exist. We may find that the average number of persons per family is 5.12, although such a number is evidently impossible.

(*d*) It cannot be located on a frequency graph when such is already in existence.

454. What are the advantages of the median?

(*a*) It is easy to determine and is exactly defined.

(*b*) It is only slightly affected by items having extreme deviation from the normal. A $1,000 check in the church collection does not affect the mode at all, and affects the median only as much as any other single item larger than the median would do, that is, the item receives the same weight as any other instance and no more. Thus it is useful whenever extreme items are of little importance.

(*c*) The median is particularly useful in groups to which a measure cannot be applied, groups of nonmathematical type.

(*d*) Its location can never depend upon a small number of items, as is sometimes the case with the mode.

(*e*) If the number of extreme items is known, their values are not needed in getting the median. The median is a position average. Merely the number of items, not their size, influences the position of the median.

(*f*) On the whole it is one of the most valuable types for practical use; and for such studies as wages and distribution of wealth it is often superior to either the mode or the mean.

455. What are the disadvantages of the median?

(*a*) It is not so readily determined by a simple mathematical process.

(*b*) We cannot obtain a total by multiplying the median by the number of items.

(*c*) It is not useful where it is desirable to give large weight to extreme variations.

(*d*) It is insensitive, which means that we can replace certain measurements or values of a given group by other values without affecting the median.

EXAMPLE: In the values 2, 4, 6, ⑧, 10, 12, 14 the median is 8, the average is

$$\frac{2 + 4 + 6 + 8 + 10 + 12 + 14}{7} = \frac{56}{7} = 8$$

Now we may replace the three values which are larger than 8 and this replacement will have no effect upon the median. Thus, the values are 2, 4, 6, ⑧, 17, 21, 34. The median is still 8. But the mean becomes

$$\frac{2 + 4 + 6 + 8 + 17 + 21 + 34}{7} = \frac{92}{7} = 13.1$$

(*e*) Unlike mode, but like arithmetic mean, it is frequently located at a point in the array at which actual items are few.

(*f*) Where there are many items of the same size as the median, the number of items larger than the median may be very different from the number of items smaller than the median, and the value of the median as an average is largely destroyed.

456. What are the advantages of the mode?

(*a*) It is useful in cases in which one desires to eliminate extreme variations which do not effect it.

(*b*) One need know only that extreme items are few in number, not their size.

(*c*) Mode may be determined with considerable accuracy from well-selected sample data.

(*d*) It is the best way to represent the group. It means more to say that the modal wage of workingmen in a locality is $16 per day, than to say that the average wage is $16.32, which no one actually receives.

457. What are the disadvantages of the mode?

(*a*) In many cases no single well-defined type actually exists. There is no such thing as a modal size city. We are likely to find several distinct modes corresponding to the various grades of labor.

(*b*) Mode is difficult to determine accurately by any method.

(*c*) It is not useful when you want to give any weight to extreme variations.

(*d*) Mode times the number of items does not equal the correct total as in arithmetic mean.

(*e*) Unless grouping is used, mode may be determined by a comparatively small number of like items in a large group of varying size. If only 4 people owned $3,000 each in a community having a great variation in wealth, this would be the modal value while the wealth of all others varied.

PROBLEMS

1. The wages of a man for six weeks are: $92, $87, $99.50, $91, $97.50, and $89. What is the average wage for these six weeks?

2. A school system had the following attendances in one week: Monday 248,585, Tuesday 248,326, Wednesday 247,963, Thursday 248,658, and Friday 248,597. What is the average daily attendance?

3. If a car travels 235 miles in 5 hours, what is the average speed?

4. What is the average mark of a group of 8 students, when their respective grades are 83, 86, 90, 92, 87, 82, 81, and 84?

5. What is the average mark when the grades are 86, 98, 100, 60, 84, 91, 77, and 89?

6. Which average describes the performance of the group better, the one in Problem 4 or the one in Problem 5?

7. If a train takes the following times between stops: 37 minutes, 44 minutes, 1 hour 2 minutes, and 31 minutes, what is the average time between stops?

8. A car travels 8 miles up a steep grade at 32 mph, and then 80 miles on a level road at 52 mph. What is its average speed?

9. If you paid an income tax of 20% on $3,200 one year and 26% on $4,400 the following year, how much did you pay altogether?

10. If a total of only 14 persons are permitted to ride in an elevator, and the average weight of a person is assumed to be 150 lb., what is the carrying capacity of this elevator?

11. If you sell 40 radios that cost $35 per set at an average profit of $33\frac{1}{3}$%, and 70 sets that cost $58 per set at an average profit of 40%, what is the total profit, assuming the percentages are based on the cost price?

12. If the weights in an examination are: Arithmetic 2, English 3, Practical 3, Oral 1, Citizenship 1, what is the average mark of a candidate whose marks are: Arithmetic 94%, English 89%, Practical 75%, Oral 80%, Citizenship 80%?

13. If the minimum passing average is 75%, what must a candidate get for the English mark in order to pass in Problem 12?

14. What is the average daily sales figure if the daily sales record is $435.89, $307.64, $396.38, $429.07, and $436.79, using the simplified method by first determining mentally the approximate average?

15. (*a*) What is the median of 9, 16, 13, 4, 19, 3, 24, 14, and 10?
(*b*) What is the median of 13, 39, 50, 19, 6, 24, 9, 12, and 31?

16. If 10 salesmen report their average weekly incomes as $370, $310, $105, $310, $560, $385, $760, $300, $260, and $385, what is the average weekly income of the group, and is this average representative of the group?

17. What is the median income of the group of Problem 16?

18. What is the mode of the weekly incomes of Problem 16?

19. What is the mode when the frequencies are grouped in Problem 16?

20. In the series 3, 5, 7, 9, 18, 22, and 35, what is the median?

21. Is the median of Problem 20 affected if 18, 22, and 35 are replaced by 11, 12, and 15?

22. A group of 50 persons contributed to a church collection in the following amounts:

\$5.00	(1)	\$.50	(12)
\$3.00	(2)	\$.25	(22)
\$1.00	(3)	\$.15	(6)
\$.75	(4)		

(*a*) How much did the group contribute?
(*b*) What was the average contribution?
(*c*) What was the median contribution?
(*d*) What was the mode?
(*e*) Which type gives the truest picture of the contributions of the group?

23. If the median grade of a class in a biology test is 81, what can be said about the grades in that test?

24. A plane covers 290 miles in the first hour of its flight, 504 miles in the next $2\frac{1}{2}$ hours of flight, and 376 miles in the final $1\frac{1}{2}$ hours of flight. What is the average speed for the entire journey?

CHAPTER XII

DENOMINATE NUMBERS

458. What is a denominate number?

It is a concrete number whose unit of value or measure has been fixed by law or custom. It is used to specify the units of measurements. When standard units are used with a stated quantity, they are commonly referred to as denominate numbers.

EXAMPLE: 3 feet, 4 yards, 8 pounds are denominate numbers.

459. What is meant by reduction of denominate numbers?

It is the process of changing a number, expressed in one denomination, to an equivalent expressed in another denomination.

EXAMPLES: 3 feet changed to inches equals 36 inches.
3 quarts changed to pints equals 6 pints.

460. What is meant by (*a*) reduction descending, (*b*) reduction ascending?

(*a*) Changing a number from a higher to a lower denominator = reduction descending.

EXAMPLE: 2 yards = 6 feet = 72 inches.

(*b*) Changing a number from a lower to a higher denomination = reduction ascending.

EXAMPLES: 200 cents = 2 dollars 36 inches = 3 feet.

Note that in reduction the expression is changed without changing the value.

461. What are the standard linear measures?

12 inches (in.)	= 1 foot (ft.)	320 rods	= 1 mile (mi.)
3 feet	= 1 yard (yd.)	1,760 yards	= 1 mile
$5\frac{1}{2}$ yards	= 1 rod (rd.)	5,280 feet	= 1 mile
$16\frac{1}{2}$ feet	= 1 rod		

Note (*a*): Marine measures are expressed in fathoms (= 6 feet), long cable lengths (= 120 fathoms), short cable lengths (= 100 fathoms), knots (= 1.15 miles), and leagues (= 3 knots).

Note (*b*): The units in the above table represent length only. They are used to measure distances, lengths, widths, or thicknesses of objects. The unit of length is the standard yard.

Note (*c*): Symbol for inches = ["] placed at upper right

$$5'' = 5 \text{ in.}$$

Symbol for feet = ['] placed at upper right

$$5' = 5 \text{ ft.}$$

462. What is the result of the reduction of the following?

(*a*) 5 ft. 5 in. to inches?	$5 \times 12'' + 5'' = 65$ in.
(*b*) 5 yd. 3 ft. to feet?	$5 \times 3' + 3' = 18$ ft.
(*c*) 5 rd. to yards?	$5 \times 5\frac{1}{2}$ yd. $= 27\frac{1}{2}$ yd.
(*d*) 108 in. to feet?	$\frac{108}{12} = 9$ ft.
(*e*) 4 mi. to rods?	4×320 rd. $= 1{,}280$ rd.
(*f*) 1 rd. to inches?	$1 \times 16\frac{1}{2}' \times 12'' = 198$ in.
(*g*) 66 ft. to yards?	$\frac{66}{3} = 22$ yd.
(*h*) 72 in. to yards?	$\frac{72}{36} = 2$ yd.
(*i*) 66 ft. to rods?	$66/16\frac{1}{2} = 4$ rd.
(*j*) 2 rd. to feet?	$2 \times 16\frac{1}{2} = 33$ ft.
(*k*) $16\frac{1}{2}$ yd. to rods?	$16\frac{1}{2}/5\frac{1}{2} = 3$ rd.
(*l*) 15,840 ft. to miles?	$15{,}840/5{,}280 = 3$ mi.

463. What is the procedure for reduction to lower denominations when the length is expressed in several denominations?

Reduce each unit to the next lower denomination in regular order.

EXAMPLE: What is the reduction to inches of 6 rd. 5 yd. 2 ft. 6 in.?

$$6 \text{ rd.} \times 5\tfrac{1}{2} \text{ yd.} = 33 \text{ yd.}$$
$$+\ 5 \text{ yd.}$$
$$38 \text{ yd.} \times 3 \text{ ft.} = 114 \text{ ft.}$$
$$+\ 2 \text{ ft.}$$
$$116 \text{ ft.} \times 12'' = 1{,}392 \text{ in.}$$
$$+\ 6 \text{ in.}$$
$$\text{Ans.} = 1{,}398 \text{ in.}$$

464. What is the procedure for reduction to higher denominations?

Reduce each unit to the next higher denomination in regular order.

EXAMPLE: What is the reduction to rods, yards, feet, and inches of 1,503 inches?

12)1,503 inches	(Reduce inches to feet. Divide by 12.)
3)125 ft. and 3 in.	(Reduce to yards. Divide by 3.)
$5\frac{1}{2}$)41 yd. and 2 ft.	(Reduce to rods. Divide by $5\frac{1}{2}$.)
7 rd. and $2\frac{1}{2}$ yd.	

Ans. = 7 rd. $2\frac{1}{2}$ yd. 2 ft. 3 in.
or, 7 rd. 2 yd. 3 ft. 9 in.

465. What are the units used in measuring the areas of surfaces (square measure)?

144	square inches (sq. in.)	= 1 square foot (sq. ft.)
9	square feet	= 1 square yard (sq. yd.)
$30\frac{1}{4}$	square yards	= 1 square rod (sq. rd.)
160	square rods	= 1 acre (A) = 43,560 sq. ft.
640	acres	= 1 square mile (sq. mi.) or section
36	square miles	= 1 township (twp.)

1 sq. mi. = 102,400 sq. rd. = 3,097,600 sq. yd. = 27,878,400 sq. ft.

Note that 12 in. × 12 in. = 144 sq. in. = 1 sq. ft.;
3 ft. × 3 ft. = 9 sq. ft. = 1 sq. yd., etc.

Note: A *square* 10′ × 10′ = 100 sq. ft. is commonly used in roofing.

466. What is the result of the reduction of the following?

(*a*) 442 sq. in. to sq. ft.? 442 ÷ 144 = 3 sq. ft.
(*b*) 45 sq. ft. to sq. yd.? 45 ÷ 9 = 5 sq. yd.
(*c*) 4 sq. yd. to sq. ft.? 4 × 9 = 36 sq. ft.
(*d*) 640 sq. rd. to acres? 640 ÷ 160 = 4 A
(*e*) 432 sq. mi. to twp.? 432 ÷ 36 = 12 twp.
(*f*) 10 sq. mi. to acres? 10 × 640 = 6,400 A
(*g*) 10 twp. to acres? 10 × 36 × 640 = 230,400 A
(*h*) 120 A, 240 sq. rd. to sq. yd.?

$$120 \times 160 \times 30\tfrac{1}{4} + 240 \times 30\tfrac{1}{4} = 580{,}800 + 7{,}260 = 588{,}060 \text{ sq. yd.}$$

(*i*) 24 sq. yd. 14 sq. ft. to sq. in.?

$$24 \times 9 \times 144 + 14 \times 144 = 31{,}104 + 2{,}016 = 33{,}120 \text{ sq. in.}$$

(*j*) 2 sections to sq. rd.? 2 × 640 × 160 = 204,800 sq. rd.
(*k*) 24,320 sq. rd. to acres? 24,320 ÷ 160 = 152 A
(*l*) 152,460 sq. ft. to acres? 152,460 ÷ 43,560 = $3\frac{1}{2}$ A

467. What are the measurements for solids (cubic measure)?

Cubic measure is used to measure the contents or capacity of bins, tanks, and the like as well as solids (volume).

1,728 cubic inches (cu. in.)	= 1 cubic foot (cu. ft.)
27 cubic feet	= 1 cubic yard (cu. yd.)
128 cubic feet	= 1 cord of wood (cd.)
$24\frac{3}{4}$ cubic feet	= 1 perch (P)

Note: A *cord* of wood is 8 ft. long × 4 ft. wide × 4 ft. high = 128 cu. ft. A *perch* (used to measure stone masonry) is $16\frac{1}{2}$ ft. long × $1\frac{1}{2}$ ft. wide × 1 ft. high = $24\frac{3}{4}$ cu. ft.

468. What are the units applicable to liquid measure?

4 gills (gi.) = 1 pint (pt.)	$31\frac{1}{2}$ gallons = 1 barrel (brl.)
2 pints = 1 quart (qt.)	63 gallons = 1 hogshead (hgs.)
4 quarts = 1 gallon (gal.)	$7\frac{1}{2}$ gallons = 1 cubic foot

231 cubic inches = 1 gallon (U.S.)
277.274 cu. in. = 1 gallon (imperial gallon of England)
A gallon of water (English gallon) weighs 10 pounds.
A gallon of water (U.S. gallon) weighs about $8\frac{1}{3}$ pounds.
A cubic foot of water weighs $62\frac{1}{2}$ pounds.
Liquid measure is used in measuring liquids except medicine.

Note: A fluid ounce is equal to $\frac{1}{16}$ of a pint, or $\frac{1}{4}$ of a gill.

469. What are the units applicable to dry measure?

2 pints (pt.) = 1 quart (qt.)	2.75 bushel = 1 barrel (bbl.)
8 quarts = 1 peck (pk.)	2,150.42 cu. in. = 1 stricken bu.
4 pecks = 1 bushel (bu.)	2,747.71 cu. in. = 1 heaped bu.
	67.2 cu. in. = 1 dry quart

Dry measure is used in measuring grains, seeds, produce, and the like.

470. How many kinds of weight are in use in the United States?

Four kinds:

(*a*) Avoirdupois weight is used in weighing heavy, coarse products such as grain, hay, coal, iron, and the like.

(*b*) Troy weight is used in weighing precious metals—minerals, gold, silver, and diamonds. It is also used by the government in weighing coins at the mint.

(*c*) Apothecaries' weight is used in weighing drugs and chemicals.

(*d*) Metric or decimal system of weights is used extensively in the United States in scientific work.

471. What constitutes the avoirdupois table of weights?

16 ounces (oz.)	= 1 pound (lb.)
100 pounds	= 1 hundredweight (cwt.)
20 hundredweight	= 1 ton (T)
2,000 pounds	= 1 ton
7,000 grains (gr.)	= 1 pound avoirdupois
437.5 grains	= 1 ounce
2,240 pounds	= 1 long ton (sometimes called gross ton)

The long ton is used by the U.S. Custom House in determining duty on merchandise taxed by the ton. It is also used when coal and iron are sold at wholesale at the mines. Unless otherwise specified, a ton is taken to be 2,000 pounds.

472. What constitutes the troy table of weights?

24 grains (gr.)	= 1 pennyweight (pwt.)
20 pennyweights	= 1 ounce (oz.)
12 ounces	= 1 pound (lb.)
5,760 grains	= 1 pound troy
480 grains	= 1 ounce

The *carat*, used in weighing precious stones, is equivalent to 3.168 grains troy, or 205.5 milligrams. The term *karat* is used to denote the fineness of gold, and means $\frac{1}{24}$ by weight of gold. For example, 24 karats fine means pure gold, 18 karats means $\frac{18}{24}$ pure gold by weight.

473. What constitutes the apothecaries' table of weights?

20 grains (gr.)	= 1 scruple (sc.) = ℈
3 scruples	= 1 dram (dr.) = ʒ
8 drams	= 1 ounce (oz.) = ℥
12 ounces	= 1 pound (lb.) = ℔
5,760 grains	= 1 pound
480 grains	= 1 ounce

Although avoirdupois weight is used in buying and selling drugs and chemicals wholesale, druggists and physicians use apothecaries' weight in compounding medicines.

Apothecaries' fluid measure

60 minims (m.)	= 1 fluid drachm, or dram (fʒ)
8 fluid drachms	= 1 fluid ounce (f℥)
16 fluid ounces	= 1 pint (O)
8 pints	= 1 gallon (Cong.)

Apothecaries' fluid measure is used by druggists in preparing medicines.

474. What are some comparisons of weights?

	Pound	*Ounce*
Troy	5,760 grains	480 grains
Apothecaries'	5,760 grains	480 grains
Avoirdupois	7,000 grains	$437\frac{1}{2}$ grains

The grain is the same in all three systems. The troy and apothecaries' pound and ounce are respectively alike.

475. What are the units for measurement of time?

The measures are based on the movements of the earth and other bodies of the solar system. One revolution of the earth on its axis is designated a day, and one complete revolution of the earth around the sun is one year. The month is derived from the revolution of the moon around the earth.

60 seconds (sec.)	=	1 minute (min.)
60 minutes	=	1 hour (hr.)
24 hours	=	1 day (da.)
7 days	=	1 week (wk.)
30 days	=	1 month (mo.) (See Note (*b*) below)
52 weeks	=	1 year (yr.)
12 months	=	1 common year (yr.)
365 days	=	1 common year
366 days	=	1 leap year (l. yr.)
10 years	=	1 decade
20 years	=	1 score
100 years	=	1 century (C.)

One revolution of the earth around the sun requires 365 days 5 hours 48 minutes and 49.7 seconds. Since the fraction is almost $\frac{1}{4}$ of a day, one entire day is added every fourth year to make a leap year. Because this does not exactly take care of the fraction, every centennial year which is not divisible by 400 is regarded as a common year.

Note (*a*): All months have 31 days except April, June, September, and November, which have 30 days, and February, which has 28 days in the common year and 29 days in the leap year.

Note (*b*): It is customary in business to regard a year as 12 months of 30 days each, or as 360 days. This practice is for convenience only in making interest calculations as explained earlier.

476. What are the measures of counting?

20 units	=	1 score
12 units	=	1 dozen
12 dozen	=	1 gross (gro.)
12 gross	=	1 great gross (gr. gro.)

477. What are the units for paper measure?

24 sheets	=	1 quire (qr.)
20 quires	=	1 ream (rm.)
2 reams	=	1 bundle (bdl.)
5 bundles	=	1 bale (bl.)

Publishers and printers estimate on a basis of 1,000 sheets, and allow 500 sheets to a ream, although there are usually 480 sheets in a ream.

478. What are some measures of value?

U.S. money

10 mills	=	1 cent (¢)
5 cents	=	1 nickel
10 cents	=	1 dime
100 cents	=	1 dollar ($)
10 dollars	=	1 eagle

German money

100 pfennig	=	1 mark

English money

12 pence (d.)	=	1 shilling (s.)
20 shillings	=	1 pound sterling (£)

French money

10 millimes	=	1 centime
10 centimes	=	1 decime
10 decimes	=	1 franc

479. What is the metric system of weights and measures?

It is a decimal system in which the fundamental unit is the *meter*, the unit of length. From this, the units of capacity (*liter*) and of weight (*gram*) were derived. Decimal subdivisions or multiples of these comprise all the other units.

One meter (= 39.37 in.) was taken to be one ten millionth of the distance from the equator to the pole. More accurate measurements later proved this to be only approximately correct.

Six numerical prefixes combine with meter, gram, and liter to form the metric tables.

The Latin prefixes are:

milli-	= one thousandth	= .001	=	$\frac{1}{1000}$
centi-	= one hundredth	= .01	=	$\frac{1}{100}$
deci-	= one tenth	= .1	=	$\frac{1}{10}$

The Greek prefixes are:

deca-	= ten	= 10
hecto-	= one hundred	= 100
kilo-	= one thousand	= 1,000

480. What is the linear measure table in the metric system?

10 millimeters (mm.)	= 1 centimeter (cm.)	=	.01	meter
10 centimeters	= 1 decimeter (dm.)	=	.1	meter
10 decimeters	= 1 meter (m.)	=	1.	meter
10 meters	= 1 decameter (Dm.)	=	10.	meters
10 decameters	= 1 hectometer (hm.)	=	100.	meters
10 hectometers	= 1 kilometer (km.)	=	1,000.	meters

Move the decimal point to the *right* to change from a higher to a lower denomination, and to the *left* to change from a lower to a higher denomination.

Ex. (*a*) Express 82.6 meters as decimeters. Higher to lower; move point to right, getting 826 decimeters.

Ex. (*b*) Express 832.34 centimeters to meters. Lower to higher; move point to left, getting 8.3234 meters.

Ex. (*c*) Express 15,283 meters in the proper denominations.

15 kilometers 2 hectometers 8 decameters 3 meters

481. What is the area measure table in the metric system?

The unit measure for small surfaces is the square meter. One hundred units of any denomination are required to make one unit of the next higher denomination.

100 sq. millimeters (sq. mm.)	= 1 sq. centimeter (sq. cm.)
	= .0001 sq. meter
100 sq. centimeters	= 1 sq. decimeter (sq. dm.)
	= .01 sq. meter
100 sq. decimeters	= 1 sq. meter (sq. m.)
	= 1 sq. meter = 1 centare
100 sq. meters	= 1 sq. decameter (sq. Dm.)
	= 100 sq. meters = 1 are
100 sq. decameters	= 1 sq. hectometer (sq. hm.)
	= 10,000 sq. meters = 1 hectare
100 sq. hectometers	= 1 sq. kilometer (sq. km.)
	= 1,000,000 sq. meters

Move decimal point to the *right* to change from a higher to a lower denomination.

Ex. (*a*) Express 82.6 sq. meters as sq. decimeters. Higher to lower; move point to right, getting 8,260 sq. decimeters.

Move point to the left to change from a lower to a higher denomination.

Ex. (*b*) Express 832.34 sq. centimeters as sq. meters. Lower to higher; move point to left, getting .083234 sq. meters.

482. What is the volume or cubic measure table in the metric system?

The cubic meter is the practical unit of measures of volume. When used in measuring wood the cubic meter is called a stere. One thousand units of any denomination are required to make one unit of the next higher denomination.

1,000 cu. millimeters (cu. mm.)	= 1 cu. centimeter (cu. cm.)
	= .000001 cu. meter
1,000 cu. centimeters	= 1 cu. decimeter (cu. dm.)
	= .001 cu. meter (= 1 liter)
1,000 cu. decimeters	= 1 cu. meter (cu. m.)
1,000 cu. meters	= 1 cu. decameter (cu. Dm.)
	= 1,000 cu. meters
1,000 cu. decameters	= 1 cu. hectometer (cu. hm.)
	= 1,000,000 cu. meters
1,000 cu. hectometers	= 1 cu. kilometer (cu. km.)
	= 1,000,000,000 cu. meters

Ex. (*a*) Express 82.6 cu. meters as cubic decimeters. Higher to lower; move point to right, getting 82,600 cu. decimeters.

Ex. (*b*) Express 83,234 cu. centimeters as cu. meters. Lower to higher; move point to left, getting .083234 cu. meters.

483. What is the table for measures of liquid and dry capacity in the metric system?

The liter, a cube the side of which is one decimeter (= $\frac{1}{10}$ meter), is the unit of capacity for both liquid and dry measures.

10 milliliters (ml.)	= 1 centiliter	= .01 liter
10 centiliters (cl.)	= 1 deciliter	= .1 liter
10 deciliters (dl.)	= 1 liter	= 1 liter
10 liters (l.)	= 1 decaliter	= 10 liters
10 decaliters (Dl.)	= 1 hectoliter	= 100 liters
10 hectoliters (hl.)	= 1 kiloliter (kl.)	= 1,000 liters

484. What is the table for measures of weight in the metric system?

The unit of weight is the gram which is the weight of a cube of distilled water having an edge $\frac{1}{100}$ meter in length. One pound = 453.5924 grams.

10 milligrams (mg.)	= 1 centigram (cg.)	= .01 gram
10 centigrams	= 1 decigram (dg.)	= .1 gram
10 decigrams	= 1 gram (g.)	= 1 gram

10 grams	= 1 decagram (Dg.)	= 10 grams
10 decagrams	= 1 hectogram (hg.)	= 100 grams
10 hectograms	= 1 kilogram (kg.)	= 1,000 grams
10 kilograms	= 1 myriagram (Mg.)	= 10,000 grams
10 myriagrams	= 1 quintal (Q)	= 100,000 grams
10 quintals	= 1 metric ton (MT)	= 1,000,000 grams

485. What are the units for circular measure?

60 seconds (″)	= 1 minute (′)
60 minutes	= 1 degree (°)
360 degrees	= 1 circle (cir.)
An angle of 90 degrees (90°)	= a right angle
$\frac{1}{4}$ of a circle (90°)	= a quadrant
$\frac{1}{6}$ of a circle (60°)	= a sextant
$\frac{1}{12}$ of a circle (30°)	= a sign

486. In reducing 4 bu. 3 pk. 5 qt. 2 pt. to pints, what is the procedure?

Reduce each denomination to pints by multiplying by the appropriate units and find the total.

$$
\begin{array}{lr}
4\text{ (bu.)} \times 4\text{ pk.} \times 8\text{ qt.} \times 2\text{ pt.} = & 256\text{ pt.} \\
3\text{ (pk.)} \times 8\text{ qt.} \times 2\text{ pt.} \quad = & 48\text{ pt.} \\
5\text{ (qt.)} \times 2 \quad = & 10\text{ pt.} \\
 & 2\text{ pt.} \\ \hline
\text{Ans.} = & 316\text{ pt.}
\end{array}
$$

487. What is the result of reducing $\frac{7}{8}$ gal. to lower denominations?

$$\frac{7}{\not 8_{2}} \times \not 4^{1}\text{ qt.} = 3\tfrac{1}{2}\text{ qt.}$$

$$\tfrac{1}{2} \times 2\text{ pt.} = 1\text{ pt.}$$

$$\therefore \tfrac{7}{8}\text{ gal.} = 3\text{ qt. } 1\text{ pt. Ans.}$$

488. What is the result of reducing 10 qt. 2 pt. to the fraction of a bushel?

$$4\text{ pk.} \times 8\text{ qt.} \times 2\text{ pt.} = 64\text{ pt. in } 1\text{ bu.}$$

$$10\text{ qt.} \times 2\text{ pt.} + 2\text{ pt.} = 22\text{ pt.}$$

$$\therefore \tfrac{22}{64} = \tfrac{11}{32}\text{ bu. Ans.}$$

489. What is the result of reducing $\frac{7}{8}$ ft. to the fraction of a rod?

$$\tfrac{7}{8} \div 3 \text{ ft.} = \tfrac{7}{8} \times \tfrac{1}{3} = \tfrac{7}{24} = \text{the fraction of a yd.}$$
$$\tfrac{7}{24} \div 5\tfrac{1}{2} \text{ yd.} = \tfrac{7}{24} \times \tfrac{2}{11} = \tfrac{7}{132} = \text{the fraction of a rd. Ans.}$$

490. What is the result of reducing 2 pk. 6 qt. $\frac{1}{2}$ pt. to a decimal of a bushel?

2 pt.	$\frac{1}{2}$ pt.	Divide $\frac{1}{2}$ pt. by 2 pt. (= 1 qt.)
	$\frac{1}{4}$ (or .25) qt. + 6 qt. = 6.25 qt.	
8 qt.	6.25 qt.	Divide 6.25 qt. by 8 qt. (= 1 pk.)
	.78125 pk. + 2 pk. = 2.78125 pk.	
4 pk.	2.78125	Divide 2.78125 pk. by 4 pk. (= 1 bu.)
	.6953125 bu. Ans.	

491. What is the result of reducing .27 lb. apothecaries' to lower denominations?

.27 lb. × 12 oz. (= 1 lb.) = 3.24 oz.
.24 oz. × 8 dr. (= 1 oz.) = 1.92 dr.
.92 dr. × 3 sc. (= 1 dr.) = 2.76 sc.
.76 sc. × 20 gr. (= 1 sc.) = 15.2 gr.

∴ 3 oz. 1 dr. 2 sc. 15.2 gr. Ans.

492. What is the result of reducing .62 gill to a decimal of a gallon?

.62 gill ÷ 4 gill (= 1 pt.) = .155 pt.
.155 pt. ÷ 2 pt. (= 1 qt.) = .0775 qt.
∴ .0775 qt. ÷ 4 qt. (= 1 gal.) = .019375 gal. Ans.

493. What is the procedure for addition of denominate numbers?

Arrange so that like units are under like units (pounds under pounds, ounces under ounces, etc.). Begin with the lowest denomination and work to the left.

Example: Add:

	6 lb.	3 oz.	10 pwt.	12 gr.
	3 lb.	8 oz.	8 pwt.	10 gr.
	12 lb.	6 oz.	15 pwt.	16 gr.
Sum =	22 lb.	6 oz.	14 pwt.	14 gr.

Sum of the gr. is 38 gr., which divide by 24 gr. (= 1 pwt.) = 1 pwt. + 14 gr., remaining.

Sum of pwt. = 33 + 1 carry = 34 pwt., which divide by 20 pwt. (= 1 oz.) = 1 oz. + 14 pwt., remaining.

Sum of oz. = 17 + 1 carry = 18 oz., which divide by 12 oz. (=1 lb.) = 1 lb. + 6 oz. remaining.

Sum of lb. = 21 + 1 carry = 22 lb.

∴ 22 lb. 6 oz. 14 pwt. 14 gr. Ans.

494. What is the procedure for subtraction of denominate numbers?

Place like units under like units. Start with the lowest denomination. Borrow from higher denomination when necessary.

EXAMPLE:

From	3 yr.	7 mo.	14 days	7 hr.	35 min.	24 sec.	
Subtract	1 yr.	8 mo.	22 days	4 hr.	50 min.	32 sec.	
	1 yr.	10 mo.	22 days	2 hr.	44 min.	52 sec.	Ans.

Borrow 1 min. = 60 sec. from 35 min., leaving 34 min.
Add 60 sec. to 24 sec. = 84 sec. and subtract 32 sec., leaving 52 sec.

Borrow 1 hr. = 60 min. from 7 hr, leaving 6 hr.
Add 60 min. to 34 min. = 94 min. and subtract 50 min., leaving 44 min.

Subtract 4 hr. from 6 hr., leaving 2 hr.

Borrow 1 mo. = 30 days from 7 mo., leaving 6 mo.
Add 30 days to 14 days = 44 days and subtract 22 days, leaving 22 days.

Borrow 1 yr. = 12 mo. from 3 yr., leaving 2 yr.
Add 12 mo. to 6 mo. = 18 mo. and subtract 8 mo., leaving 10 mo.

Subtract 1 yr. from 2 yr, leaving 1 yr.

∴ 1 yr. 10 mo. 22 days. 2 hr. 44 min. 52 sec. Ans.

495. What is the result of multiplying 26 sq. rd. 10 sq. yd. 5 sq. ft. 34 sq. in. by 8?

Multiply each denomination by 8 and place results in position.

sq. rd.	sq. yd.		sq. ft.	sq. in.
26	10		5	34
				×8
208_2	80_4		40_1	272
210	23		5	128
		($\frac{1}{2}$ sq. yd. =)	4	72
210	23		9_1	200
210	24		1	56

272 sq. in. = 1 sq. ft. + 128 sq. in remaining.
40 sq. ft. + 1 sq. ft. = 41 sq. ft. = 4 sq. yd. + 5 sq. ft. remaining.
80 sq. yd. + 4 sq. yd. = 84 sq. yd. = 2 sq. rd. + $23\frac{1}{2}$ sq. yd. remaining.
208 sq. rd. + 2 sq. rd. = 210 sq. rd.
Change $\frac{1}{2}$ sq. yd. to 4 sq. ft. 72 sq. in. and adjust the result by adding.
72 + 128 sq. in. = 200 sq. in. = 1 sq. ft. + 56 sq. in. remaining.
4 + 5 + 1 sq. ft. = 10 sq. ft. = 1 sq. yd. + 1 sq. ft. remaining.
23 sq. yd. + 1 sq. yd. = 24 sq. yd.

∴ 210 sq. rd. 24 sq. yd. 1 sq. ft. 56 sq. in. Ans.

496. What is the result of dividing 18 A, 142 sq. rd. 24 sq. yd. by 7?

In division start at the left with the highest denomination, and divide each in turn.

	A	sq. rd.	sq. yd.	sq. ft.	sq. in.
7)	18	142	24		
	2	111	25	0	$46\frac{2}{7}$ (or 46.3)

$$\frac{18\text{ A}}{7} = 2\text{ A} + 4\text{ A remaining,}$$

4 A × 160 sq. rd. + 142 sq. rd. = 782 sq. rd.

$$\frac{782\text{ sq. rd.}}{7} = 111\text{ sq. rd.} + 5\text{ sq. rd. remaining,}$$

5 sq. rd. × $30\frac{1}{4}$ sq. yd. + 24 sq. yd. = $175\frac{1}{4}$ sq. yd.

$$\frac{175\frac{1}{4}\text{ sq. yd.}}{7} = 25\text{ sq. yd.} + \frac{1}{4}\text{ sq. yd. remaining,}$$

$\frac{1}{4}$ × 9 sq. ft. = $2\frac{1}{4}$ sq. ft. $\quad$ $2\frac{1}{4}$ × 144 sq. in. = 324 sq. in.

$$\frac{324\text{ sq. in.}}{7} = 46\tfrac{2}{7}\text{ or }46.3\text{ sq. in.}$$

∴ 2 A, 111 sq. rd. 25 sq. yd. $46\frac{2}{7}$ sq. in. Ans.

497. How many pounds of avoirdupois are 25 pounds troy weight?

$$\frac{\overset{1}{\cancel{25}} \times \overset{144}{\cancel{5{,}760}}}{\underset{\underset{7}{\cancel{280}}}{\cancel{7{,}000}}} = \frac{144}{7} = 20\tfrac{4}{7}\text{ lb. Ans.}$$

There are 5,760 gr. in the pound troy, and 7,000 gr. in the pound avoirdupois.

498. How can we reduce 6 km. 4 hm. 3 m. 5 dm. 9 mm. to meters?

Insert a decimal point after the measure required, first making sure to insert a zero whenever any unit is omitted.

6 km. 4 hm. 0 Dm. 3 m. 5 dm. 0 cm. 9 mm.

Here meters are called for. Insert a decimal point after meters, getting

6,403.509 meters. Ans.

499. How can we reduce 53,276.98 dm. to km.?

Here the 6 represents whole dm.
the 7 represents whole m.
the 2 represents whole Dm.
the 3 represents whole hm.
the 5 represents whole km.

As km. are called for, put the decimal point after the 5, getting 5.327698 km.

Another way of doing it is to note that from the table of units $10 \times 10 \times 10 \times 10$ or 10,000 dm. = 1 km.

Therefore, divide 53,276.98 dm. by 10,000 or move the decimal point 4 places to the left, getting

5.327698 km. Ans.

500. What is the result of adding 4.8 m. 284 cm. and 5 Dm. 2 dm., with the answer expressed in meters?

Write down each quantity in meters, keeping the points underneath each other.

m.	
4.8	
2.84	
50.2	
57.84	m. Ans.

501. How many centimeters remain when from a pipe 2.83 m. long, 16.7 cm. is cut off?

cm.	
283.	
− 16.7	
266.3	cm. remain. Ans.

502. What is the total weight in kg. of 3,450 cartons when each carton weighs 3,600 g.?

3,600 g. = 3.6 kg.

$\therefore$ 3.6 × 3,450 = 12.420 kg. Ans.

PROBLEMS

1. Express:

(*a*) 3 ft. 3 in. in inches
(*b*) 3 yd. 3 ft. in feet
(*c*) 6 rd. in yards
(*d*) 112 in. in feet
(*e*) 5 mi. in rods
(*f*) 2 rd. in inches
(*g*) 88 ft. in yards
(*h*) 96 in. in yards
(*i*) 92 ft. in rods
(*j*) 3 rd. in feet
(*k*) 34,860 ft. in miles
(*l*) 6 mi. in feet

2. Reduce 5 rd. 4 yd. 4 ft. 7 in. to inches.

3. What is the reduction of 1,608 inches to rods, yards, feet, and inches?

4. What is the result of the reduction of:

(*a*) 562 sq. in. to sq. ft.?
(*b*) 36 sq. ft. to sq. yd.?
(*c*) 6 sq. yd. to sq. ft.?
(*d*) 860 sq. rd. to acres?
(*e*) 362 sq. mi. to twp.?
(*f*) 12 sq. mi. to acres?
(*g*) 8 twp. to acres?
(*h*) 80 A 120 sq. rd. to sq. yd.?
(*i*) 12 sq. yd. 10 sq. ft. to sq. in.?
(*j*) 3 sections to sq. rd.?
(*k*) 12,460 sq. rd. to acres?
(*l*) 174,240 sq. ft. to acres?

5. How many cubic inches are there in a bar of metal $4\frac{1}{2}$ inches long, 3 inches wide, and $1\frac{1}{2}$ inches thick?

6. At $2.25 a cubic yard, what would be the cost of excavating a basement 25 feet 9 inches by 34 feet 6 inches by 9 feet deep?

7. A tank is 40 ft. 6 in. high and 5 ft. 9 in. square. How much will this tank full of water weigh, assuming water weighs 62.5 lb. per cubic foot?

8. Express:

(*a*) 4 gallons in pints
(*b*) 96 pints in bushels
(*c*) 3 pints in gills
(*d*) 6 quarts in pints
(*e*) 2 bushels in pints
(*f*) 12 pecks in bushels
(*g*) 3 pecks in pints
(*h*) 8 quarts in gills
(*i*) 2 bushels in pecks
(*j*) 692 cu. in. in gallons
(*k*) 4 bushels in quarts
(*l*) 12 gills in pints
(*m*) 12 pints in quarts
(*n*) 24 quarts in pecks
(*o*) 32 pints in pecks
(*p*) $22\frac{1}{2}$ gallons in cubic feet

9. A bin holds 832 bushels of grain. What is its capacity in barrels?

10. Express:

(*a*) 4,000 pounds in tons
(*b*) 4 pounds in ounces (troy)
(*c*) 3 pennyweights in grains
(*d*) 5 lb. in ounces (avoirdupois)
(*e*) 60 pennyweights in ounces
(*f*) 48 ounces in pounds
(*g*) 60 hundredweight in tons
(*h*) 3 carats in grains

11. What is the value of a gold nugget which weighs 6 ounces 4 pennyweights 12 grains, at $35 an ounce?

12. How many pounds are there in 103,680 grains?

13. Express 22 long tons in pounds.

14. Express:

(*a*) 2 weeks in hours
(*b*) 4 hours in minutes
(*c*) 3 days in hours
(*d*) 6 hours in minutes
(*e*) 3 common years in days
(*f*) $4\frac{1}{2}$ minutes in seconds
(*g*) 72 hours in days
(*h*) 300 seconds in minutes
(*i*) 7 days in minutes
(*j*) 4,000 years in centuries
(*k*) 414,720 seconds in days
(*l*) 1 day in seconds

15. How many days will the year 2000 have?

16. Express 18 hours 38 minutes 20 seconds in seconds.

17. How many days in the second six months of a common year?

18. How many hours are there in the month of September?

19. A crate contains 504 eggs—how many dozen?

20. A single card contains 24 hooks and eyes. How many gross are there on 48 cards?

21. How many years are there in three decades?

22. How many years are there in three score years and ten?

23. How many sheets are there in:

(*a*) 12 quires?
(*b*) 3 reams?
(*c*) 2 bundles?
(*d*) 3 bales?

24. How many mills in $5\frac{1}{2}$ cents?

25. Express:

(*a*) 73.5 meters as decimeters
(*b*) 741.26 centimeters as meters

26. Express 18,362 meters in the proper denominations.

27. Express:

(*a*) 482.61 sq. centimeters as sq. meters
(*b*) 74.8 sq. meters as sq. decimeters

28. Express:

(*a*) 74.6 cu. meters as cubic decimeters
(*b*) 94,364 cu. centimeters as cu. meters

29. Express 8 pounds and 10 ounces in decigrams.

30. Express 3 kilograms in ounces.

31. How many grains are there in $4\frac{1}{2}$ hectograms?

32. Express 3 pounds and 6 ounces in milligrams.

33. Express 2 pounds and 4 ounces in centigrams.

34. How many grains are there in 4.5 grams?

35. Express 4 centigrams and $3\frac{1}{4}$ milligrams in grains.

36. How many minutes in 10°? 12°? 28°?

37. How many seconds in 4′? 6′? 2°?

38. Reduce:

(*a*) 50° 40′ 30″ to seconds (*b*) 43,200″ to degrees

39. Reduce 5 bu. 4 pk. 3 qt. 2 pt. to pints.

40. Reduce $\frac{5}{8}$ gal. to lower denominations.

41. Reduce 12 qt. 2 pt. to the fraction of a bushel.

42. Reduce $\frac{5}{6}$ ft. to the fraction of a rod.

43. Reduce 3 pk. 4 qt. $\frac{1}{2}$ pt. to a decimal of a bushel.

44. Reduce .38 lb. apothecaries' to lower denominations.

45. Reduce .58 gill to a decimal of a gallon.

46. Add:

5 lb.	4 oz.	8 pwt.	10 gr.
4 lb.	7 oz.	6 pwt.	8 gr.
14 lb.	8 oz.	16 pwt.	18 gr.

47. From 4 yr. 8 mo. 12 da. 8 hr. 30 min. 22 sec. subtract 2 yr. 10 mo. 24 da. 3 hr. 45 min. 30 sec.

48. Multiply 24 sq. rd. 8 sq. yd. 4 sq. ft. 28 sq. in. by 9.

49. Divide 20 A, 138 sq. rd. 22 sq. yd. by 6.

50. How many pounds avoirdupois are 22 pounds troy weight?

51. Reduce 7 km. 5 hm. 4 m. 6 dm. 10 mm. to meters.

52. Reduce 654,389.79 dm. to km.

53. Add and express result in meters: 5.6 m. 376 cm. and 7 Dm. 4 dm.

54. How many centimeters remain when from a pipe 3.69 m. long, 26.8 cm. are cut off?

55. What is the total weight in kg. of 4,860 cartons, when each carton weighs 2,400 g.?

CHAPTER XIII

POWER — ROOTS — RADICALS

503. How can we show that the square of a number is the product of a number with itself?

In the figure there are three units on each side of the square. There are 9 square units in a square of 3 units on a side. Therefore, 9 is said to be the square of 3. Similarly, a square with 5 units on each side has a total of 25 square units. $5 \times 5 = 25$.

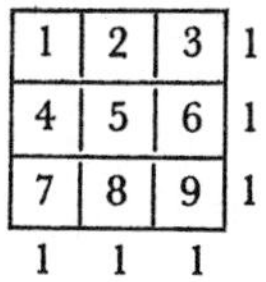

504. How can we show that the cube of a number is the product of the number taken 3 times as a factor?

In the figure there are three units on each edge. There are 27 cubic units in a cube with 3 units on an edge. Therefore, 27 is said to be the cube of 3. Similarly a cube with 5 units on an edge has a total of 125 cubic units, $5 \times 5 \times 5 = 125$.

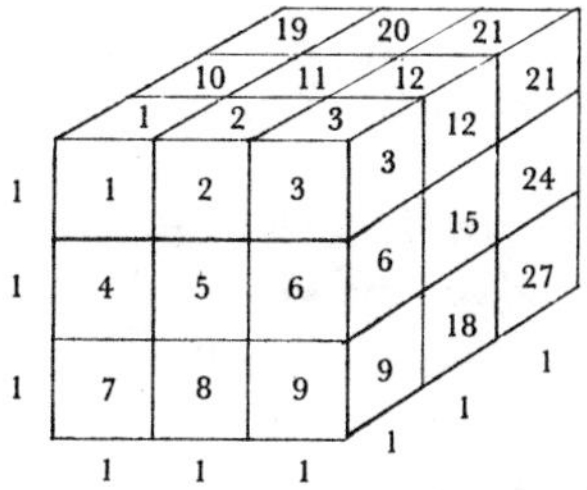

505. What is meant by raising a number to a power?

A number multiplied by itself is said to be raised to a power.

EXAMPLES:

$$2 \times 2 = \text{square or second power of } 2 = 4$$
$$2 \times 2 \times 2 = \text{cube or third power of } 2 = 8$$
$$2 \times 2 \times 2 \times 2 = \text{fourth power of } 2 = 16$$

Other higher powers are denoted by numbers indicating the number of times the factor is used.

$$3 \times 3 \times 3 \times 3 \times 3 = \text{fifth power of } 3 = 243$$

506. What is meant by (*a*) an exponent, (*b*) a base?

(*a*) The exponent is a small figure written to the upper right of a number to be raised to a power and indicates the power taken (or how many times the number is multiplied by itself).

EXAMPLES:

3^2 means the square of 3 ($3 \times 3 = 9$), exponent is 2
4^3 means the cube of 4 ($4 \times 4 \times 4 = 64$), exponent is 3
7^8 means the eighth power of 7
($7 \times 7 \times 7 \times 7 \times 7 \times 7 \times 7 \times 7 = 5{,}764{,}801$), exponent is 8

(*b*) The factor to be raised to a power is called the base.

EXAMPLE: In 7^8, 7 is the base and 8 is the exponent.

507. How do we raise an algebraic symbol to a power?

By the use of an exponent which denotes the number of times the symbol is used.

Ex. (*a*) x squared $= x \cdot x = x^2$, which means that two equal quantities x have been multiplied together.

Ex. (*b*) x cubed $= x \cdot x \cdot x = x^3$, which means that three equal quantities x have been multiplied together.

Ex. (*c*) ($3x$) squared means 3 squared multiplied by x squared; or, $3 \cdot 3 \cdot x \cdot x = 3^2x^2 = (3x)^2 = 9x^2$, which is read "9 ($x$ squared)."

Ex. (*d*) $3x$ raised to the fourth power $= 3 \cdot 3 \cdot 3 \cdot 3 \cdot x \cdot x \cdot x \cdot x = 3^4x^4 = 81x^4$.

508. What is the operation of raising quantities or terms to given powers called?

The process is called *involution.*

509. How can we show that the square of the sum of any two numbers is the square of the first plus the square of the second plus twice the product of the two numbers?

The square of a number is the number multiplied by itself. The square of 26 is $26 \times 26 = 676$.

We may write this multiplication as

		2 tens + 6 units	
		× 2 tens + 6 units	
	2 tens × 6 units	(6 units)^2	← multiplying by 6 units
(2 tens)^2	2 tens × 6 units		← multiplying by 2 tens
(2 tens)^2 +	2 (2 tens × 6 units) +	(6 units)^2	= sum of partial products
400 +	240 +	36	= 676

Since any number greater than 10 may be considered as the sum of two numbers, the square of the sum of any two numbers = the square of the first + square of the second + twice the product of the two numbers.

510. How can the above be shown graphically?

Cut the lines of the sides into 20 units and 6 units to represent 2 tens + 6 units. The whole square of 26 consists of the large square = 20^2 + 2 times the rectangle 20×6 + small square 6^2, or

$$400 + 240 + 36 = 676$$

If $20 = a$ and $6 = b$ we get the formula

$$(a + b)^2 = a^2 + 2ab + b^2$$

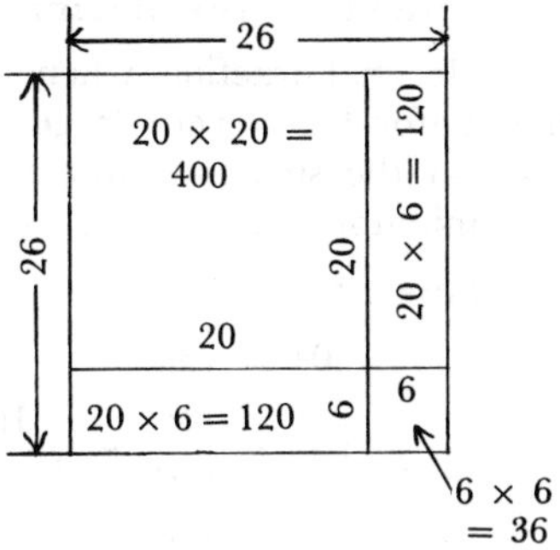

511. How do we find the number of square units in the surface of any plane figure or flat surface?

Multiply the unit of length by the unit of width of the same denomination (inches by inches, feet by feet, etc.).

Here there are 8 units of length and 6 units of width, therefore, $8 \times 6 = 48$ square units. This may be thought of as 6 rows of 8 square units per row.

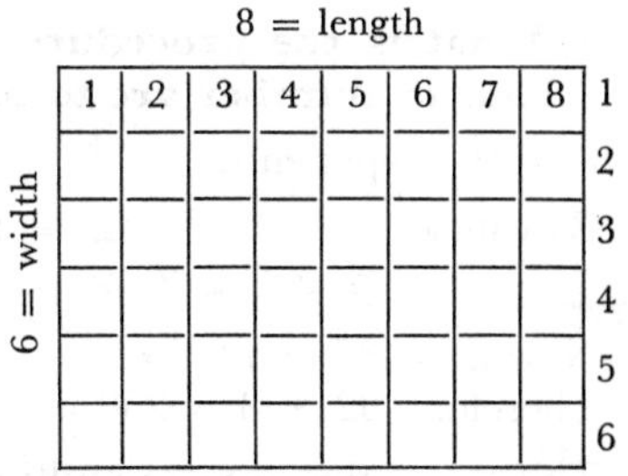

512. How do we calculate a higher power of a common fraction?

Raise the numerator to the power required.
Raise the denominator to the power required.
Express the powers as a fraction.

EXAMPLE: Find the fourth power of $\frac{5}{8}$.

$$\left(\frac{5}{8}\right)^4 = \frac{5\cdot5\cdot5\cdot5}{8\cdot8\cdot8\cdot8} = \frac{625}{4{,}096}$$

513. What are the rules affecting the powers of decimal fractions?

(*a*) The square of a decimal fraction must have at least two decimal places.

EXAMPLES: $.2^2 = .04$
$.5^2 = .25$

(*b*) There must be an even number of decimal places.

EXAMPLES: $.\overline{12}^2 = .0144$
$\overline{.123}^2 = .015129$

514. Why is a decimal fraction raised to a power of a smaller value than the original fraction?

A decimal fraction when converted to a numerator and a denominator has a very large decimal denominator. In raising the fraction the smaller numerator is divided by a larger and larger denominator as the power to which the fraction is raised increases.

EXAMPLE:

$$.04 = \tfrac{4}{100}$$

$$.\overline{04}^2 = \frac{4^2}{100^2} = \frac{16}{10{,}000} = .0016$$

$$.\overline{04}^3 = \left(\frac{4}{100}\right)^3 = \frac{4^3}{100^3} = \frac{64}{1{,}000{,}000} = .000064$$

515. What is the procedure when two powers of the same base or number are to be multiplied?

Add the exponents.

EXAMPLE: $2^5 \times 2^3 = 2^{5+3} = 2^8$

$$(2\cdot2\cdot2\cdot2\cdot2) \times (2\cdot2\cdot2) = 2^8$$

Now $2^5 = 32$ and $2^3 = 8$.
Therefore, $32 \times 8 = 256 = 2^5 \times 2^3 = 2^{5+3} = 2^8$.

This shows that we can multiply 32×8 by means of exponents.

516. What is the procedure when two powers of the same base or number are to be divided?

Subtract the exponent of the divisor (or denominator) from the exponent of the dividend (or numerator).

Ex. (*a*) Divide 32 by 8.

$$32 = 2^5 \quad \text{and} \quad 8 = 2^3 \quad \left(\frac{\not2\cdot\not2\cdot\not2\cdot2\cdot2}{\not2\cdot\not2\cdot\not2} = 2^2\right)$$

$$\therefore \frac{32}{8} = \frac{2^5}{2^3} = 2^{5-3} = 2^2 = 4$$

Ex. (*b*) Divide 243 by 9.

$$243 = 3^5 \quad \text{and} \quad 9 = 3^2 \quad \left(\frac{\not3\cdot\not3\cdot3\cdot3\cdot3}{\not3\cdot\not3} = 3^3\right)$$

$$\therefore \frac{243}{9} = \frac{3^5}{3^2} = 3^{5-2} = 3^3 = 27$$

This shows that division can be performed by means of exponents.

517. What limits the above processes?

They are only good for division and multiplication of exact powers of 2, 3, or exact powers of any other numbers or bases for which you have built up tables.

518. What is the procedure when the power of a number is itself to be raised to a power?

Multiply the exponents.

EXAMPLE: Find the third power of 4^2.

$$(4^2)^3 = 4^2 \cdot 4^2 \cdot 4^2 = 4^{2\times 3} = 4^6 = 4{,}096$$

Multiply exponent 2 by exponent 3 to get exponent 6.

519. How can we show that any number or base to the zero power equals 1?

Any quantity or base raised to the first power is represented by the quantity or base itself. Thus 2 raised to the first power is $2^1 = 2$. x to the first power is written $x^1 = x$.

Ex. (*a*) $2 \div 2 = 1$ But $\frac{2^1}{2^1} = 2^{1-1} = 2^0$ $\therefore\ 2^0 = 1$

(*b*) $5 \div 5 = 1$ But $\frac{5^1}{5^1} = 5^{1-1} = 5^0$ $\therefore\ 5^0 = 1$

(*c*) $10 \div 10 = 1$ But $\frac{10^1}{10^1} = 10^{1-1} = 10^0$ $\therefore\ 10^0 = 1$

The same procedure can be followed for any base or number.

$$\therefore\ 1 = \text{Any number to the zero power.}$$

520. How can we show that the sign of an exponent may be changed by changing the position of the number from one side of the denominator line to the other?

When a factor does not appear, its exponent is zero and the value 1 can be substituted for it.

EXAMPLE: In $3 \times 5 = 15$, 7 is not used as a factor, which means its exponent is zero or the factor is used zero times. This may be written

$$3 \times 5 \times 7^0 = 3 \times 5 \times 1 = 3 \times 5$$

Now

$$7^5 \div 7^8 = \frac{\not{7}\cdot\not{7}\cdot\not{7}\cdot\not{7}\cdot\not{7}}{\not{7}\cdot\not{7}\cdot\not{7}\cdot\not{7}\cdot\not{7}\cdot 7\cdot 7\cdot 7} = \frac{7^0}{7^3} = \frac{1}{7^3}$$

But, subtracting exponents,

$$\frac{7^5}{7^8} = 7^{5-8} = 7^{-3} \quad \text{(as in division)}$$

$$\therefore\ 7^{-3} = \frac{1}{7^3}$$

The sign of the exponent may be changed by changing the position of the number from one side of the denominator line to the other.

Thus a negative exponent means division of 1 by the number with the same positive exponent.

EXAMPLES: 2^{-3} means $\frac{1}{2^3} = \frac{1}{8}$

$$x^{-3} = \frac{1}{x^3} \quad \text{where } x \text{ is any number}$$

521. Why is a decimal fraction raised to a negative power of greater value than the original decimal fraction?

In negative powers the very large denominator becomes the numerator, which increases the value of the fraction.

Ex. (*a*) $(.04)^{-2} = \left(\frac{4}{100}\right)^{-2} = \frac{1}{(\frac{4}{100})^2} = \frac{1}{\frac{4^2}{100^2}} = \frac{100^2}{4^2}$

$$= \frac{10{,}000}{16} = 625$$

(*b*) $(.04)^{-3} = \left(\frac{4}{100}\right)^{-3} = \frac{1}{(\frac{4}{100})^3} = \frac{1}{\frac{4^3}{100^3}} = \frac{100^3}{4^3}$

$$= \frac{1{,}000{,}000}{64} = 15{,}625$$

522. Why are the negative powers of whole numbers smaller than the original numbers?

A negative power makes a fraction of a whole number and reduces its value.

Ex. (*a*) $4^{-2} = \frac{1}{4^2} = \frac{1}{16} = .0625$

(*b*) $4^{-3} = \frac{1}{4^3} = \frac{1}{64} = .015625$

Higher negative powers make the results smaller and smaller.

523. How can we simplify the raising of a number to a power that can be factored?

(*a*) Factor the power.
(*b*) Raise the number to the power of one of the factors.
(*c*) Raise this result to the power of the next factor and so on until all the factors are used up.

Ex. (*a*) Raise 3 to the eighth power.
Factor exponent 8 into $2 \times 2 \times 2$ (3 factors)
Raise $3^2 = 9$ Then $9^2 = 81$ Then $81^2 = 6{,}561 = 3^8$

Ex. (*b*) Raise 5 to the twelfth power.
Factor exponent 12 into $2 \times 2 \times 3$
Raise $5^2 = 25$ Then $25^2 = 625$ Then $625^3 = 244{,}140{,}625 = 5^{12}$

524. What is the basis for a short method of squaring a number from 1 to 100?

We know from algebra that $(a - b)(a + b) = a^2 - b^2$. The product of the sum and difference of two numbers is the same as the difference of their squares.

EXAMPLE: If we want to square 29, we set up

$$(29 + 1)(29 - 1) = (29^2 - 1)$$

or

$$30 \times 28 = 840 = (29^2 - 1)$$

$$\therefore\ 29^2 = 840 + 1 = 841$$

525. What then is the procedure for a short method of squaring a number from 1 to 100?

(*a*) Add or subtract a number to make one of the multipliers a decimal number.
(*b*) Subtract the same number from the original.
(*c*) Multiply the above and add the square of the number added or subtracted.

Ex. (*a*) $\overline{68}^2$ Add and subtract 2, getting

$$(68 + 2)(68 - 2) = 70 \times 66 = 4{,}620$$
$$\text{Add } 2^2 = 4$$
$$4{,}624 \text{ Ans.}$$

(*b*) $\overline{74}^2$ Subtract and add 4, getting

$$(74 - 4)(74 + 4) = 70 \times 78 = 5{,}460$$
$$\text{Add } 4^2 = 16$$
$$5{,}476 \text{ Ans.}$$

526. How does the procedure of Question 509 compare with the above as a short method of squaring a number from 1 to 100?

$$(a + b)^2 = a^2 + b^2 + 2ab$$

EXAMPLE: $\overline{74}^2 = (70 + 4)^2 = 70^2 + 4^2 + 2 \times 70 \times 4$
$= 4{,}900 + 16 + 560 = 5{,}476$ Ans.

For numbers between 1 to 100, the procedure of Question 525 would appear to be somewhat simpler.

527. How can we apply the procedure of Question 509 to mixed numbers as $6\frac{1}{3}$, $8\frac{1}{2}$, $7\frac{1}{5}$, etc.?

$$(a + b)^2 = a^2 + 2ab + b^2 = a(a + 2b) + b^2$$

In this case a is made an integral number and b is made the fraction.

Add twice the fraction to the integral number and multiply this by the integral number. Then add the square of the fraction.

Ex. (*a*) $(6\frac{1}{3})^2 = (6 + \frac{1}{3})^2 = 6(6 + 2 \times \frac{1}{3}) + (\frac{1}{3})^2 = 40\frac{1}{9}$
(*b*) $(8\frac{1}{2})^2 = (8 + \frac{1}{2})^2 = 8(8 + 2 \times \frac{1}{2}) + (\frac{1}{2})^2 = 72\frac{1}{4}$
(*c*) $(7\frac{1}{5})^2 = (7 + \frac{1}{5})^2 = 7(7 + 2 \times \frac{1}{5}) + (\frac{1}{5})^2$
$= 51\frac{4}{5} + \frac{1}{25} = 51\frac{21}{25}$

528. How may aliquot parts be applied to the above method?

Convert the number to a mixed number theoretically.

EXAMPLE: To square 825, convert to $(8\frac{1}{4})^2$ theoretically, and apply above rule.

$$(8\tfrac{1}{4})^2 = (8 + \tfrac{1}{4})^2 = 8(8 + 2 \times \tfrac{1}{4}) + (\tfrac{1}{4})^2 = 68\tfrac{1}{16}$$

Now $\frac{1}{16} = .0625$, and since the original number has no decimal the answer is 680625.

529. How is the squaring of a number that is divisible by factor 2, 3, or 5 made simpler?

Divide by the factor; square the quotient, and multiply by the factor squared.

Ex. (*a*) To square 36, divide 36 by 3, getting 12 as the quotient. Square 12, getting 144, which multiply by 3^2, getting 1,296.

(*b*) Square 35. $\frac{35}{5} = 7$ $7^2 = 49$ $49 \times 5^2 = 1{,}225$
(*c*) $\overline{14}^2$ $\frac{14}{2} = 7$ $7^2 = 49$ $49 \times 2^2 = 196$
(*d*) $\overline{18}^2$ $\frac{18}{2} = 9$ $9^2 = 81$ $81 \times 2^2 = 324$

530. What is the procedure for getting the square of the mean between two numbers?

Multiply the two numbers and add the square of half their difference.

Ex. (*a*) What is the square of the mean of 12 and 16, or 14?

$12 \times 16 = 192 \qquad 16 - 12 = 4 \qquad \frac{4}{2} = 2$

$(\frac{4}{2})^2 = 4$

$\overline{196}$ Ans. $(= \overline{14^2})$

(*b*) What is the square of the mean of 30 and 40, or 35?

$30 \times 40 = 1{,}200 \qquad 40 - 30 = 10 \qquad \frac{10}{2} = 5$

$5^2 = 25$

$\overline{1{,}225}$ Ans. $(= \overline{35^2})$

(*c*) What is the square of the mean of 24 and 25, or $24\frac{1}{2}$?

$24 \times 25 = 600 \qquad 25 - 24 = 1 \qquad \frac{1}{2}$

$(\frac{1}{2})^2 = \frac{1}{4}$

$\overline{600\frac{1}{4}}$ Ans. $(= \overline{24\frac{1}{2}^2})$

531. What is an easy way of squaring a number ending in $\frac{1}{2}$?

Multiply the integral by the next higher integral and add $\frac{1}{4}$. (This is similar to Example (*c*) of Question 530.)

Ex. (*a*) $(7\frac{1}{2})^2 = 7 \times 8 + \frac{1}{4} = 56\frac{1}{4}$

(*b*) $(12\frac{1}{2})^2 = 12 \times 13 + \frac{1}{4} = 156\frac{1}{4}$

532. What is the procedure when the number ends in 5 instead of $\frac{1}{2}$?

The 5 is taken as representing the $\frac{1}{2}$ of the above.

Ex. (*a*) $(7\frac{1}{2})^2 = 7 \times 8 + \frac{1}{4}$ is similar to

$(75)^2 = 70 \times 80 + 25 = 5{,}625$

(*b*) $(125)^2 = 120 \times 130 + 25 = 15{,}600 + 25 = 15{,}625$

The proof of the above when the number ends in $\frac{1}{2}$ is

$$(a + \tfrac{1}{2})^2 = a^2 + a + \tfrac{1}{4} = a(a + 1) + \tfrac{1}{4}$$

533. What is the procedure for squaring a number consisting of 9's?

Place 1 as the right-hand figure
Then zeros, one less than the number of 9's
Then figure 8
Then 9's, one less than the number of 9's

Ex. (*a*) $\underbrace{\overline{99999}}_{\text{five 9's}}{}^2 = \underbrace{9\,9\,9\,9}_{\text{four 9's}}\,8\,\underbrace{0\,0\,0\,0}_{\text{four zeros}}\,1$ Ans.

(*b*) $\overline{999}^2 = 998001$ Ans.

534. What does the exponent of any power of 10 indicate?

It indicates the number of zeros after the 1 in representing the result.

$$10^0 = 1 \qquad 10^1 = 10 \qquad 10^2 = 100 \qquad 10^3 = 1{,}000$$

$$10^4 = 10{,}000 \qquad 10^{⑤} = \underbrace{100{,}000}_{\text{five zeros}} \qquad 10^{⑥} = 1{,}\underbrace{000{,}000}_{\text{six zeros}}, \text{ etc.}$$

Each power adds one more zero successively.

The reverse also holds; that is, if the result is 10,000,000, you count the zeros to get the number of factors of 10, or the exponent of 10, which in this case is 10^7.

How can large numbers be expressed conveniently in terms of powers of 10?

Ex. (*a*) $3{,}900 = 39 \times 100 = 39 \times 10^2$

(*b*) $4{,}000{,}000 = 4 \times 1{,}000{,}000 = 4 \times 10^6$

(*c*) $36{,}300{,}000 = 3.63 \times 10{,}000{,}000 = 3.63 \times 10^7$

535. Does the above apply to negative exponents of base 10?

Yes.

$$10^{-7} = \frac{1}{10^7} = \frac{1}{10{,}000{,}000} = .0000001$$

Exponent = 7 factors = 7 zeros = 7

536. How can we express decimals as powers of 10?

$$1 \div 10 = \frac{1}{10} = \frac{10^0}{10^1} = 10^{0-1} = 10^{-1} = .1$$

$$1 \div 100 = \frac{1}{100} = \frac{10^0}{10^2} = 10^{0-2} = 10^{-2} = .01$$

$$1 \div 1{,}000 = \frac{1}{1{,}000} = \frac{10^0}{10^3} = 10^{0-3} = 10^{-3} = .001$$

Negative power of 10 = a decimal

How can decimals be expressed conveniently in terms of negative powers of 10?

Ex. (*a*) $0.03 = 3 \times 0.01 = 3 \times 10^{-2}$

(*b*) $0.0021 = 21 \times 0.0001 = 21 \times 10^{-4}$

(*c*) $0.0000462 = 462 \times 0.00001 = 4.62 \times 10^{-5}$

537. What is done with the exponents in multiplying powers of 10?

The exponents are added algebraically.

Ex. (*a*) $10^6 \times 10^2 = 10^{6+2} = 10^8$
(*b*) $10^3 \times 10^2 = 10^{3+2} = 10^5$
(*c*) $10^6 \times 10^{-4} = 10^{6-4} = 10^2$
(*d*) $10^{12} \times 10^{-8} = 10^{12-8} = 10^4$
(*e*) $10^2 \times 10^{-7} = 10^{2-7} = 10^{-5}$

538. What is done with the powers of 10 in division?

Subtract the exponent of the denominator from the exponent of the numerator. The same thing is obtained by changing the sign of the exponent of the denominator.

Ex. (*a*) $\dfrac{10^3}{10^4} = 10^{3-4} = 10^{-1}$

(*b*) $\dfrac{10^6}{10^4} = 10^{6-4} = 10^2$

(*c*) $\dfrac{10^2}{10^{-5}} = 10^{2+5} = 10^7$

539. What is meant by a root of a number or power?

If a given number or term can be produced by multiplying together two or more equal numbers or terms, then each of the equal numbers or terms is said to be a root of that product.

Ex. (*a*) $9 = 3 \times 3$; then 3 is a root of 9.
(*b*) $125 = 5 \times 5 \times 5$; then 5 is the cube root of 125.
(*c*) $81 = 3 \times 3 \times 3 \times 3$; then 3 is the fourth root of 81.
(*d*) $x^3 = x \cdot x \cdot x$; then x is a root of x^3.

The root of a number is always one of the equal factors of that number.

540. What is meant by evolution?

It is the inverse process of involution. In evolution the problem is to determine one of a given number of equal factors when their product alone is given. The factors so found are called square root, cube root, fourth root, etc., depending upon the number of factors involved.

541. What is the symbol of evolution?

The symbol is $\sqrt{}$, which is an abbreviation: *r*, for root, followed by a line. This symbol is known as the *radical sign*, and indicates that a root is to be taken of the expression before which it stands. A small number called an *index* is written over the radical sign and indicates the root to be taken, except for a square root when it is usually omitted. The quantity or expression within the radical sign is known as the *radicand*. In $\sqrt[4]{81}$, 81 is the radicand and 4 is the index.

Ex. (*a*) $\sqrt{9}$ indicates that the square root is to be extracted.
(*b*) $\sqrt[3]{27}$ indicates that the cube root is to be extracted.
(*c*) $\sqrt[4]{81}$ indicates that the fourth root is to be extracted.

542. What is meant by (*a*) a perfect power, (*b*) an imperfect power?

(*a*) A number is a perfect power when its root can be extracted without leaving a remainder.

(*b*) A number is an imperfect power when its root cannot be extracted exactly.

Ex. (*a*) 81 is a perfect power because $\sqrt[4]{81} = 3$.
(*b*) 87 is an imperfect power because its root cannot be extracted exactly.

543. What is the simplest method of extracting a root?

Divide the number by its lowest prime factor and continue the process.

EXAMPLE: Find the cube root of 216. There are three factors 2, and three factors 3, or $\sqrt[3]{216} = \sqrt[3]{2^3 \times 3^3} = 2 \times 3$. Then $2 \times 3 = 6 =$ the cube root of 216.

```
2)216
2)108
 2)54
 3)27
  3)9
    3
```

544. What is the rule for extracting the required root of a quantity?

Divide the exponent of the quantity by the index of the root and then perform indicated operations when possible.

Ex. (*a*) $\sqrt[4]{1{,}296} = \sqrt[4]{2^4 \times 3^4} = 2 \times 3 = 6$
(*b*) $\sqrt[4]{81b^{12}} = \sqrt[4]{3^4 \times b^{12}} = 3b^3$
(*c*) $\sqrt{a^4b^2} = a^{4/2} \cdot b^{2/2} = a^2b$
(*d*) $\sqrt[3]{x^9y^6} = x^{9/3} \cdot y^{6/3} = x^3y^2$
(*e*) $\sqrt[3]{\dfrac{5^6}{9^9}} = \dfrac{5^{6/3}}{9^{9/3}} = \dfrac{5^2}{9^3} = \dfrac{25}{729}$
(*f*) $\sqrt[3]{7^2} = 7^{2/3}$

545. What is the rule for fractional exponents?

The numerator indicates the power to which the base is to be raised, and the denominator, the root which is to be extracted of that power.

Ex. (*a*) $\sqrt[3]{7^2} = 7^{2/3}$ (Question 544*f*)

We see that in the fractional exponent $\frac{2}{3}$ of the base 7, the denominator 3 is the index of the root and the numerator 2 is the exponent of the base or quantity.

Ex. (*b*) $32^{2/5} = \sqrt[5]{32^2} = \sqrt[5]{1{,}024} = 4$

(*c*) $x^{3/5} = \sqrt[5]{x^3}$

(*d*) $b^{r/s} = \sqrt[s]{b^r}$

(*e*) $y^{1/7} = \sqrt[7]{y}$

(*f*) $(1 + i)^{1/n} = \sqrt[n]{1 + i}$

(*g*) $\left(1 + \frac{k}{t}\right)^{p/q} = \sqrt[q]{\left(1 + \frac{k}{t}\right)^p}$

546. When are radicals similar?

When they have the same indices and the same radicands.

Example: $5\sqrt[3]{6}$ and $\frac{3}{7}\sqrt[3]{6}$ are similar radicals.

547. When may a factor of the radicand be removed from under the radical sign?

When the factor is an exact power of the indicated order.

Ex. (*a*) $\sqrt{75} = \sqrt{25 \times 3} = \sqrt{5^2 \times 3} = 5\sqrt{3}$

(*b*) $\sqrt[3]{81} = \sqrt[3]{27 \times 3} = \sqrt[3]{3^3 \times 3} = 3\sqrt[3]{3}$

548. How may a factor in the coefficient of a radical be introduced under the radical sign?

By raising the factor to the power of the index.

Ex. (*a*) $3\sqrt{5} = \sqrt{3^2 \times 5} = \sqrt{9 \times 5} = \sqrt{45}$

(*b*) $2\sqrt[3]{4} = \sqrt[3]{2^3 \times 4} = \sqrt[3]{8 \times 4} = \sqrt[3]{32}$

549. How may a fraction with a radical in the denominator be reduced to a fraction with a rational denominator?

Multiply numerator and denominator by the same radical expression which would make the denominator rational.

Ex. (*a*) $\frac{1}{\sqrt{5}} = \frac{1\sqrt{5}}{\sqrt{5} \times \sqrt{5}} = \frac{\sqrt{5}}{\sqrt{25}} = \frac{\sqrt{5}}{5}$

(*b*) $\frac{1}{\sqrt{5} - \sqrt{3}} = \frac{\sqrt{5} + \sqrt{3}}{(\sqrt{5} - \sqrt{3})(\sqrt{5} + \sqrt{3})}$

$$= \frac{\sqrt{5} + \sqrt{3}}{(\sqrt{5})^2 - (\sqrt{3})^2} = \frac{\sqrt{5} + \sqrt{3}}{5 - 3} = \frac{\sqrt{5} + \sqrt{3}}{2}$$

550. How may a radical with a fractional radicand be reduced to a fraction whose denominator has no radical?

Multiply the numerator and denominator by the same number which will make the denominator a rational number.

EXAMPLE: $\sqrt{\frac{3}{7}} = \sqrt{\frac{3}{7} \times \frac{7}{7}} = \sqrt{\frac{21}{49}} = \frac{\sqrt{21}}{\sqrt{49}} = \frac{\sqrt{21}}{7}$

551. How may a radical be changed to one of a higher order with an index that is a multiple of the original index?

Multiply the numerator and denominator of the fractional exponent of the base by the same number.

EXAMPLE: $\sqrt[6]{3} = 3^{\frac{1}{6}} = 3^{\frac{3}{18}} = \sqrt[18]{3^3} = \sqrt[18]{27}$

552. When may a radical be reduced to a radical of a lower order?

When the exponent of the radicand is a factor of the index of the radical.

EXAMPLE: $\sqrt[12]{81} = 81^{\frac{1}{12}} = (3^4)^{\frac{1}{12}} = 3^{\frac{4}{12}} = 3^{\frac{1}{3}} = \sqrt[3]{3}$

553. When is a radical expression said to be in simplest form?

When:

(*a*) the index is as small as possible
(*b*) the radicand has no fractions
(*c*) the denominator of the expression has no radical
(*d*) every factor of the radicand has an exponent less than the index.

554. What is the result of $\sqrt[4]{16b^4y^3}$ reduced to its simplest form?

$$\sqrt[4]{16b^4y^3} = \sqrt[4]{2^4b^4y^3} = 2b\sqrt[4]{y^3} \text{ Ans.}$$

This is the simplest form, as the index 4 is as small as possible, the radicand has no fraction, there is no radical in the denominator of the expression, and the radicand y^3 has no factor which is a fourth power of y.

555. What is the result of reducing (*a*) $\sqrt[8]{16b^{12}}$, (*b*) $\sqrt{\frac{5}{9}}$ to the simplest form?

(*a*) $\sqrt[8]{16b^{12}} = \sqrt[8]{2^4 \cdot b^8 \cdot b^4} = 2^{\frac{4}{8}} \cdot b^{\frac{8}{8}} \cdot b^{\frac{4}{8}} = b\sqrt{2b}$

(*b*) $\sqrt{\frac{5}{9}} = \sqrt{\frac{5}{9} \times \frac{9}{9}} = \sqrt{\frac{45}{81}} = \frac{\sqrt{45}}{9}$

556. How many figures does it take to express the square root of a number of (*a*) 1 or 2 figures, (*b*) 3 or 4 figures, (*c*) 5 or 6 figures?

(*a*) When a number has 1 or 2 figures, the square root has 1 figure

$$\sqrt{9} = 3 \qquad \sqrt{16} = 4$$

(*b*) When a number has 3 or 4 figures, the square root has 2 figures

$$\sqrt{144} = 12 \qquad \sqrt{1{,}024} = 32$$

(*c*) When a number has 5 or 6 figures, the square root has 3 figures

$$\sqrt{10{,}000} = 100 \qquad \sqrt{105{,}625} = 325$$

If a whole number be divided into groups of 2 figures each, beginning at the units place, the number of groups will equal the number of figures in the root.

557. (*a*) What is the relation of the number of decimal places in the square of a decimal to that of the decimal itself, and (*b*) what is the relation of the number of decimal places in the square root of a decimal to that of the decimal itself?

(*a*) The square of a decimal has twice as many decimal places as does the decimal itself.

EXAMPLES: $\overline{.09}^2 = .0081 \qquad \overline{.25}^2 = .0625$

$\overline{.75}^2 = .5625 \qquad 1.25 = 1.5625$

In each case 2 places in the decimal produce 4 places in the square.

(*b*) The square root of a decimal has half as many decimal places as does the decimal itself.

EXAMPLES: $\sqrt{.0081} = .09 \qquad \sqrt{.0625} = .25$

$\sqrt{.5625} = .75 \qquad \sqrt{1.5625} = 1.25$

In each case 4 places in the decimal produce 2 places in the square root.

To get the square root of a decimal, there must be an even number of figures. Annex a zero, if need be. If a decimal number be divided into groups of 2 figures each, beginning at the decimal point, the number of groups will equal the number of figures in the root.

558. What is the square root of 676?

Divide the number into groups of two figures starting from the units figure and going to the left, getting $\overline{6}$, $\overline{76}$. There are 2 groups and the root will have 2 figures, one of tens and one of units.

From Questions 509 and 510, we know the basic formula for the square of the sum of two numbers is

$$(a + b)^2 = a^2 + 2ab + b^2 = a^2 + (2a + b)b$$

If $a = 2$ tens and $b = 6$ units, we get by substitution

$$\begin{aligned}(2 \text{ tens} + 6)^2 &= (2 \text{ tens})^2 + (2 \times 2 \text{ tens} + 6)6 \\ &= 20^2 + (2 \times 20 + 6)6 \\ &= 400 + (240 + 36) = 400 + 276 = 676\end{aligned}$$

We may start with 676 and work back to get the square root.

(*a*) The square of the tens of the root is contained in the second group $\overline{6}$, which represents 600, and the greatest square in 600 is 400. The square root of 400 is $20 = 2$ tens $= a$. Put 2 in the root.

	$a^2 + (2a + b)b$	$a + b$
	$\overline{6}$ $\overline{76}$	)2 (tens) + 6 (units) ←root
$2a + b$	4 00	$(= a^2)$
4 6	)2 76	(= remainder)
×6	2 76	$[= (2a + b)b]$
	0	(= remainder)

(*b*) Subtract 400 from 676, getting 276 = remainder of the number. We have now accounted for the a^2 part of the formula.

(*c*) To account for the remainder $(2a + b)b$, get a trial value of b by dividing 4 $(= 2a)$ into 27 of the remainder and getting 6 $(= b)$. Put 6 in the root.

(*d*) Add the 6 to the 4 tens, getting 46 $(= 2a + b)$, and multiply by 6, getting 276 $(= 46 \times 6) = (2a + b)b$.

(*e*) Subtract this 276 from the remainder 276, getting zero.

$$\therefore \sqrt{676} = 26$$

To prove a square root multiply the square root by itself.

559. What is the rule for the extraction of a square root?

(*a*) Separate the number into groups of 2 figures, going to the left from the decimal point for the whole part of the number and to the right for the decimal part.

(*b*) Determine the greatest square in the farthest left group. Get its root and put this in the root.

(*c*) Subtract the square of this root from this left group and bring down the next group to the remainder.

(*d*) Divide the remainder by twice the root already found, considered as tens, as a trial divisor, getting the next figure of the root.

(*e*) To the trial divisor add the new figure of the root, then multiply by the last figure found, and subtract this product from the last remainder.

(*f*) Bring down the next group to the remainder and continue as before.

If the number is not a perfect square or if you want more decimal places in the root, add zeros to the number and continue the process.

560. What is the square root of 702.25?

$$
\begin{array}{rrrl}
 & \overline{7}\ \ \overline{02}\,.\,\overline{25} &)\underline{26\,.\,5} & \leftarrow \text{root} \\
2a+b & -4\qquad\downarrow & & \\
(\text{Trial divisor } 2\times 2 = 4)\rightarrow 4\ \ 6 &)3\ \ 02 & \leftarrow & \text{remainder} \\
\underline{\times 6} & 2\ \ 76 & \leftarrow & (2a+b)b \\
2a+b & & & \\
(\text{Trial divisor } 2\times 26 = 52)\rightarrow 52\ \ 5 &)\ 26\ \ 25 & \leftarrow & \text{remainder} \\
\underline{\times 5} & \underline{26\ \ 25} & \leftarrow & (2a+b)b \\
 & 0 & \leftarrow & \text{remainder}
\end{array}
$$

(*a*) The root will have 2 whole figures and 1 decimal figure.

(*b*) The greatest square in $\overline{7}$ (or 700) is 4 (or 400), whose root is 2 (=2 tens) = *a*. Put 2 in the root.

(*c*) Subtract 4 from 7, getting 3, and bring down the next group, getting 302 = remainder.

(*d*) Divide twice the root already found, or $2 \times 2 = 4$, as a trial divisor into 30 of the remainder, getting 6 (7 would be too large), the next figure of the root. Put 6 in the root.

(*e*) Add 6 to the trial divisor 4 (as tens), getting 46, and multiply by 6 (the last figure found), getting 276. Subtract 276 from the last remainder 302, getting 26.

(*f*) Bring down the next group, $\overline{25}$, getting 2,625 = remainder.

(*g*) Divide twice the root already found, or $2 \times 26 = 52$, as a trial divisor into 262 of the remainder, getting 5, the next figure of the root. Put 5 in the root.

(*h*) Add 5 to the trial divisor 52 (as tens), getting 525, and multiply by 5 (the last root figure found), getting 2,625. Subtract 2,625 from the last remainder 2,625, getting zero.

$$\therefore \sqrt{702.25} = 26.5$$

561. What is the square root of 704.3716?

(*a*) The root will have 2 whole figures and 2 decimal figures.

```
                                        7  04 . 37  16)26.54←root
                             2a + b−4    ↓      |    |
(Trial divisor 2 × 2 = 4) →4     6 )3   04      |    |
                                ×6  2   76      |    |
                            2a + b              ↓    |
(Trial divisor 2 × 26 = 52) → 52  5     )28    37    |
                                 ×5      26    25    |
                             2a                      ↓
(Trial divisor 2×265 = 530) → 530  4     )2    12   16
                                  ×4      2    12   16
                                                     0
```

(*b*) Greatest square in 7 is 4, whose root is 2. Put 2 in root.

(*c*) Subtract 4 from 7, getting 3. Bring down next group, getting 304 = remainder.

(*d*) Divide twice root already found, or 2 × 2 = 4, as a trial divisor into 30 of remainder, getting 6, the next figure of root. Put 6 in the root.

(*e*) Add 6 to trial divisor 4 (as tens), getting 46, and multiply by 6 (the last figure found), getting 276. Subtract 276 from last remainder 304, getting 28.

(*f*) Bring down the next group 37, getting 2,837 = remainder.

(*g*) Divide twice root already found, or 2 × 26 = 52, as a trial divisor into 283 of remainder, getting 5, the next figure of root. Put 5 in the root.

(*h*) Add 5 to the trial divisor 52 (as tens), getting 525, and multiply by 5 (the last figure found), getting 2,625. Subtract 2,625 from last remainder 2,837, getting 212.

(*i*) Bring down the next group 16, getting 21,216 = remainder.

(*j*) Divide twice root already found, or 2 × 265 = 530, as a trial divisor into 2,121 of remainder, getting 4, the next figure of root. Put 4 in the root.

(*k*) Add 4 to the trial divisor 530 (as tens), getting 5,304, and multiply by 4 (the last figure found), getting 21,216. Subtract 21,216 from last remainder 21,216, getting zero.

$$\therefore \sqrt{704.3716} = 26.54 \text{ Ans.}$$

Note: In each step you consider the part of the root already found as tens in relation to the next figure.

562. What is the square root of 94,864?

(*a*) The root will have 3 whole figures.
(*b*) The greatest square in 9 is 9, whose root is 3. Put 3 in root.

```
           _   __   __
           9   48   64 )308 ← root
          −9    ↓    ↓
   6  08  )0   48   64
      ×8       48   64
                     0
```

(*c*) Subtract 9 from 9, getting zero. Bring down the next group, getting 48 = remainder.

(*d*) Divide twice root already found, or 2 × 3 = 6, as a trial divisor into 4 of remainder, which results in zero. Put a zero in the root and in the divisor, and bring down the next group, getting 4,864 = remainder.

(*e*) Divide the new trial divisor 60 into 486 of the remainder, getting 8, the next figure of root. Put 8 in the root.

(*f*) Add 8 to the trial divisor 60 (as tens), getting 608, and multiply by 8 (the last figure found), getting 4,864. Subtract 4,864 from last remainder 4,864, getting zero.

$$\therefore \sqrt{94{,}864} = 308 \text{ Ans.}$$

563. What is the square root of 69,284.7642?

```
                                          _   __   __   __   __
                                          6   92   84 . 76   42 )263.22 ← root
                                         −4    ↓    |    |    |
(Trial divisor 2 × 2 = 4)  →      4   6  )2   92    |    |    |
                                     ×6   2   76    ↓    |    |
(Trial divisor 2 × 26 = 52) →    52   3  )16  84         |    |
                                     ×3   15  69         ↓    |
(Trial divisor
      2 × 263 = 526) →          526   2  )1   15   76         |
                                     ×2   1   05   24         ↓
(Trial divisor
      2 × 2,632 = 5,264) →      5264  2  )10  52   42
                                     ×2   10  52   84
```

The final subtraction cannot be made, as 105,284 is a little larger than the remainder 105,242, but is close to it, so that the root is

$$\therefore 263.22 \text{ Ans. (approx.)}$$

564. How do we get the root of a fraction?

Extract the root of both the numerator and denominator separately.

Ex. (*a*) $\sqrt{\frac{49}{64}} = \frac{\sqrt{49}}{\sqrt{64}} = \frac{7}{8}$

(*b*) $\sqrt{\frac{225}{256}} = \frac{\sqrt{225}}{\sqrt{256}} = \frac{\sqrt{3^2 \times 5^2}}{\sqrt{2^2 \times 8^2}} = \frac{15}{16}$

565. What is the rule for the extraction of the cube root?

(*a*) Separate the number into groups of 3 figures each to left of decimal point for whole numbers, and to right for decimal portion.

(*b*) Find greatest cube contained in farthest left-hand group. Put its cube root in the root.

(*c*) Subtract this cube from the first group and bring down the next group to get the remainder.

(*d*) Divide remainder by 3 times the square of the root already found, considered as tens, as a trial divisor, to get the next figure of the root. Put this figure in the root.

(*e*) To trial divisor add 3 times the product of the two parts of the root plus the square of the second part of the root to make the complete divisor.

(*f*) Multiply the complete divisor by the second figure of the root. Subtract and bring down the next group.

(*g*) Continue in this manner until all groups have been used.

566. What is the cube root of 245,314,376?

		$\overline{245}\ \overline{314}\ \overline{376}$)626	← root
Cube of first figure of root →		−216 ↓	
1st trial divisor $3 \times \overline{60}^2 =$	10,800	29 314	
$3 \times (60 \times 2) =$	360	22 328	← 1st complete divisor × 2nd figure of root (11,164 × 2)
$\overline{2}^2 =$	4		
1st complete divisor →	11,164		
2nd trial divisor $3 \times \overline{620}^2 =$	1,153,200	6 986 376	
$3 \times (620 \times 6) =$	11,160	6 986 376	2nd complete divisor × 3rd figure of root (1,164,396 × 6)
$\overline{6}^2 =$	36	0	
2nd complete divisor →	1,164,396		

(*a*) Separate into groups.

(*b*) The cube of 6 is the largest cube contained in the first group $\overline{245}$. Put 6 in the root.

(*c*) Subtract $\overline{6}^3 = 216$ from 245, getting 29. Bring down the next group, getting 29,314 = remainder.

(*d*) The root already found, considered as tens, is 60 and $\overline{60}^2 =$ 3,600. 3 × 3,600 = 10,800 = first trial divisor. This is contained in 29,314 twice. Put 2 as the next figure of the root.

(*e*) The two parts of the root already found are 60 and 2. 60 × 2 = 120 and 3 × 120 = 360. The square of the last figure found is 4. Adding 360 + 4 to the trial divisor, we get the complete divisor = 11,164.

(*f*) Multiply complete divisor by the second figure of the required root, 11,164 × 2 = 22,328, and subtract from the remainder, getting 6,986. Bring down the next group, getting 6,986,376 = remainder.

(*g*) The root already found is 62, or, considered as tens, 620. 620^2 = 384,400, and 3 × 384,400 = 1,153,200 = second trial divisor. This trial divisor is contained 6 times in the remainder. Put 6 as the next figure of the root.

(*h*) The two parts of the root already found are 620 and 6. 620 × 6 = 3,720, and 3 × 3,720 = 11,160. The square of the last number of the root is 6^2 = 36. Adding 11,160 and 36 to 1,153,200, we get 1,164,396 = second complete divisor.

(*i*) Multiply complete divisor by the third figure of the root,

1,164,396 × 6 = 6,986,376, and subtract from remainder 6,986,376, getting zero.

$$\therefore \sqrt[3]{245{,}314{,}376} = 626 \text{ Ans.}$$

Note: There are as many decimal places in a cube root of a decimal as there are periods of 3 figures each in the decimal. If the number is not a perfect cube, annex zeros and continue the process to as many places as you desire.

The cube root of a fraction is found by taking the cube root of its numerator and of its denominator, or by reducing the fraction to a decimal and then extracting the root.

567. In summary, what are the principles applying to exponents?

(*a*) *Multiplication* $a^m \cdot a^n = a^{m+n}$ $a^{\frac{3}{8}} \cdot a^{\frac{3}{4}} = a^{\frac{3}{8}+\frac{3}{4}} = a^{\frac{9}{8}}$

(*b*) *Division* $a^m \div a^n = a^{m-n}$

(*c*) *Raising to a power* $(a^m)^n = a^{mn}$ $\left(\frac{a}{b}\right)^m = \frac{a^m}{b^m}$ $(a^2b)^m = a^{2m} \cdot b^m$

(*d*) *Extracting a root* $\sqrt[m]{a^n} = a^{\frac{n}{m}}$

(*e*) *Negative exponent* $a^{-m} = 1/a^m$

(*f*) *Fractional exponent* $a^{\frac{1}{n}} = \sqrt[n]{a}$ $a^{\frac{n}{m}} = \sqrt[m]{a^n} = [(\sqrt[m]{a})^n]$

(*g*) *Zero exponent* $a^0 = 1$ $y^0 = 1$ $\left[\frac{159}{(a+b)^2}\right]^0 = 1$

PROBLEMS

1. Find:

(*a*) 5^2	(*b*) 8^2	(*c*) 20^2	(*d*) 1^4
(*e*) 11^2	(*f*) 1^9	(*g*) 10^3	(*h*) 3^4
(*i*) 25^2	(*j*) 17^3	(*k*) 83^3	(*l*) 1.25^3
(*m*) $.\overline{75}^2$	(*n*) $(\frac{3}{4})^2$	(*o*) $(1\frac{1}{4})^3$	(*p*) $(\frac{7}{8})^3$
(*q*) $x \cdot x \cdot x \cdot x$	(*r*) $(4x)$ squared	(*s*) $(2b)$ cubed	(*t*) $(12.5)^3$

2. Find the square of the following by the formula $(a + b)^2 = a^2 + 2ab + b^2$:

(*a*) 64	(*b*) 89	(*c*) 36	(*d*) 72
(*e*) 93	(*f*) 783	(*g*) 209	

3. How many square feet are there in a lot 40′ × 100′?

4. How many acres are there in a field 140 rd. square?

5. How many square yards are there in the floor of a room 24 feet long and 18 feet wide?

6. What is the square of: (*a*) .3? (*b*) .6? (*c*) .14? (*d*) .134? (*e*) .07?

7. What is the value of:

(*a*) $2^6 \times 2^2$?	(*b*) $3^5 \times 3^4$?	(*c*) $\frac{2^6}{2^2}$?
(*d*) $3^5 \div 3^4$?	(*e*) $a^x \div a^y$?	(*f*) $a^x \times a^y$?
(*g*) $(4^3)^2$?	(*h*) $(5^2)^3$?	(*i*) 7^0?
(*j*) a^0?	(*k*) $\left(\frac{12x^3y^6}{29 \times 13}\right)^0$?	(*l*) $4 \times 6 \times 8^0$?
(*m*) $5^3 \div 5^8$?	(*n*) 2^{-3}?	(*o*) 4^{-4}?

8. Raise 4 to the 8th power by factoring-the-power method.

9. Raise 6 to the 12th power by factoring-the-power method.

10. Reduce the following to equivalent expressions free from zero, and negative exponents:

(*a*) $3^{-3} \times 2^0$	(*b*) $a^0 \cdot a^{-1}$	(*c*) $\frac{a^{-3}}{a^{-4}}$
(*d*) 3×4^{-1}	(*e*) $(.05)^{-2}$	(*f*) $(a^{-m})^{-n}$

11. Square the following by the short method as indicated by the formula $(a - b)(a + b) = a^2 - b^2$:

(*a*) 28 (*b*) 67 (*c*) 76 (*d*) 89

12. Apply $(a + b)^2 = a(a + 2b) + b^2$ to squaring:

(*a*) $7\frac{1}{4}$ (*b*) $9\frac{1}{2}$ (*c*) $8\frac{1}{5}$

13. Square 975 by aliquot part method and $a(a + 2b) + b^2$.

14. Square the following by first dividing by 2, 3, or 5:

(*a*) 16 (*b*) 45 (*c*) 24 (*d*) 24

15. What is the square of the mean between the two numbers in:

(*a*) 14 and 18? (*b*) 40 and 50? (*c*) 25 and 26?

16. Square the following by the simple method:

(*a*) $8\frac{1}{2}$ (*b*) $13\frac{1}{2}$ (*c*) $11\frac{1}{2}$ (*d*) 65 (*e*) 225

17. What is the square of: (*a*) 9999? (*b*) 99? (*c*) 999,999?

18. What is the value of:

(*a*) 10^7? (*b*) 10^9? (*c*) 10^{-6}? (*d*) 10^{-4}?
(*e*) $10^5 \times 10^2$? (*f*) $10^4 \times 10^3$? (*g*) $10^5 \times 10^{-2}$?
(*h*) $10^{14} \times 10^{-6}$? (*i*) $10^4 \div 10^6$? (*j*) $10^3 \div 10^{-6}$?

19. What is the value of:

(*a*) $\sqrt[3]{1{,}728}$? (*b*) $\sqrt[4]{256b^{16}}$? (*c*) $\sqrt{a^6b^3}$?

(*d*) $\sqrt[3]{x^6y^{12}}$? (*e*) $\sqrt[3]{\dfrac{5^9}{8^{12}}}$? (*f*) $\sqrt[5]{\dfrac{1}{3125}}$?

(*g*) $\sqrt[3]{8^3}$ (*h*) $\sqrt[4]{72}$? (*i*) $\sqrt[9]{93}$?

20. Express in radical form:

(*a*) $28^{\frac{2}{5}}$ (*b*) $x^{\frac{4}{5}}$ (*c*) $b^{\frac{s}{r}}$

(*d*) $y^{\frac{2}{7}}$ (*e*) $(1 + n)^{\frac{1}{r}}$ (*f*) $\left(1 + \dfrac{p}{q}\right)^{\frac{m}{n}}$

21. Remove a factor of the radicand from under the radical sign:

(*a*) $\sqrt[3]{600}$ (*b*) $\sqrt{144}$ (*c*) $\sqrt[4]{768}$

22. Introduce the coefficient of the radical under the radical sign:

(*a*) $4\sqrt{6}$ (*b*) $3\sqrt[3]{7}$ (*c*) $2\sqrt[4]{9}$

23. Make the denominator rational:

(*a*) $\dfrac{1}{\sqrt{7}}$ (*b*) $\dfrac{1}{\sqrt{7} - \sqrt{3}}$ (*c*) $\dfrac{1}{\sqrt{a}}$

24. Reduce to a fraction whose denominator has no radical:

(*a*) $\sqrt{\frac{5}{7}}$ (*b*) $\frac{5}{8}\sqrt{\frac{3}{5}}$ (*c*) $y\sqrt[3]{\dfrac{1}{y}}$

25. Change to a higher order with an index that is a multiple of the original index:

(*a*) $\sqrt[5]{2}$ (*b*) $\sqrt[4]{3}$ (*c*) $\sqrt[6]{a}$

26. Reduce to a radical of lower order:

(*a*) $\sqrt[15]{3{,}125}$ (*b*) $\sqrt[9]{729}$ (*c*) $\sqrt[8]{256b^4}$

27. Reduce to simplest form:

(*a*) $\sqrt[4]{256a^8y^5}$ (*b*) $\sqrt[6]{8x^9}$ (*c*) $\sqrt{\frac{7}{8}}$

28. A square room contains 784 sq. ft. What is the length of one side?

29. If there are 6,084 sq. rd. in the area of a square park, what is the length of one side?

30. If there are 2,916 sq. in. in a square table top, what is the length in feet of one side?

31. Find the square root of 39.864 to three decimal places.

32. What is the square root of 167,321.9025?

33. Find the square root of:

(*a*) $\frac{121}{144}$ (*b*) $\frac{49}{64}$ (*c*) $\frac{16}{25}$ (*d*) $\frac{1}{2}$

(*e*) $\frac{3}{4}$ (*f*) .0178 (*g*) $\frac{3025}{3249}$ (*h*) .9

(*i*) $6\frac{19}{25}$ (*j*) $94\frac{3}{11}$ (*k*) .00065

34. Extract the cube root of 242,970,624.

35. What is the value of $\sqrt[3]{\frac{27}{64}}$?

CHAPTER XIV

LOGARITHMS

568. What is meant by: (*a*) logarithm (abbreviated "log"), (*b*) exponent, (*c*) base?

A *logarithm* is an *exponent.*

A quantity raised to an exponent equals a number (power of the quantity).

The word "logarithm" may be substituted for "exponent."

Then $(\text{quantity})^{logarithm} = \text{a number}$.

Now the quantity to be raised to a power is called the *base.*

Thus $\text{base}^{logarithm} = \text{a number}$.

Ex. (*a*) $(\text{base})8^{2\,(log)} = 64$ (number).

Here, exponent *2* is the log of the number *64.* Or the *log* of 64 to the base 8 is *2.*

Ex. (*b*) $4^3 = 64$.

Here, exponent *3* is the log of the number *64.* Or the *log* of 64 to the base 4 is *3.*

We see that the same number may have a different log, depending upon the base used.

Note carefully that when we *raise* a base, or a quantity, to a certain *power,* we apply an exponent to the base, and the *number* obtained as a result of this process is called the power of the base.

569. What are the two forms of expressing the relationship between the base, the power, and the exponent?

(*a*) $8^2 = 64$ = exponential form.
(*b*) $\log_8 64 = 2$ = logarithmic form.

$$\text{logarithm} = \text{exponent}$$

Note: In the logarithmic form the question arises, "To what exponent must the base 8 be raised to produce 64?" Always ask yourself this question when you see this form. However, any value may be chosen as the base of a system of logarithms (or exponents), except the base 1.

570. What two systems of logarithms are in general use?

(*a*) The Napierian, or the natural system. Here the base is ϵ = epsilon, which denotes the irrational number 2.7182+. (An irrational number is one which cannot be expressed as the quotient of two whole numbers.) It is used principally in theoretical mathematics, engineering, and advanced statistics.

(*b*) The Briggs, or the common system. Here the base is 10, which is most applicable to our decimal number system.

571. To what exponent (logarithm) must the base 10 be raised to produce a number between 1 and 10?

We can readily get the logs of the following numbers:

$$10^1 = 10.000 = \text{number}$$
$$10^{\frac{2}{3}} = \sqrt[3]{100} = 4.642 = \text{number}$$
$$10^{\frac{1}{2}} = \sqrt{10} = 3.162 = \text{number}$$
$$10^{\frac{1}{3}} = \sqrt[3]{10} = 2.154 = \text{number}$$
$$10^0 = 1.000 = \text{number}$$

In logarithmic form these are written as:

$$\log_{10} 10 = 1.0000 = \text{exponent}$$
$$\log_{10} 4.642 = \tfrac{2}{3} = .6667 = \text{fractional exponent}$$
$$\log_{10} 3.162 = \tfrac{1}{2} = .5000 = \text{fractional exponent}$$
$$\log_{10} 2.154 = \tfrac{1}{3} = .3333 = \text{fractional exponent}$$
$$\log_{10} 1 = .0000 = \text{exponent}$$

We see that the log (exponent) of a number between 1 and 10 is a decimal fraction.

From now on we shall omit writing the base 10 which will be understood, thus $\log 10 = 1$ will mean $\log_{10} 10 = 1$.

572. To what exponent (log) must the base 10 be raised to produce a number between 10 and 100?

$$10^2 = 100$$
$$10^{1\frac{2}{3}} = 10^{\frac{5}{3}} = 10 \times 10^{\frac{2}{3}} = 10\sqrt[3]{100} = 10 \times 4.642 = 46.42$$
$$10^{1\frac{1}{2}} = 10^{\frac{3}{2}} = 10 \times 10^{\frac{1}{2}} = 10\sqrt{10} = 10 \times 3.162 = 31.62$$
$$10^{1\frac{1}{3}} = 10^{\frac{4}{3}} = 10 \times 10^{\frac{1}{3}} = 10\sqrt[3]{10} = 10 \times 2.154 = 21.54$$
$$10^1 = 10$$

In logarithmic form these are written as:

$$\log 100 = 2.0000 = \text{exponent}$$
$$\log 46.42 = 1\tfrac{2}{3} = 1.6667 = \text{exponent}$$
$$\log 31.62 = 1\tfrac{1}{2} = 1.5000 = \text{exponent}$$
$$\log 21.54 = 1\tfrac{1}{3} = 1.3333 = \text{exponent}$$
$$\log 10 = 1.0000 = \text{exponent}$$

We see that the log (exponent) of a number between 10 and 100 is 1 + a fraction.

Note that the digit sequence of the numbers whose logs are required is the same as for Question 571, and the decimal part of the log is the same in each case. The only difference is in the position of the decimal point in the number, which produces a corresponding whole number value of the log.

573. To what exponent (log) must the base 10 be raised to produce a number between 100 and 1,000?

$$10^3 = 1{,}000$$
$$10^{2\frac{2}{3}} = 10^2 \times 10^{\frac{2}{3}} = 100\sqrt[3]{100} = 100 \times 4.642 = 464.2$$
$$10^{2\frac{1}{2}} = 10^2 \times 10^{\frac{1}{2}} = 100\sqrt{10} = 100 \times 3.162 = 316.2$$
$$10^{2\frac{1}{3}} = 10^2 \times 10^{\frac{1}{3}} = 100\sqrt[3]{10} = 100 \times 2.154 = 215.4$$
$$10^2 = 100$$

In logarithmic form these are written as:

$$\log 1{,}000 = 3.0000 = \text{exponent}$$
$$\log 464.2 = 2\tfrac{2}{3} = 2.6667 = \text{exponent}$$
$$\log 316.2 = 2\tfrac{1}{2} = 2.5000 = \text{exponent}$$
$$\log 215.4 = 2\tfrac{1}{3} = 2.3333 = \text{exponent}$$
$$\log 100 = 2.0000 = \text{exponent}$$

For a number between 100 and 1,000 the log is 2 + a fraction. The fractional parts of the logs are the same as before for the same sequence of digits. The whole part of the log is affected only by the position of the decimal point in the number.

574. How does this condition apply to higher powers of 10 for any number you may want to produce?

Ex. (*a*) $4{,}642 = 1{,}000 \times 4.642$

$$\log 1{,}000 = 3$$
$$\log 4.642 = .6667 \quad \text{(Question 571)}$$
$$\therefore \log 4{,}642 = 3.6667 \text{ Ans.}$$

Ex. (*b*) $46{,}420 = 10{,}000 \times 4.642$

$$\log 10{,}000 = 4$$
$$\log 4.642 = .6667$$
$$\therefore \log 46{,}420 = 4.6667, \text{ etc. Ans.}$$

575. Why is the log of a number between 1 and .1 expressed as −1 plus the same positive decimal fraction as for Question 571, with the same sequence of digits in the number?

$$10^0 = 1 \qquad 10^{-1} = .1$$

$$.4642 = .1 \times 4.642 \quad (\log .1 = -1 \quad \text{and} \quad \log 4.642 = .6667)$$

$$\therefore \log .4642 = -1 + .6667$$

$$.3162 = .1 \times 3.162 \quad (\log .1 = -1 \quad \text{and} \quad \log 3.162 = .5000)$$

$$\therefore \log .3162 = -1 + .5000$$

$$.2154 = .1 \times 2.154 \quad (\log .1 = -1 \quad \text{and} \quad \log 2.154 = .3333)$$

$$\therefore \log .2154 = -1 + .3333$$

The positive fractional part of the log is the same as in Question 571, for the same sequence of the digits of the number in each case.

576. How does this apply to finding the log of still smaller decimal fractions?

EXAMPLES:

(*a*) $.04642 = .01 \times 4.642$ ($\log .01 = -2$ and $\log 4.642 = .6667$)

$$\therefore \log .04642 = -2 + .6667$$

(*b*) $.004642 = .001 \times 4.642$ ($\log .001 = -3$ and $\log 4.642 = .6667$)

$$\therefore \log .004642 = -3 + .6667, \text{ etc.}$$

577. Why may numbers between 1 and 10 be considered as basic numbers for a system of logs having 10 as a base?

3.0, 6.2, 1.643, 8.769, and 9.37482 are called *basic* numbers.

Logarithms of all numbers having 10 for a base can be obtained from the logs of the basic numbers.

4.642 is a basic number.
$\log 4.642 = .6667$ (Question 571).

$$\begin{aligned} 46.42 &= 4.642 \times 10^1 & \therefore \log 46.42 &= 1.6667 \\ 464.2 &= 4.642 \times 10^2 & \therefore \log 464.2 &= 2.6667 \\ 4642. &= 4.642 \times 10^3 & \therefore \log 4642 &= 3.6667 \\ 46420 &= 4.642 \times 10^4 & \therefore \log 46420 &= 4.6667, \text{ etc.} \end{aligned}$$

$$\begin{aligned} .4642 &= 4.642 \times 10^{-1} & \therefore \log .4642 &= -1 + .6667 \\ .04642 &= 4.642 \times 10^{-2} & \therefore \log .04642 &= -2 + .6667 \\ .004642 &= 4.642 \times 10^{-3} & \therefore \log .004642 &= -3 + .6667 \end{aligned}$$

578. What is meant by the characteristic of a logarithm?

The logarithm of a basic number is a decimal fraction. For other numbers, a positive or negative integer must be added to the fraction to get the logarithm of the number. This integral part or integer is called the *characteristic* of the logarithm.

EXAMPLE: In $\log 46.42 = 1.6667$, 1 is the characteristic.

579. What is meant by the mantissa of a logarithm?

The *decimal* part of the logarithm is the *mantissa*.

EXAMPLE: In $\log .004642 = -3 + .6667$; -3 is the characteristic; .6667 is the mantissa.

The mantissa depends only on the sequence of the digits of the number and not on the position of the decimal point.

580. What is the rule for finding the characteristic of the logarithm of a number?

Count the number of digits in the integral part of the number. The characteristic is one less than that number. This follows from the fact that a basic number has *one integral digit* and its logarithm has no characteristic.

Ex. (*a*)

Number	*Integral digits*	*Characteristic*
1 to 9	1	$1 - 1 = 0$
10 to 99	2	$2 - 1 = 1$
100 to 999	3	$3 - 1 = 2$ etc.

Ex. (*b*) The characteristic of the log of 86537.94 is 4 which is one less than the number of integral digits.

581. What is the rule for finding the characteristic of a purely decimal number?

Count the number of places the decimal point must be moved to make the number basic. The negative characteristic is that number.

Ex. (*a*) What is the negative characteristic of the log of .000865? Move decimal point 4 places to get 8.65, which is a basic number. Then -4 is the characteristic.

$$\log .000865 = -4 + .9370$$

Ex. (*b*) What is the negative characteristic of the log of .00427? Move decimal point 3 places to get 4.27, which is a basic number. Then -3 is the characteristic.

$$\log .00427 = -3 + .6304$$

582. Why is a negative characteristic kept distinct from the mantissa of a logarithm?

In computation it is advantageous to have the mantissa positive in every case and to keep it equal to the mantissa of the log of the basic number. The log of a purely decimal number then consists of a negative integer plus a positive decimal.

583. How are negative characteristics generally expressed?

(*a*) With a minus sign over the characteristic. This indicates that it alone is negative.

Ex. (*a*) $\log .000865 = \bar{4}.9370$
(*b*) $\log .00427 = \bar{3}.6304$

(*b*) By adding and subtracting 10.

Ex. (*a*) $\log .000865 = \bar{4} + 10 + .9730 - 10 = 6.9370 - 10$
(*b*) $\log .00427 = \bar{3} + 10 + .6304 - 10 = 7.6304 - 10$

584. May a negative characteristic be expressed in other ways?

It may sometimes be found useful to add and subtract a number other than 10.

EXAMPLE: $\log .00427 = \bar{3}.6304$ may be written as:

$$\bar{3} + 8 + .6304 - 8 = 5.6304 - 8$$

or

$$\bar{3} + 30 + .6304 - 30 = 27.6304 - 30$$

Any combination may be used as long as the net result is the original $\bar{3}$.

However, the form $9.\cdots\cdot - 10$ is most convenient for operations of addition and subtraction of logs, and these operations are quite common.

585. What is a table of common logarithms?

A table of logs is a table of mantissas. It is a table of the exponents of 10 corresponding to basic numbers. It answers the question: "What is the power of 10 required to give a certain basic number?" Finding the exponent is finding the log. The differences between successive logs are not the same because they form an exponential scale of powers of 10. See Table 3, Appendix B.

The same sequence of numbers gives the same log independent of the position of the decimal point.

586. How do we look up a log in a table?

Look at the left of the table to get the sequence of digits in the number as far as it will go, and then go to the top for the next digit in the sequence. When the number has more than three significant figures, add to the log reading the proportional part of the number between the two adjacent logs in the table. For less than three significant figures add zeros.

Ex. (*a*) Find the log of 42. Look up 420 figures. Enter 42 at left and 0 column on top and get 62325 for the mantissa.* Then add 1 as the characteristic.

$$\therefore \log 42 = 1.62325$$

Ex. (*b*) For the sequence of figures 420, the mantissa is the same but the characteristic is one less than the number of digits.

$$\therefore \log 420 = 2.62325$$

Ex. (*c*)
$$\begin{aligned} \log 4.2 &= .62325 \\ \log .42 &= \bar{1}.62325 \quad \text{or} \quad 9.62325 - 10 \\ \log .042 &= \bar{2}.62325 \quad \text{or} \quad 8.62325 - 10 \\ \log .0042 &= \bar{3}.62325 \quad \text{or} \quad 7.62325 - 10 \end{aligned}$$

587. What is meant by a proportional part of a log?

The proportional part of the difference between two adjacent logs represented by the required log is known as the proportional part of the log.

Ex. (*a*) Find the log of 681.6.

Enter 68 at left and move right until you reach column 1 at the top of table. Read .83315.

The next adjacent log is of 682. Read .83378.

Difference is $83378 - 83315 = 63$.

Now .6 of this difference is $.6 \times 63 = 37.8$ or 38 to nearest digit.

Then
$$\begin{array}{r} .83315 \\ +.00038 \\ \hline .83353 \end{array}$$

Characteristic of 681.6 is 2.

$$\therefore \log 681.6 = 2.83353 \text{ Ans.}$$

Ex. (*b*) Find log of 764.52.

$$\begin{array}{r} \log 765 = .88366 \\ \log 764 = .88309 \\ \hline \text{Difference} = 1 = \quad 57 \end{array}$$

$$.52 \times 57 = 29.64 = 30 \text{ to nearest digit.}$$

$$\begin{array}{r} \log 764 = 2.88309 \\ + \quad .00030 \\ \hline \therefore \log 764.52 = 2.88339 \text{ Ans.} \end{array}$$

* Calculations here are shown to five places. Because of limitations of space it has not been possible to include a table of five-place logarithms. A table of four-place logarithms, however, may be found on pp. 424-425 (Table 3, Appendix B).

588. What is meant by an antilogarithm?

An antilogarithm is the number corresponding to a given logarithm. When the exponent is given and the number is required, the process is called finding the antilogarithm. It is the reverse of finding the logarithm.

EXAMPLE: In the above, 2.88339 is the log and 764.52 is the antilog.

589. How do we obtain an antilog, or number, from a table of logs?

(*a*) Find the number corresponding to the two mantissas between which the desired mantissa is located.

(*b*) Get their difference. Find the difference between the lower mantissa and the desired one.

(*c*) Find the proportional part and add this to the number.

EXAMPLE: Find the antilog of $8.61768 - 10 = \bar{2}.61768$.

	Mantissa 61805 produces number	41500 sequence
	Mantissa 61700 produces number	41400 sequence
Difference	= 105	100

Desired mantissa = 61768
Lower mantissa = 61700
Difference = 68

$$\frac{68}{105} = \frac{x}{100} \qquad x = 65$$

Add $41400 + 65 = 41465$.

$\therefore$ Antilog of $\bar{2}.61768$ is .041465 Ans.

590. Upon what laws do computations with logs depend?

Upon the laws of exponents. The essential laws of exponents are:

(*a*) To multiply, add the exponents algebraically.

$$10^5 \times 10^{-\frac{1}{3}} \times 10^{-\frac{2}{3}} = 10^{(5-\frac{1}{3}-\frac{2}{3})} = 10^4 = 10{,}000$$

(*b*) To divide, subtract the exponents algebraically.

$$10^{-5} \div 10^{-8} = 10^{-5-(-8)} = 10^3 = 1{,}000$$

(*c*) To raise to a power, multiply the exponents.

$$(10^{-3})^{-2} = 10^{(-3\times-2)} = 10^6 = 1{,}000{,}000$$

(*d*) To extract a root, divide the exponents.

$$\sqrt[3]{10^9} = 10^{9/3} = 10^3 = 1{,}000$$

591. What is the procedure for multiplying two or more quantities by logs?

A number can be expressed in exponential form to any base or to base 10.

EXAMPLE: $160 = 10^{2.2041}$ $236 = 10^{2.3729}$ $28 = 10^{1.4472}$

Now, by the laws of exponents, to multiply we add the exponents. But exponents are logs. So to multiply, add the logs. Thus

$$160 \times 236 \times 28 = 10^{2.2041} \times 10^{2.3729} \times 10^{1.4472}$$
$$= 10^{2.2041+2.3729+1.4472} = 10^{6.0242}$$

$$\therefore \log (160 \times 236 \times 28) = 6.0242 \text{ (characteristic} = 6, \text{ mantissa} = .0242)$$

$$\text{and antilog} = 1057000 = \text{product}$$

The procedure may be stated in logarithmic form as:

$$\begin{aligned} \log 160 &= 2.2041 \\ +\log 236 &= 2.3729 \\ +\log 28 &= \underline{1.4472} \\ \log (160 \times 236 \times 28) &= 6.0242 = \text{sum} = \text{log of product} \end{aligned}$$

$$\therefore \text{antilog} = 1057000 = \text{product Ans.}$$

592. What is the procedure for getting the quotient of two numbers by logs?

By the laws of exponents, to divide, subtract the exponents. Thus, the log of a quotient is the log of the numerator minus the log of the denominator.

EXAMPLE: $135.834 = 10^{2.13301}$ $8.96 = 10^{.95230}$

Thus

$$\frac{135.834}{8.96} = \frac{10^{2.13301}}{10^{.95230}} = 10^{2.13301-.95230} = 10^{1.18071}$$

$$\therefore \log \frac{135.834}{8.96} = 1.18071 \text{ (characteristic} = 1, \text{ mantissa} = .18071)$$

The antilog is 15.1605.
This procedure may be expressed in logarithmic form as:

$$\begin{aligned} \log 135.834 &= 2.13301 \\ -\log 8.96 &= \underline{0.95230} \\ \log \left(\frac{135.834}{8.96}\right) &= 1.18070 = \text{difference} = \text{log of quotient} \end{aligned}$$

$$\therefore \text{antilog} = 15.1605 = \text{quotient Ans.}$$

593. What is the procedure for raising a number to a power by logs?

By the law of exponents, to raise to a power, multiply the exponents.

EXAMPLE: $37.4 = 10^{1.5729}$

This means that exponent 1.5729 is the log of 37.4. Now, $(37.4)^3 = (10^{1.5729})^3 = 10^{1.5729 \times 3}$.

$$\therefore \log (37.4)^3 = 3 \times 1.5729$$

This means, multiply the log of the number by the power.

In logarithmic form this is stated as:

$$\log 37.4 = 1.5729$$

$$\log (37.4)^3 = 3 \times 1.5729 = 4.7187 \quad \text{(characteristic = 4, mantissa = .7187)}$$

$$\therefore \text{antilog} = 52320 \text{ Ans.}$$

594. What is the procedure for getting the root of a number by logs?

By the law of exponents, to extract a root, divide the exponents.

EXAMPLE: $37.4 = 10^{1.5729}$

$$\text{and } \sqrt[3]{37.4} = (10^{1.5729})^{\frac{1}{3}} = 10^{1.5729 \times \frac{1}{3}}$$

$$\therefore \log \sqrt[3]{37.4} = \frac{1.5729}{3}$$

This means, divide the log of the number by the root.

In logarithmic form this is expressed as:

$$\log 37.4 = 1.5729$$

$$\log \sqrt[3]{37.4} = \frac{1.5729}{3} = .5243$$

$$\therefore \text{antilog} = 3.344 \text{ Ans.}$$

595. How can we express the log of 75 in terms of the log of 5 and the log of 3?

$$75 = 5^2 \times 3$$

Then, $\log 75 = \log (5^2 \times 3) = \log 5^2 + \log 3 = 2 \log 5 + \log 3$.

596. How can we express $12^{\frac{3}{4}}/5\sqrt{52}$ as an algebraic sum of logs?

$$\log \frac{12^{\frac{3}{4}}}{5\sqrt{52}} = \log^{\frac{3}{4}} 12 - \log 5\sqrt{52} = \tfrac{3}{4} \log 12 - [\log 5 + \log \sqrt{52}]$$

$$= \tfrac{3}{4} \log 12 - \log 5 - \tfrac{1}{2} \log 52$$

597. How can we reduce log 7 + 3 log 5 to the log of a single number?

$$\begin{aligned}\log 7 + 3\log 5 &= \log 7 + \log 5^3\\ &= \log 7 + \log 125\\ &= \log 7 \times 125 = \log 875\end{aligned}$$

598. What is the log of 1 to any base?

We know that

$$1 = \frac{a^1}{a^1} = a^{1-1} = a^0 \quad (a = \text{any number except } 1)$$

$$\therefore\ \log 1 = 0$$

$0 =$ the exponent $=$ the log of 1 to any base a. Ans.

599. What is the log of the base itself in any system?

We know that $a = a^1$

$$\therefore\ \log a = 1$$

$1 =$ the a exponent $=$ log of a to base a. Ans.

600. What is the log of 0 in any system whose base is greater than 1?

We know that

$$a^{-\infty} = \frac{1}{a^{\infty}} = \frac{1}{\infty} = 0 \quad (\infty = \text{infinity})$$

$$\therefore\ \log 0 = -\infty$$

$-\infty =$ the exponent $=$ log of 0 to any base greater than 1. Ans.

Thus log 0 is negative = numerically greater than any assigned number, however great.

601. How can we find the log of a number to a new base, when the logs of numbers to a particular base are given?

Divide the log of the number to the particular base by the log of the new base referred to the particular base.

EXAMPLE: We have a table of logs (exponents) to base 10 and we want to get the log of 47.25 to a new base $\epsilon = 2.718$.

$$\log_{\epsilon} 47.25 = \frac{\log_{10} 47.25}{\log_{10} 2.718} = \frac{\log_{10} 47.25}{.4343} = 2.3026 \log_{10} 47.25$$

$$= 2.3026 \times 1.9199 = 4.421 = \text{exponent Ans.}$$

602. How are natural and common logs related, as seen from the above?

(*a*) To get the natural log of a number multiply its common log by 2.3026.

EXAMPLE:

$$\log_\varepsilon 100 = 2.3026 \times \log_{10} 100 = 2.3026 \times 2 = 4.6052$$

(*b*) To get the common log of a number multiply the natural log by .4343.

EXAMPLE:

$$\log_{10} 100 = .4343 \log_\varepsilon 100 = .4343 \times 4.6052 = 2$$

603. What is meant by the cologarithm of a number?

The cologarithm of a number is the logarithm of the reciprocal of the number.

EXAMPLE: If a is a given number, then

$$\text{colog } a = \log \frac{1}{a}$$

But

$$\log \frac{1}{a} = \log 1 - \log a$$

$$\therefore \text{colog } a = 0 - \log a$$

This may be written as:

$$\text{colog } a = (10 - 10) - \log a$$

604. What is the rule for obtaining the colog of a number to base 10?

Subtract the logarithm of the number from (10 − 10).

EXAMPLE: If the log of a number is 7.15625 − 10, then the colog is:

$$\begin{array}{r} 10.00000 - 10 \\ -(7.15625 - 10) \\ \hline 2.84375 = \text{colog} \end{array}$$

605. When are cologs used to advantage?

In finding the log of a fraction or quotient.

Instead of subtracting the log of the denominator, add the colog of the denominator to the log of the numerator. In a series of multiplication and division use cologs for the denominators, or the terms by which you have to divide. This enables you to combine the log value in one operation of addition.

EXAMPLE: What is the value of $\frac{87.56}{34.45}$?

$$\log\left(\frac{87.56}{34.45}\right) = \log 87.56 + \text{colog } 34.45$$

$$\begin{aligned} \log 87.56 &= 1.9423 \\ \text{colog } 34.45 &= \underline{8.4628 - 10} \\ & 10.4045 - 10 \end{aligned} \qquad \begin{aligned} & 10.0000 - 10 \\ \log 34.45 &= \underline{1.5372} \\ \text{colog } 34.45 &= 8.4628 - 10 \end{aligned}$$

$$\log\left(\frac{87.56}{34.45}\right) = .4045$$

$$\frac{87.56}{34.45} = 2.538 = \text{antilog Ans.}$$

606. What is the result of .005864 × 272.6 × 84.65?

$$\begin{aligned} \log .005864 &= 7.7683 - 10 \\ \log 272.6 &= 2.4356 \\ \log 84.65 &= \underline{1.9277} \\ \text{Sum} &= 12.1316 - 10 = 2.1316 \end{aligned}$$

(characteristic = 2; mantissa = .1316)

$\therefore$ antilog = 135.4 Ans.

607. What is the result of $(262)^4$?

$$\log (262)^4 = 4 \times \log 262 = 4 \times 2.4183 = 9.6732$$

(characteristic = 9; mantissa = .6732)

$\therefore$ antilog = 4,712,000,000 Ans.

608. What is the result of $\frac{(4.687)^3}{.8476}$?

$$\begin{aligned} 3 \times \log 4.687 = 3 \times .6709 &= 2.0127 \\ \log .8476 &= \underline{\bar{1}.9282} \\ \text{Difference} &= 2.0845 \end{aligned}$$

(characteristic = 2; mantissa = .0845)

$\therefore$ antilog = 121.5 Ans.

609. What is the value of $(18.34)^{-3}$?

$$\log (18.34)^{-3} = -3 \times \log 18.34 = -3 \times 1.2634 = -(3.7902)$$

Here the entire number, including the decimal part, is negative. To obtain a positive mantissa for use in the table of logs, change the form of this log by adding and subtracting 10.

$$\begin{array}{rl} & 10.0000 - 10 \\ & -3.7902 \\ \log (18.34)^{-3} = & 6.2098 - 10 = \bar{4}.2098 \end{array}$$

(characteristic = $\bar{4}$; mantissa = .2098)

$\therefore$ antilog = .0001621 Ans.

This problem may be solved by using the colog method, because

$$(18.34)^{-3} = \frac{1}{(18.34)^3} = \left(\frac{1}{18.34}\right)^3$$

Then

$$\log \left(\frac{1}{18.34}\right)^3 = 3 \times \log \frac{1}{18.34} = 3 \times \text{colog } 18.34$$

$$\begin{array}{rl} & 10.0000 - 10 \\ \log 18.34 = & 1.2634 \\ \text{colog } 18.34 = & 8.7366 - 10 \end{array}$$

and

$$\begin{aligned} 3 \times \text{colog } 18.34 = 3(8.7366 - 10) &= (26.2098 - 30) \\ &= \bar{4}.2098 \end{aligned}$$

$\therefore$ antilog = .0001621 Ans.

610. What is the value of $(2.718)^{-14}$?

$$(2.718)^{-14} = \frac{1}{(2.718)^{14}} = \left(\frac{1}{2.718}\right)^{14}$$

$$\log \left(\frac{1}{2.718}\right)^{14} = 14 \times \text{colog } 2.718$$

$$\begin{array}{rl} & 10.0000 - 10 \\ \log 2.718 = & .4343 \\ \text{colog } 2.718 = & 9.5657 - 10 \end{array}$$

$$\begin{aligned} 14 \times \text{colog } 2.718 = 14(9.5657 - 10) &= (133.9198 - 140) \\ &= \bar{7}.9198 \end{aligned}$$

(characteristic = $\bar{7}$; mantissa = .9198)

$\therefore$ antilog = .0000008314 = $8{,}314 \times 10^{-10}$ Ans.

611. What is the result of $\sqrt[7]{-0.7982}$?

$$\sqrt[7]{-0.7982} = (-0.7982)^{\frac{1}{7}}$$

The log of a negative number is not defined in real numbers. However, this problem may be solved by considering the base as a positive number, and prefixing a minus sign to the result.

$$\log 0.7982 = \bar{1}.9021 = 9.9021 - 10 = 69.9021 - 70$$
$$\tfrac{1}{7}\log 0.7982 = \tfrac{1}{7}(69.9021 - 70) = 9.9860 - 10 = \bar{1}.9860$$
$$(\text{characteristic} = \bar{1};\ \text{mantissa} = 0.9860)$$
$$\text{antilog} = 0.9683$$

Prefix minus sign to result,

$$\therefore\ -0.9683 \text{ Ans.}$$

Note: Since *even* powers can never be negative (see Question 628), it is impossible to express an *even* root of a negative quantity by the "real" system of numbers. In higher mathematics such *even* roots are called "imaginary" numbers.

612. What is the result of $(.462)^{\frac{3}{7}}$?

$$\log .462 = \bar{1}.6646 = (9.6646 - 10) = (69.6646 - 70)$$
$$\tfrac{3}{7}\log .462 = \tfrac{3}{7}(69.6646 - 70) = \tfrac{3}{7} \times 69.6646 - \tfrac{3}{7} \times 70$$
$$= \frac{208.9938}{7} - 30 = 29.8562 - 30 = 9.8562 - 10$$
$$(\text{characteristic} = \bar{1};\ \text{mantissa} = .8562)$$
$$\therefore\ \text{antilog} = .7181 \text{ Ans.}$$

613. What is the result of $(.08746)^{-\frac{3}{5}}$?

$$\log .08746 = \bar{2}.9418 = 8.9418 - 10$$
$$-\tfrac{3}{5}\log .08746 = -\tfrac{3}{5}(8.9418 - 10) = -\tfrac{3}{5}(48.9418 - 50)$$
$$= -\tfrac{3}{5} \times 48.9418 - \tfrac{3}{5} \times (-50) = -\frac{146.8254}{5} + 30$$
$$= -29.3651 + 30 = \left(\begin{array}{r} 30.0000 \\ -29.3651 \\ \hline .6349 \end{array}\right)$$
$$(\text{characteristic} = 0;\ \text{mantissa} = .6349)$$
$$\therefore\ \text{antilog} = 4.314 \text{ Ans.}$$

Using the colog procedure,

$$(.08746)^{-\frac{3}{5}} = \frac{1}{(.08746)^{\frac{3}{5}}}$$

$$\begin{array}{rr} & 10.0000 - 10 \\ \log .08746 = & 8.9418 - 10 \\ \hline \text{colog } .08746 = & 1.0582 \end{array}$$

$$\tfrac{3}{5}\,\text{colog } .08746 = \tfrac{3}{5} \times 1.0582 = \frac{3.1746}{5} = .6349$$

$$\therefore\ \text{antilog} = 4.314 \text{ Ans.}$$

Here the colog procedure is simpler.

614. What is the result of $(.04782)^{1.64}$?

$$\log .04782 = \bar{2}.3007 = 8.3007 - 10$$

$$1.64 \times \log .04782 = 1.64(8.3007 - 10) = 1.64(98.3007 - 100)$$

[(98 − 100) is used to simplify the multiplication]

$$= 161.2132 - 164 = 9.2132 - 10$$

(characteristic = $\bar{1}$; mantissa = .2132)

$\therefore$ antilog = .1634 Ans.

615. What is the result of $(.3846)^{-1.6}$?

$$\log .3846 = \bar{1}.5850 = 9.5850 - 10$$

$$-1.6 \log .3846 = -1.6(9.5850 - 10)$$

$$= -1.6 \times 9.5850 - 1.6 \times (-10)$$

$$= -15.3360 + 16 = \left(\begin{array}{r} 16.0000 \\ -15.3360 \\ \hline .6640 \end{array}\right)$$

(characteristic = 0; mantissa = .6640)

$\therefore$ antilog = 4.613 Ans.

616. What is the result of $(.42)^{.71} \times (7.6)^{-.62} \times (432 - 69)$?

Perform (432 − 69) first, getting 363.

$$\log .42 = \bar{1}.6232 = 9.6232 - 10$$

$$.71 \log .42 = .71(9.6232 - 10) = 6.8325 - 7.1 \quad \text{(First factor)}$$

$$\log 7.6 = .8808$$

$$-.62 \log 7.6 = -.62 \times .8808 = -.5461$$

Change the negative number −.5461 to a positive mantissa by adding and subtracting 10.

$$\begin{array}{r} 10.0000 - 10 \\ .5461 \quad\;\; \\ \hline 9.4539 - 10 \end{array} \quad \text{(Second factor)}$$

$$\log 363 = 2.5599 \quad \text{(Third factor)}$$

Now add all the factors

$$\begin{array}{lrl} \text{First factor} & = & 6.8325 - 7.1 \\ \text{Second factor} & = & 9.4539 - 10 \\ \text{Third factor} & = & 2.5599 \\ \hline & & 18.8463 - 17.1 \end{array} \qquad \begin{array}{r} 18.8463 \\ -17.1000 \\ \hline 1.7463 \end{array}$$

(characteristic = 1; mantissa = .7463)

$\therefore$ antilog = 55.75 Ans.

617. What is the result of $\dfrac{24.6 \times 288 \times \log \frac{7}{8}}{.794}$ **?**

$$\log \tfrac{7}{8} = \log .875 = 9.9420 - 10$$

Now, since log .875 is to be used as a number and not as a log, evaluate it by getting the difference between 9.9420 and −10.

$$\begin{array}{r} -10.0000 \\ 9.9420 \\ \hline -\quad .0580 \end{array}$$

$$\text{Now } \log \left[\frac{24.6 \times 288 \times (-.058)}{.794}\right]$$
$$= \log 24.6 + \log 288 + \log .058 + \text{colog } .794$$

Disregard the negative sign of .058 during calculation and prefix it to the result.

$$\begin{array}{rl} \log 24.6 = & 1.3909 \\ \log 288 = & 2.4594 \\ \log .058 = & 8.7634 - 10 \\ \text{colog } .794 = & .1002 \\ \hline & 12.7139 - 10 = 2.7139 \end{array}
\qquad
\begin{array}{rr} & 10.0000 - 10 \\ \log .794 = & 9.8998 - 10 \\ \hline \text{colog } .794 = & .1002 \end{array}$$

(characteristic = 2; mantissa = .7139)

$\therefore$ antilog = −517.5 Ans.

618. What is the result of $\sqrt[4]{\dfrac{.00008436 \times \sqrt[6]{946.38}}{(8.58)^4 \times (.006439)^{5/9}}}$ **?**

$$\begin{array}{rl} \log .00008436 = & 5.9258 - 10 \\ \frac{1}{6} \log 946.38 = & .4960 \\ 4 \text{ colog } 8.58 = & 4.2660 - 10 \\ \frac{5}{9} \text{ colog } .006439 = & 1.2173 \\ \hline & 4)\overline{11.9051} - 20 \\ \text{or} & 4)\overline{31.9051} - 40 \\ \hline & 7.9763 - 10 \end{array}
\qquad
\begin{array}{rl} \log 946.38 = & 2.9761 \\ \frac{1}{6} \text{ of this} = & .4960 \end{array}$$

$$\begin{array}{rl} 10.0000 - 10 & \\ 7.8088 - 10 & = \log .006439 \\ \hline 2.1912 & = \text{colog } .006439 \end{array}$$

$$\frac{10.9560}{9} = 1.2173 \qquad = \tfrac{5}{9} \text{ colog } .006439$$

$$\begin{array}{rl} 10.0000 - 10 & \\ .9335 & = \log 8.58 \\ \hline 9.0665 - 10 & = \text{colog } 8.58 \end{array}$$

$$4(9.0665 - 10) = (36.2660 - 40) = 4.2660 - 10 = 4 \text{ colog } 8.58$$

$\therefore$ antilog = .00947 Ans.

In getting $\frac{5}{9}$ of colog of .006439, first multiply by 5 and then divide by 9 to eliminate any error that would result from inexact division, an error that would be multiplied 5 times.

619. What is the result of $(.58)^y = 5.67$?

Take the logs of both sides.

$$y \log .58 = \log 5.67$$

or

$$y = \frac{\log 5.67}{\log .58} = \frac{.7537}{9.7634 - 10}$$

Carry out the indicated subtraction in the denominator.

$$\begin{array}{r} -10.0000 \\ 9.7634 \\ \hline -\ \ .2366 \end{array}$$

Then

$$y = -\frac{.7536}{.2366}$$

$$\begin{array}{r} \log .7536 = 9.8772 - 10 \\ \log .2366 = 9.3740 - 10 \\ \hline .5032 \end{array}$$

(characteristic = 0; mantissa = .5032)

$\therefore$ antilog $= -3.186$ Ans.

620. How accurate are results of numerical computations by logs?

Results obtained by logarithmic computations are approximate.

A log of a number cannot, in general, be found exactly, but only approximately to four, five, or any desired number of decimal places. Therefore, the results of numerical computations by means of logs are not, in any case, correct beyond the four, five, or other number of decimal places in the logs used to make the computations.

PROBLEMS

1. Give the log and write the log form of:

(*a*) $5^3 = 125$ (*b*) $10^6 = 1{,}000{,}000$ (*c*) $(\frac{1}{3})^5 = \frac{1}{243}$
(*d*) $9^{\sqrt{2}} = 22.35$ (*e*) $3^4 = 81$ (*f*) $2^{-2} = \frac{1}{4}$

2. Write the log form of:

(*a*) $4^3 = 64$ (*b*) $36^{-\frac{3}{2}} = \frac{1}{216}$ (*c*) $10^d = 600$
(*d*) $p^t = n$ (*e*) $(.01)^4 = .00000001$ (*f*) $2^{-4} = \frac{1}{16}$

3. Express in exponential form:

(*a*) $\log_4 256 = 4$ (*b*) $\log_x a = b$ (*c*) $\log_b 1 = 0$

(*d*) $\log_{10} .000001 = -6$ (*e*) $\log_{10} 10{,}000 = 4$ (*f*) $\log_6 1{,}296 = 4$

4. If the logs to the base 4 are 0, 1, 2, 3, 4, -1, -2, $\frac{1}{2}$, what are the numbers?

5. If the base is 5 what are the logs of the following numbers: 1, 5, 25, 125, 625, $\frac{1}{5}$, $\frac{1}{25}$, $\frac{1}{125}$, $\frac{1}{625}$?

6. If the base is 10 what are the logs of the following numbers: 0, 10, 100, 1,000, 10,000, 100,000, 0.1, 0.01, 0.001, 0.0001, 0.00001?

7. Find the value of x in each of the following:

(*a*) $\log_{10} x = 3$ (*b*) $\log_{16} x = \frac{3}{2}$ (*c*) $x = \log_{\frac{1}{3}} 243$

(*d*) $\log_x 64 = \frac{2}{3}$ (*e*) $\log_5 x = -5$ (*f*) $\log_x 10{,}000 = 4$

(*g*) $2 \log_{25} x = -3$ (*h*) $x = \log_{100} 1{,}000$ (*i*) $\log_x 49 = 2$

8. Are the following true statements?

(*a*) $\log_{10} 10{,}000 - \log_{10} 1{,}000 + \log_{10} 100 + \log_{10} 10 + \log_{10} 1 = 4$

(*b*) $\log_{10} 0.0001 + \log_{10} 0.001 - \log_{10} 0.01 - \log_{10} 0.1 = -4$

(*c*) $3 \log_3 3 + 4 \log_3 \frac{1}{3} + \log_3 1 = -1$

(*d*) $3 \log_5 \sqrt{.008} + 3 \log_{10} \dfrac{\sqrt[3]{10}}{10} = -3\frac{1}{2}$

9. What is the characteristic of the logs of each of the following numbers?

(*a*) 9.854 (*b*) 9854 (*c*) .9854

(*d*) 9.854×10^6 (*e*) 98.5 (*f*) .000098

(*g*) 985.41 (*h*) 985,000,000 (*i*) .0098541

(*j*) 98.5413 (*k*) 4.62915 (*l*) 3.1416

(*m*) 2.718×10^{-14} (*n*) $.00054 \times 10^{-4}$ (*o*) 3,755,000

(*p*) .4343

10. If the mantissa of the log of a number is .4064, where should the decimal point be for each of the following characteristics?

(*a*) 2 (*b*) -1 (*c*) 0 (*d*) -3

(*e*) 5 (*f*) -4 (*g*) 1 (*h*) 3

(*i*) 6 (*j*) $(3 - 1)$ (*k*) $(11 - 10)$ (*l*) $(10 - 10)$

(*m*) $(2 - 3)$ (*n*) $(8 - 10)$ (*o*) $(27 - 30)$ (*p*) $(34 - 38)$

11. Find the log of each of the following numbers:

(*a*) 5.9433 (*b*) 971.4 (*c*) .0642

(*d*) .008793 (*e*) 37.93 (*f*) 1.379

(*g*) .0306 (*h*) .00006794 (*i*) 5.674×10^{-5}

(*j*) $.00638 \times 10^4$

12. Find the antilogs of the following logs:

(*a*) .9954 (*b*) 3.4789 (*c*) $\bar{1}.9572$

(*d*) 3.0358 (*e*) $\bar{4}.3762$ (*f*) $7.8617 - 10$

(*g*) $18.6742 - 20$ (*h*) $2.4169 - 5$ (*i*) 3.1606

(*j*) $1.2168 - 0.7$ (*k*) $.5464 - \frac{6}{5}$ (*l*) $-.3649$

13. Express 196 in terms of the log of 7 and the log of 4.

14. Express $\frac{24\frac{5}{8}}{6\sqrt[3]{78}}$ as an algebraic sum of logs.

15. Express log 9 + 3 log 6 as a log of a single number.

16. Express each of the following as the sum or difference of logs:

(*a*) $83 \times 92 \times 28$ (*b*) $\frac{942.6}{168.4}$

(*c*) $\frac{38.4 \times .6592}{7.431}$ (*d*) $\frac{(3.742)^3 \times (82)^2}{\sqrt{462.1}}$

17. Express in expanded form:

(*a*) $\log \frac{4.32 \times 7.48}{5.66}$ (*b*) $\log \sqrt{\frac{9.48}{72.4 \times .069}}$

18. Find the value of each of the following:

(*a*) $\log (.01)^3 + \log \sqrt[4]{.0001}$ (*b*) $\log \sqrt{10{,}000} + \log \sqrt{.001}$

(*c*) $\log \sqrt{\frac{1}{100}} + \log \sqrt{100}$ (*d*) $\log \sqrt[5]{1{,}000} + \log (.001)^2$

(*e*) $\log (.001)^5 - \log (100)^2 + \log \sqrt[5]{.01}$

(*f*) $\log_5 \sqrt{625} + \log \sqrt[3]{512}$

19. Contract each of the following expressions:

(*a*) $4 \log 6 + \frac{1}{6} \log 5 - 7 \log 8$

(*b*) $\frac{9}{5} \log 25 - \frac{7}{9} \log 10 - \frac{1}{7} \log 5 + \log 9$

(*c*) $\frac{3}{5}[6 \log 2 + 6 \log 5 - \frac{1}{3} \log 6 - \frac{3}{8} \log 7]$

(*d*) $3 \log 2 + \log 3 - \frac{1}{9} \log 4$

20. Evaluate each of the following, given that log 2 = .3010 and log 3 = .4771:

(*a*) log 8 (*b*) log 6 (*c*) log 12
(*d*) log 27 (*e*) log 1.5 (*f*) log 432

21. Find the result of each of the following:

(*a*) $\log_2 9$ (*b*) $\log_6 112$ (*c*) $\log_5 11$
(*d*) $\log_8 9$ (*e*) $\log_6 122$ (*f*) $\log_5 \sqrt{.0001}$
(*g*) $\log_4 .1$ (*h*) $\log_4 10$ (*i*) $\log_4 3$
(*j*) $\log_7 6$ (*k*) $\log_5 .01$ (*l*) $\log_5 100$

22. Find the natural log of each of the following numbers:

(*a*) 872.1 (*b*) .782 (*c*) 6,928
(*d*) .0432 (*e*) 1.872 (*f*) .000496

23. Find the common log if the natural logs are as given by each of the following:

(*a*) .782 (*b*) 847.2 (*c*) .0083
(*d*) 9,248 (*e*) .00062 (*f*) 3.78

24. Evaluate the following, using logs:

(*a*) $.006943 \times 342.2 \times 82.43$ (*b*) $(358)^4$ (*c*) $\frac{(5.792)^3}{.7931}$

(*d*) $(21.12)^{-3}$ (*e*) $(2.718)^{-12}$ (*f*) $\sqrt[5]{-.8279}$

(*g*) $(.06493)^{-\frac{3}{8}}$ (*h*) $(.5937)^{-1.3}$

(*i*) $(.36)^{.69} \times (5.3)^{-.58} \times (238 - 43)$ (*j*) $\dfrac{25.8 \times 369 \times \log \frac{5}{9}}{.642}$

(*k*) $\sqrt[3]{\dfrac{.00006384 \times \sqrt[5]{839.17}}{(9.42)^{3.} \times (.007986)^{\frac{3}{7}}}}$ (*l*) $(.42)^x = 6.49$

(*m*) $\dfrac{1}{3\sqrt[5]{-.064}}$ (*n*) $\sqrt[9]{2} \times \sqrt[7]{3} \times \sqrt[5]{.03}$

CHAPTER XV

POSITIVE AND NEGATIVE NUMBERS

621. What is meant by "signed" numbers?

Numbers preceded by a plus sign, or a minus sign, are called signed numbers. Such numbers show the amount and direction of change, and may thus denote quality as well as quantity.

Ex. (*a*) If $+32°$ represents $32°$ above zero, then $-32°$ represents $32°$ below zero.

Ex. (*b*) If $+8$ miles represents 8 miles to the east, then -8 represents 8 miles to the west.

Ex. (*c*) If $+\$5$ represents a credit of \$5, then $-\$5$ represents a debit of \$5.

Ex. (*d*) If $+100$ represents a distance above sea-level, then -100 represents a distance below sea-level.

622. What is meant by "positive" and "negative" numbers?

Numbers preceded by a plus [+] sign, or by no sign at all, are called positive numbers, as 32, $+5$, $+.7$, 11, $+\frac{5}{8}$.

Numbers preceded by a minus [−] sign are called negative numbers, as -7, $-.14$, $-\frac{5}{6}$, $-28\frac{3}{8}$, -23.

623. What is meant by the absolute value of a number?

The absolute value is the value of the number without the sign.

EXAMPLES: The absolute value of $+32$ is 32
The absolute value of -8 is 8

624. How can the relations between the plus numbers, the minus numbers, and zero be shown by the number scale?

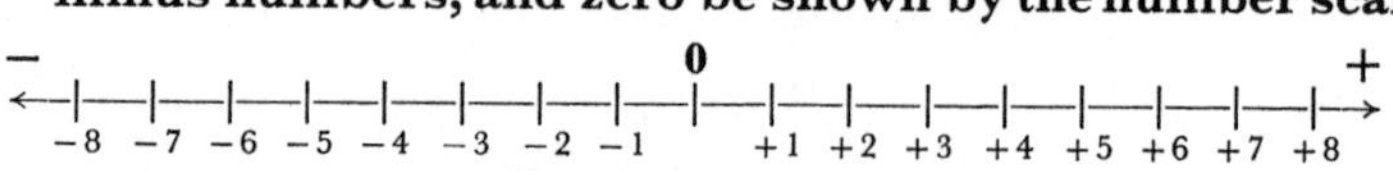

The ordinary numbers of arithmetic are positive numbers and are greater than zero. These are shown to the right of zero. Negative numbers are to the left of zero. Corresponding to $+4$ we have -4 which is as much below zero as $+4$ is greater than zero.

EXAMPLE: -6 is less than -5, or -2, or 0, or $+1$, or $+6$. Numbers increase as you go to the right and decrease as you go to the left.

625. What are the two meanings of plus and minus signs?

The plus sign [+] may direct us to add, or it may indicate the quality of the number, as a positive number.

The minus sign [−] may direct us to subtract, or it may indicate a negative number opposite in quality or sense to a positive number.

To distinguish the sign of operation from the sign of quality (positive or negative), the quality sign is enclosed in parentheses.

EXAMPLES: (*a*) $18 + (+3)$ (*b*) $18 - (+3)$
(*c*) $18 + (-3)$ (*d*) $18 - (-3)$

For the sake of brevity, (*a*) and (*b*) may be written as $18 + 3$ and $18 - 3$, since a plus sign is not necessary in front of a positive number.

626. What is the procedure for addition of positive and negative numbers?

If the numbers have the same signs, add the numbers and prefix the common (or same) sign.

If the numbers have unlike signs, find the difference and use the sign of the larger number.

EXAMPLES:

(*a*) $(+7) + (+5) = 7 + 5 = 12$ (like [+] signs) + result

(*b*) $(-7) + (-5) = 7 + 5 = -12$ (like [−] signs) − result

(*c*) $(+7) + (-5) = 7 - 5 = +2$ (unlike signs) {use sign of larger

(*d*) $(-7) + (+5) = 7 - 5 = -2$ (unlike signs) {use sign of larger

627. What is the procedure for subtraction of positive and negative numbers?

Change the sign of the number being subtracted and add as in addition (Question 626).

EXAMPLES:

(*a*) $(-5) - (-7)$. Change the sign of (-7) and add to (-5) or,
$$-5 + (+7) = 2$$

(*b*) $(-5) - (+7)$. Change the sign of $(+7)$ and add to (-5) or,
$$-5 + (-7) = -12$$

(*c*) $(+5) - (-7)$. Change the sign of (-7) and add to $(+5)$ or,
$$5 + (+7) = 12$$

(*d*) $(+5) - (+7)$. Change the sign of $(+7)$ and add to $(+5)$ or,
$$5 + (-7) = -2$$

628. What is the procedure for multiplication of positive and negative numbers?

The product is positive when the two numbers have the same sign whether both are (+) or both are (−).

The product is negative when the two numbers have opposite signs.

EXAMPLES:

(*a*) $(+12) \times (+8) = +96 = 96$ same sign
(*b*) $(-12) \times (-8) = +96 = 96$ same sign
(*c*) $(-12) \times (+8) = -96$ opposite signs
(*d*) $(+12) \times (-8) = -96$ opposite signs

629. What is the procedure for division of positive and negative numbers?

The quotient is positive when the dividend and the divisor have the same sign.

The quotient is negative when the dividend and the divisor have opposite signs.

EXAMPLES:

(*a*) $(+96) \div (+8) = +12 = 12$ same sign
(*b*) $(-96) \div (-8) = +12 = 12$ same sign
(*c*) $(+96) \div (-8) = -12$ opposite signs
(*d*) $(-96) \div (+8) = -12$ opposite signs

PROBLEMS

1. How would you represent the following:

(*a*) 20 miles east and 25 miles west?
(*b*) 200 feet above sea-level and 200 feet below?
(*c*) 15° above zero and 15° below zero?
(*d*) A gain of $25 and a loss of $25?

2. Answer the following:

(*a*) Is -12 greater or less than -8?
(*b*) Which is larger, $+3$ or -6?
(*c*) Which is larger, -50 or $+1$?

3. What is the absolute value of:

(*a*) $+12$? (*b*) -6? (*c*) $+\frac{5}{8}$? (*d*) -1.6? (*e*) 350?

4. What is the result of:

(*a*) $(+3) + (+14)$? (*b*) $(-16) - (-72)$?
(*c*) $(-203.04) - (-12.3)$? (*d*) $(-186.04) + 16$?

(*e*) $\begin{array}{r} +14 \\ -\ 8 \\ \hline \end{array}$? (*f*) $\begin{array}{r} -13 \\ -\ 5 \\ \hline \end{array}$? (*g*) $\begin{array}{r} -12 \\ +\ 7 \\ \hline \end{array}$? (*h*) $\begin{array}{r} -\ 6 \\ +11 \\ \hline \end{array}$?

5. What is the result of:

(*a*) $(-12.2) \times (-12)$?
(*b*) $(-7) \times (-9) \times (-6)$?
(*c*) $(-2\frac{1}{3}) \times (\frac{7}{10}) \times (-\frac{7}{15})$?
(*d*) $(-6) \times (-1\frac{1}{2}) \times (1\frac{1}{3})$?
(*e*) $(-14) \times (-6)$?
(*f*) $(-14) \times (+6)$?

6. What is the result of:

(*a*) $108 \div 12$?
(*b*) $(-108) \div (-12)$?
(*c*) $(-3.68) \div (-4.6)$?
(*d*) $1{,}330 \div .38$?
(*e*) $\dfrac{383.98}{-26.3}$?
(*f*) $\dfrac{-.6096}{76.2}$?

CHAPTER XVI

PROGRESSIONS — SERIES

630. What is a series?

A succession of terms so related that each may be derived from one or more of the preceding terms in accordance with some fixed rule or order.

631. What is an arithmetic progression?

A series of numbers each of which is increased or decreased by the same number in a definite order.

Ex. (*a*) 2, 4, 6, 8, 10, 12, etc.

Each number is *increased* by 2 in an ascending order.

Ex. (*b*) 24, 20, 16, 12, 8, 4, 0, −4, −8, etc.

Each number is *decreased* by 4 in a descending order.

632. What is a geometric progression?

One in which each term is divided or multiplied by the same number to get the next term. This constant multiplier or divider is called the *ratio*.

Ex. (*a*) 2, 8, 32, 128, 512, 2048, etc.

$$\frac{8}{2} = \frac{32}{8} = \frac{128}{32} = 4 = \text{ratio.}$$

Each term is *multiplied* by 4 to get the next term. This is called an ascending series or progression.

Ex. (*b*) 2048, 512, 128, 32, 8, 2.

$$\frac{2048}{512} = \frac{512}{128} = \frac{128}{32} = 4 = \text{ratio}$$

Each term is *divided* by 4 to get the next term in a descending series.

633. What is a harmonic progression?

A series of terms whose reciprocals form an arithmetic progression.

EXAMPLE: 1, $\frac{1}{3}$, $\frac{1}{5}$, $\frac{1}{7}$, $\frac{1}{9}$ is a harmonic progression because the reciprocals of the terms, 1, 3, 5, 7, 9, etc., form an arithmetic progression.

634. What is known as a miscellaneous series?

Any pattern or combination of patterns may constitute a miscellaneous series.

Ex. (*a*) 3, 5, 8, 10, 13, 15, 18, 20, 23 ...

To get the terms, 2, then 3, then 2, then 3 are added.

Ex. (*b*) 2–2, 4–4, 6–6, 8–8, etc.

The numbers are paired off in intervals. The next pair would be 10–10.

(*a*) and (*b*) are examples of miscellaneous arithmetic series.

Ex. (*c*) $2^2, 2^3, 2^4, 2^5, 2^6$ is a varied geometric series.

Ex. (*d*) $2, 2^2, 2^4, 2^8, 2^{16}$ is a varied geometric series.

In (*d*) each term is the square of the preceding term.

635. What is the procedure for solving an ascending arithmetic progression?

(*a*) Subtract the first term from the second term to get the common difference.

(*b*) Add the difference to the last term to find the term that follows.

EXAMPLE: 1, 3, 5, 7, 9, ?

$(3 - 1) = 2 =$ difference $\quad \therefore 2 + 9 = 11 =$ next term

636. What is the procedure for solving a descending arithmetic progression?

(*a*) Subtract the second term from the first term to get the common difference.

(*b*) Subtract this difference from the last term to get the term that follows.

EXAMPLE: 25, 21, 17, 13, ?

$(25 - 21) = 4 =$ difference $\quad \therefore (13 - 4) = 9 =$ next term

637. How can we obtain a general formula for solving an arithmetic progression?

Let $a =$ the first term
$d =$ the common difference
$n =$ the number of terms (given)
$l =$ the last term (to be found)

The progression can then be stated as:

First term	Second term	Third term	Fourth term	
a	$a + d$	$a + 2d$	$a + 3d$	...

Note that the coefficient or multiplier of d in any term is 1 less than the number of the term. This means that the multiplier of d for the nth or last term is $(n - 1)$.

$$\therefore\ l = \text{last term} = a + (n - 1)d$$

Ex. (*a*) To find the last term (the twenty-seventh term) of the progression 14, 11, 8, 5, 2, −1, −4 . . . to 27 terms.

Here $a = 14$, $d = 11 - 14 = -3$, and $n = 27$. Then

$$l = a + (n - 1)d = 14 + (27 - 1) \times (-3) = 14 + [26 \times (-3)]$$
$$= 14 - 78 = -64 = \text{twenty-seventh term Ans.}$$

Ex. (*b*) Find the seventeenth term of 5, 8, 11,. 14, 17. . . .

Here $d = 8 - 5 = 3$, $a = 5$, and $n = 17$. Then

$$l = a + (n - 1)d = 5 + (17 - 1) \times 3$$
$$= 5 + 16 \times 3 = 5 + 48 = 53 \text{ Ans.}$$

638. How can we find an expression for the sum of the terms of an arithmetic progression?

Let a = the first term
l = the last term
n = the number of terms
S = the sum of the terms
d = the difference between terms (common)

Then

$$S = a + (a + d) + (a + 2d) + \cdots + (l - d) + l$$

Now, writing the terms in the reverse order, we get

$$S = l + (l - d) + (l - 2d) + \cdots + (a + d) + a$$

Add these equations term by term and get

$$2S = (a + l) + (a + l) + (a + l) + \cdots + (a + l) + (a + l)$$
$$= n(a + l)$$

$$\therefore\ S = \frac{n}{2}(a + l)$$

which is the expression required. Add the first term to the last term and multiply this by the number of terms divided by 2.

Also, we have found previously that $l = a + (n - 1)d$. Thus

$$S = \frac{n}{2}[a + a + (n - 1)d] = \frac{n}{2}[2a + (n - 1)d]$$

which is another form for the expression required.

639. What is the sum of the first twenty-seven terms of 14, 11, 8, 5, 2, −1, −4...?

Here $a = 14$, $d = 14 - 11 = -3$, and $n = 27$. Then

$$S = \frac{n}{2}[2a + (n - 1)d] = \frac{27}{2}[2 \times 14 + (27 - 1) \times (-3)]$$

$$= \tfrac{27}{2}[28 - 78] = \tfrac{27}{2} \times (-50) = -657 \text{ Ans.}$$

As a check, we know from Example (*a*) of Question 637, that $l = -64$. Then

$$S = \frac{n}{2}(a + l) = \frac{27}{2}(14 - 64) = \frac{27}{2} \times (-50) = -675 \text{ Ans.}$$

640. When any three of the five elements of an arithmetic progression are given, how are the other two found?

Given any three of the elements a, d, n, l, and S, to find the remaining two elements, substitute in:

$$l = a + (n - 1)d \quad \text{and} \quad S = \frac{n}{2}(a + l)$$

Ex. (*a*) Given: $a = -\tfrac{5}{4}$, $n = 10$, and $S = -\tfrac{75}{16}$

Find: d and l

$$S = \frac{n}{2}(a + l)$$

Then

$$-\tfrac{75}{16} = \tfrac{10}{2}(-\tfrac{5}{4} + l)$$

$$-\tfrac{75}{16} = -\tfrac{25}{4} + 5l$$

$$-\tfrac{75}{16} + \tfrac{25}{4} = 5l$$

$$\tfrac{25}{16} = 5l$$

$$\therefore\ l = \tfrac{5}{16} \text{ Ans.}$$

Now

$$l = a + (n - 1)d$$

$$\tfrac{5}{16} = -\tfrac{5}{4} + (10 - 1)d$$

$$\tfrac{5}{16} + \tfrac{5}{4} = 9d$$

$$\tfrac{25}{16} = 9d$$

$$\therefore\ d = \tfrac{25}{144} \text{ Ans.}$$

Ex. (*b*) Given: $d = -4$, $l = -48$, and $S = -288$

Find: a and n

Now

$$l = a + (n - 1)d$$

$$-48 = a + (n - 1) \times (-4) = a - 4n + 4$$

and

$$a = 4n - 52 \qquad (1)$$

$$S = \frac{n}{2}(a + l)$$

$$-288 = \frac{n}{2}(4n - 52 - 48) = \frac{n}{2}(4n - 100) = 2n^2 - 50n$$

or

$$n^2 - 25n + 144 = 0$$

Factoring, we get $(n - 9)(n - 16) = 0$, and $n = 9$ or $n = 16$. Substituting in (1),

$$a = 4 \times 9 - 52 = -16 \quad \text{for } n = 9$$

$$a = 4 \times 16 - 52 = 12 \quad \text{for } n = 16$$

There are two progressions as an answer.

If $a = -16$ and $n = 9$, the progression is:

$$-16, -20, -24, -28, -32, -36, -40, -44, -48$$

If $a = 12$ and $n = 16$, the progression is:

$$12, 8, 4, 0, -4, -8, -12, -16, -20, -24, -28, -32, -36, -40, -44, -48$$

In each case the sum is -288.

641. How can we insert any number of arithmetic means between two given terms?

Use $l = a + (n - 1)d$ to find the common difference d and then form the series.

EXAMPLE: Insert five arithmetic means between 4 and -6. This means that we are to find an arithmetic progression of seven terms, with the first term of 4 and the last term of -6. Then

$$-6 = 4 + (7 - 1)d = 4 + 6d$$

$$6d = -10 \quad \text{or} \quad d = -\tfrac{10}{6} = -\tfrac{5}{3}$$

Thus the series is:

$$4, \tfrac{7}{3}, \tfrac{2}{3}, -1, -\tfrac{8}{3}, -\tfrac{13}{3}, -6$$

642. How can we show that the arithmetic mean between two quantities is equal to one half their sum?

If x = the arithmetic mean between terms a and b, then, by the nature of the progression,

$$x - a = b - x$$

or

$$2x = a + b$$

and

$$x = \frac{a + b}{2} = \text{half their sum}$$

EXAMPLE: What is the arithmetic mean between $\frac{5}{12}$ and $-\frac{3}{20}$?

$$x = \frac{\frac{5}{12} + \left(-\frac{3}{20}\right)}{2} = \frac{\frac{25}{60} - \frac{9}{60}}{2} = \frac{1}{2} \times \frac{16}{60} = \frac{2}{15} \text{ Ans.}$$

($\frac{5}{12}$, $\frac{2}{15}$, $-\frac{3}{20}$ is the series)

643. How can we find an expression for the last term l of a geometric progression when given the first term a, the ratio r, and the number of terms n?

The progression is $a, ar, ar^2, ar^3 \ldots.$

Note that the exponent of r in any term is 1 less than the number of the term. This means that in the nth term (last) the exponent of r is $(n - 1)$.

$$\therefore\ l = ar^{n-1}$$

EXAMPLE: Find the last term for the progression:

$$4, 1, \tfrac{1}{4} \ldots \text{ to 7 terms}$$

Here $a = 4$, $r = \frac{1}{4}$, and $n = 7$. Then

$$l = 4 \times \left(\frac{1}{4}\right)^{(7-1)} = 4 \times \left(\frac{1}{4}\right)^{6} = \frac{1}{4^5} = \frac{1}{1024} \text{ Ans.}$$

To get the ratio, divide the second term by the first, or any term by the next preceding term.

644. How can we find an expression for the sum S of a geometric progression when given the first term a, the last term l, and the ratio r?

$$S = a + ar + ar^2 + \cdots + ar^{n-3} + ar^{n-2} + ar^{n-1} \quad (1)$$

Now multiply each term by r, getting

$$rS = ar + ar^2 + ar^3 + \cdots + ar^{n-2} + ar^{n-1} + ar^n \quad (2)$$

Subtract (1) from (2), getting

$$rS - S = ar^n - a$$

All the other terms cancel out.

$$S(r - 1) = ar^n - a$$

Then

$$S = \frac{ar^n - a}{r - 1}$$

But

$$l = ar^{n-1} \quad \text{or} \quad rl = ar^n$$

$$\therefore\ S = \frac{rl - a}{r - 1} = \text{expression desired.}$$

EXAMPLE: Find the sum of the series of Question 643.

There, $l = \frac{1}{1024}$, $r = \frac{1}{4}$, and $a = 4$.

$$S = \frac{(\frac{1}{4} \times \frac{1}{1024} - 4)}{(\frac{1}{4} - 1)} = \frac{\frac{1}{4096} - 4}{-\frac{3}{4}} = \frac{1 - \frac{16384}{4096}}{-\frac{3}{4}}$$

$$= -\frac{\overset{5,461}{\cancel{16,383}}}{\underset{1,024}{\cancel{4,096}}} \times \left(-\frac{\cancel{4}}{\cancel{3}}\right) = \frac{5,461}{1,024}$$

645. How can we find two of the five elements of a geometric progression when any three are given?

Substitute in

$$l = ar^{n-1} \quad \text{and} \quad S = \frac{rl - a}{r - 1}$$

EXAMPLE: Given $a = -2$, $n = 5$, and $l = -32$

Find r and S

$$(l = ar^{n-1}) \quad -32 = -2r^{5-1}$$

$$2r^4 = 32 \quad \text{or} \quad r^4 = 16 \quad \text{and} \quad r = \pm 2$$

$$\left(S = \frac{rl - a}{r - 1}\right)$$

For $r = 2$ $\quad S = \frac{2(-32) - (-2)}{2 - 1} = -64 + 2 = -62$

For $r = -2$ $\quad S = \frac{(-2)(-32) - (-2)}{-2 - 1} = \frac{64 + 2}{-3} = -22$

$\therefore r = 2$ and $S = -62$ or $r = -2$ and $S = -22$ Ans.

For $r = 2$ the progression is:

$$-2, -4, -8, -16, -32 \quad (S = -62)$$

For $r = -2$ the progression is:

$$-2, 4, -8, 16, -32 \quad (S = -22)$$

646. (*a*) What do we call the limit to which the sum of the terms of a decreasing geometric progression approaches when the number of terms is indefinitely increased? (*b*) How can we find an expression for this limit?

(*a*) This limit is called the *sum of the series to infinity*.

(*b*) We have already found that

$$S = \frac{rl - a}{r - 1}$$

This can be written as:

$$S = \frac{-1 \times (rl - a)}{-1 \times (r - 1)} = \frac{a - rl}{1 - r}$$

Now, when we continue a decreasing geometric progression, the last term may be made numerically less than any assigned number, however small. Thus, when the number of terms is indefinitely increased, l, and therefore, rl, approaches the limit 0.

Then, the fraction $\frac{a - rl}{1 - r}$ approaches the limit $\frac{a}{1 - r}$

$$\therefore\ S = \frac{a}{1 - r} = \text{the sum to infinity}$$

EXAMPLE: Find the sum of the series 3, $-\frac{9}{4}$, $\frac{27}{16}\ldots$ to infinity.

Here $a = 3$ and $r = \frac{-\frac{9}{4}}{3} = -\frac{3}{4}$.

$$\therefore\ S = \frac{3}{1 + \frac{3}{4}} = \frac{3}{\frac{7}{4}} = \frac{12}{7} = \text{sum to infinity. Ans.}$$

647. How can we find the value of a repeating decimal by the use of the sum of a series to infinity?

EXAMPLE: Find the value of $.7\dot{6}3\dot{6}3\ldots$.

Now

$$.7\dot{6}3\dot{6}3\cdots = .7 + .063 + .00063 + \cdots$$

The terms after the first constitute a decreasing geometric progression in which:

$$a = .063 \quad \text{and} \quad r = \frac{.00063}{.0063} = \frac{1}{100} = .01$$

Then

$$S = \frac{a}{1 - r} = \frac{.063}{1 - .01} = \frac{.063}{.99} = \frac{63}{990} = \frac{21}{330}$$

The value of the given decimal is:

$$\tfrac{7}{10} + \tfrac{21}{100} = \tfrac{252}{330} = \tfrac{126}{165}\ \text{Ans.}$$

648. What is the procedure for inserting any number of geometric means between two given terms?

Use $l = ar^{n-1}$.

EXAMPLE: Insert four geometric means between 3 and 729.

This means that we must find a geometric progression of six terms with 3 as a first term and 729 as a last term.

Here $a = 3$ and $l = 729 = ar^{n-1}$.

$$729 = 3r^{(6-1)} = 3r^5 \qquad r^5 = 243 \qquad r = 3$$

$\therefore$ 3, 9, 27, 81, 243, 729 is the progression.

649. How can we show that the geometric mean between two quantities is equal to the square root of their product?

Put x between a and b terms as $a, x, b, \ldots$. Then, by the nature of the progression:

$$\frac{x}{a} = \frac{b}{x} \quad \text{or} \quad x^2 = ab \quad \text{(by cross multiplication)}$$

Hence

$$x = \sqrt{ab}$$

EXAMPLE: Find the geometric mean between $1\frac{13}{15}$ and $2\frac{11}{12}$.

$$x = \sqrt{\frac{\overset{7}{\cancel{28}}}{\underset{3}{\cancel{15}}} \times \frac{\overset{7}{\cancel{35}}}{\underset{3}{\cancel{12}}}} = \sqrt{\frac{7^2}{3^2}} = \frac{7}{3} \text{ Ans.}$$

650. What is the procedure for solving a harmonic progression?

Take the reciprocals of the terms and apply the procedures and formulae of arithmetic progression.

There is, however, no general method for finding the sum of the terms of a harmonic progression.

EXAMPLE: Find the last term of the progression $3, \frac{3}{4}, \frac{3}{7} \ldots$ to twelve terms.

Take the reciprocals to get an arithmetic progression $\frac{1}{3}, \frac{4}{3}, \frac{7}{3} \ldots$.

Here $a = \frac{1}{3}$, $d = 1$, and $n = 12$. Now

$$l = a + (n - 1)d = \tfrac{1}{3} + (12 - 1) \times 1 = 11\tfrac{1}{3} = \tfrac{34}{3}$$

Take reciprocal of $\frac{34}{3}$ to get $\frac{3}{34}$ = last term of the given harmonic progression.

651. How can we insert six harmonic means between 2 and $-\frac{10}{9}$?

This means we have to insert six arithmetic means between $\frac{1}{2}$ and $-\frac{9}{10}$.

Here $a = \frac{1}{2}$, $l = -\frac{9}{10}$, and $n = 8$. Then

$$l = a + (n - 1)d$$

$$-\tfrac{9}{10} = \tfrac{1}{2} + (8 - 1)d = \tfrac{1}{2} + 7d$$

$$-\tfrac{14}{10} = 7d \quad \text{and} \quad d = -\tfrac{2}{10} = -\tfrac{1}{5}$$

Then the arithmetic progression is:

$$\tfrac{1}{2}, \tfrac{3}{10}, \tfrac{1}{10}, -\tfrac{1}{10}, -\tfrac{3}{10}, -\tfrac{5}{10}, -\tfrac{7}{10}, -\tfrac{9}{10}$$

The required harmonic progression is:

$$2, \tfrac{10}{3}, 10, -10, -\tfrac{10}{3}, -2, -\tfrac{10}{7}, -\tfrac{10}{9}$$

652. How can we find an expression for the harmonic mean between two terms?

Let x = harmonic mean between terms a and b. Then

$$\frac{1}{x} = \text{arithmetic mean between } \frac{1}{a} \text{ and } \frac{1}{b}$$

and

$$\frac{1}{x} = \frac{\frac{1}{a} + \frac{1}{b}}{2} = \frac{b + \frac{a}{ab}}{2} = \frac{a + b}{2ab}$$

$$x = \frac{2ab}{a + b} = \text{harmonic mean}$$

EXAMPLE: What is the harmonic mean between 3 and 6?

$$x = \frac{2 \times 3 \times 6}{3 + 6} = \frac{36}{9} = 4 \text{ Ans.}$$

The harmonic series is then

$$3, 4, 6 \ldots$$

The arithmetic series is:

$$\tfrac{1}{3}, \tfrac{1}{4}, \tfrac{1}{6} \ldots \quad \text{or} \quad \tfrac{4}{12}, \tfrac{3}{12}, \tfrac{2}{12} \ldots$$

653. How is the sum of an arithmetic series applied in certain installment purchase problems?

EXAMPLE: A hutch cabinet is advertised for \$1,000 cash or on time payments of \$20 per week, plus $\frac{15}{100}$ of 1% on each weekly unpaid balance for 50 weeks. What would be the total amount paid on the weekly installment basis?

The first unpaid balance is \$1,000 and .0015 × \$1,000 = \$1.50 (.15 × .01 = .0015)

The second unpaid balance is \$980 and .0015 × \$980 = \$1.47

The third unpaid balance is \$960 and .0015 × \$960 = \$1.44

Thus, the series of the carrying charges becomes \$1.50, \$1,47, \$1.44. . . .

Here, $a = \$1.50$ = first term, $d = \$1.50 - \$1.47 = -\$.03$ = common difference, and $n = 50$ = number of terms.

$$S = \frac{n}{2}[2a + (n - 1)d] = \frac{50}{2}[2 \times 1.50 + (50 - 1) \times (-.03)]$$

$$= 25[3 - 1.47]$$
$$= 25 \times 1.53 = \$38.25 = \text{sum of carrying charges.}$$

Then

$$\$1{,}000 + \$38.25 = \$1{,}038.25 = \text{total amount paid.}$$

PROBLEMS

1. What term comes next in:

(*a*) 6, 9, 12, 15, 18, 21 . . . ?
(*b*) 10, 1, 8, 1, 6, 1 . . . ?
(*c*) 4, 5, 8, 9, 12, 13 . . . ?
(*d*) 11, 11, 9, 9, 7, 7 . . . ?
(*e*) 81, 27, 9, 3, 1, $\frac{1}{3}$ . . . ?
(*f*) 2, 8, 18, 32, 50, 72 . . . ?
(*g*) 1, 2, 4, 8, 16, 32 . . . ?
(*h*) 40, 34, 30, 28, 22, 18 . . . ?

2. Find the last term and the sum of the terms of:

(*a*) $-\frac{6}{5}, -\frac{3}{2}, -\frac{9}{5}$ . . . to 12 terms
(*b*) 3, 9, 15 . . . to 8 terms
(*c*) $-\frac{1}{3}, \frac{1}{2}, \frac{4}{3}$ . . . to 14 terms
(*d*) $-7, -12, -17$ . . . to 10 terms
(*e*) $\frac{3}{4}, -\frac{3}{8}, -\frac{3}{2}$ . . . to 14 terms

3. Given $d = 4$, $l = 71$, and $n = 15$, find a and S.

4. Given $a = -7$, $n = 12$, and $l = 56$, find d and S.

5. Insert six arithmetic means between 3 and 8.

6. Insert five arithmetic means between $-\frac{4}{3}$ and 1.

7. Find the arithmetic mean between $\frac{5}{7}$ and $-\frac{3}{8}$.

8. Find the sum of all the integers beginning with 1 and ending with 100.

9. Find the sum of all the even integers beginning with 2 and ending with 1,000.

10. Find the last term and the sum of the terms of the progression 3, 1, $\frac{1}{3}$ . . . to seven terms.

11. Find the last term and the sum of the terms of the progression $-2, 6, -18$ . . . to ten terms.

12. Given $a = -3$, $n = 4$, and $l = -45$, find r and S.

13. Find the sum of the series 4, $-\frac{8}{3}, \frac{16}{9}$ . . . to infinity.

14. Find the sum to infinity of 16, -4, 1. . . .

15. Find the value of the repeating decimal .85151.

16. Find the value of .296296.

17. Insert five geometric means between 2 and $\frac{128}{729}$.

18. Insert five geometric means between 2 and 128.

19. Find the geometric mean between 9 and 25.

20. Find the last term of the progression 2, $\frac{2}{3}$, $\frac{2}{5}$... to twenty terms.

21. Insert five harmonic means between 2 and -3.

22. What is the harmonic mean between 4 and 8?

23. A TV set is sold for \$675 cash, or for \$150 cash and \$52.50 a month, plus 1% of each monthly unpaid balance for 10 months. What would it cost to buy it on the time payment basis?

CHAPTER XVII

GRAPHS — CHARTS

654. What are graphs?

When you have statistical facts in science, sociology, business, economics, or any other relationships, you can present them graphically to advantage in a variety of forms. The pictorial relationships that are thus shown in true proportions are called graphs. They may represent the relation between two units of measure, as quantity with time, or cost with quantity, parts with reference to the whole and with reference to each other, etc.

655. What are the advantages of graphs?

(*a*) At a glance they may show information that would usually require much verbal description.

(*b*) They may stimulate the mind in a more direct, descriptive, and dramatic manner than statistics expressed in numbers.

(*c*) They may enable us to understand the facts better and help us to learn new facts more easily.

(*d*) They may save us time and work in making computations, and enable us to draw conclusions in a comparative way.

For example, in aeronautics, they may show experimental or test data, and calibration of instruments. In business, they may show changes of cost with time. In sociology, they may show growth of population with time, use of water resources with population, etc.

656. What are the disadvantages of graphs?

(*a*) They are necessarily less accurate than the figures on which they are based. However, in many cases this is of no great importance.

(*b*) They can sometimes mislead us into wrong interpretations when we are not careful. We must thus examine closely the reliability of the source and the method of presentation.

657. What questions should we ask about graphs?

(*a*) What idea is the graph trying to convey?

(*b*) What quantities are being compared—time, money, people, speed, etc.?

(*c*) What measurements are used—feet, dollars, per cent, years, weights?

(*d*) Exactly how much information does the graph supply? Where does our interpretation begin?

(*e*) Is the information reliable? How were the data obtained? Is the graph planted or honestly presented by a reliable organization?

658. What types of graphs are commonly used?

(*a*) Bar graphs (horizontal and vertical).
(*b*) Block graphs.
(*c*) Rectangle graphs (divided-bar charts).
(*d*) Circle or pie graphs.
(*e*) Broken-line graphs.
(*f*) Curved graphs (smooth-line or curve).
(*g*) Frequency distribution graphs (staircase diagrams).
(*h*) Statistical maps.
(*i*) Pictographs.

659. What are horizontal bar graphs and when are they used?

They are graphs that show a comparison of data.

They are used when the data are comparable but separate (discrete), as when you compare heights (same aspect) of different people (separate data).

They may be used to compare amounts of different kinds of things or of the same thing at two or more different times or places.

They may show the production or consumption of an item for several periods, or the amounts of several items during a single period. They are simple and convenient.

660. How is a bar graph constructed?

It is constructed on ordinary graph paper. The graph has a title, description of each bar, a horizontal scale, and, when necessary, a vertical scale. Bars are made of the same width and are placed equally far apart.

Ex. (*a*) Draw a horizontal bar graph to show the comparative sales of a business concern for the months of January and February when the sales for January were \$208,600 and for February \$276,500.

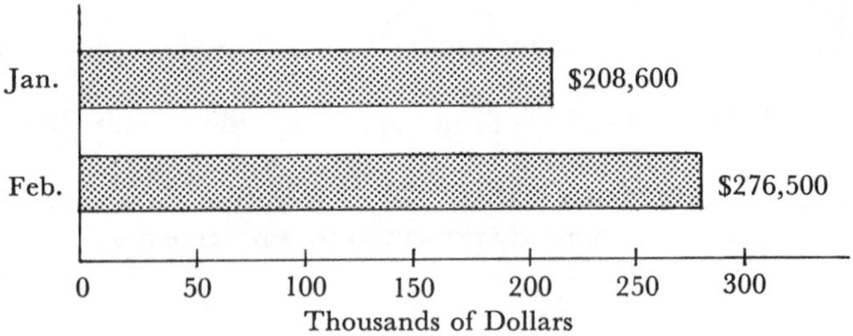

Ex. (*b*) Show with a bar graph the range of incomes of the employees of a certain company when the statistics are as follows:

Income range	*Number of employees*
\$4,000–\$4,999	12,400
\$5,000–\$5,999	10,200
\$6,000–\$6,999	8,100
\$7,000–\$7,999	3,040
\$8,000–\$8,999	2,200
\$9,000–\$9,999	1,160
\$10,000 and over	208

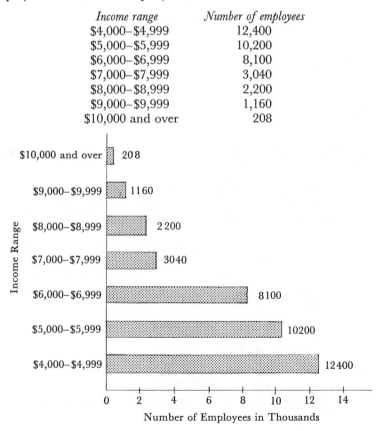

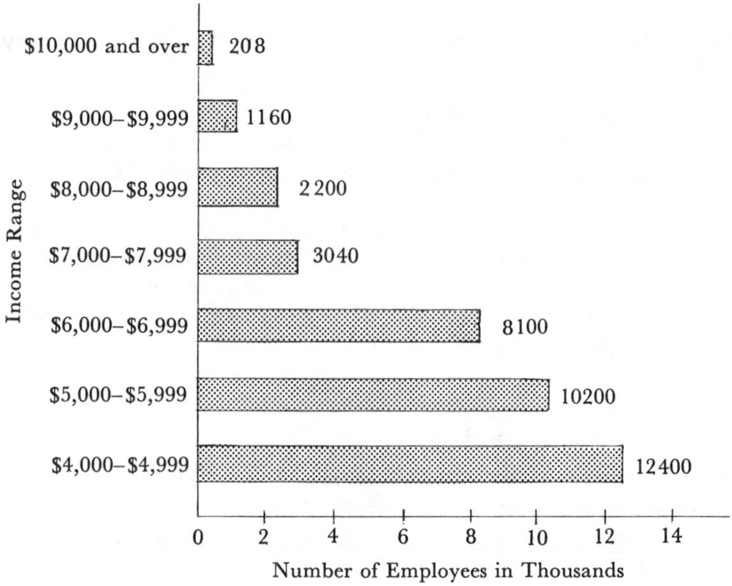

Ex. (*c*) Show with a bar graph the stopping or braking distance of a car in relation to speed of vehicle traveling on a hard, dry surface. Distance is measured from the instant the brakes are applied.

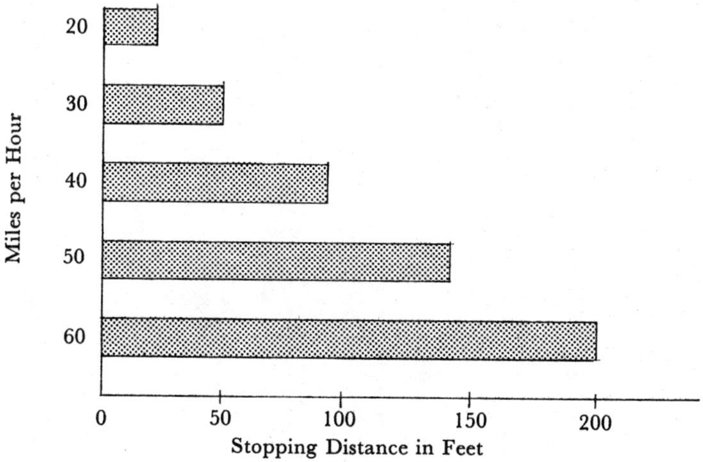

661. What are vertical bar graphs and when are they used?

When bars are drawn from bottom to top, the drawing is a vertical bar graph. The spacing between consecutive bars should be uniform and should be arranged in order of size or according to sequence of time.

Vertical bar graphs are commonly used to represent quantities or amounts at various times, and are then known as historical bar graphs. The horizontal scale is always used to represent the time and the vertical scale to represent quantities or amounts at various times. The heights of any two adjacent bars compare the increase or decrease from one time to another.

EXAMPLE: Show with a vertical bar graph the comparison of a firm's sales for 7 years, when the statistics are:

Year	*Sales*
1954	$38,260,000
1955	$47,840,000
1956	$43,190,000
1957	$45,000,000
1958	$39,080,000
1959	$47,040,000
1960	$51,000,000

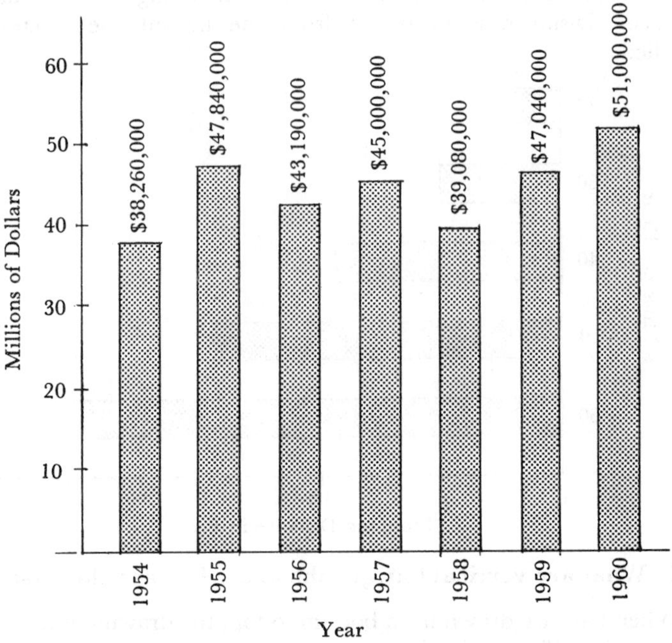

662. What types of charts or graphs are used to show the relation of the parts to the whole of an item and which type is preferred?

(*a*) The 100% bar chart.
(*b*) The divided bar chart (or rectangle graph).
(*c*) The circle graph or pie chart.

These are usually expressed in terms of per cents but not necessarily so. It is often desirable that both the actual figures and the per cents be stated directly on the chart or graph.

EXAMPLE: Statistics show that out of 100 accidents, 65 are due to falls, 25 due to burns, bruises, and blows, and 10 due to all other causes. Show this information with a 100% bar chart, divided bar chart, and circle graph.

The divided bar chart is to be preferred. Mental comparison of sectors having different central angles is not so simple to make, view, and interpret.

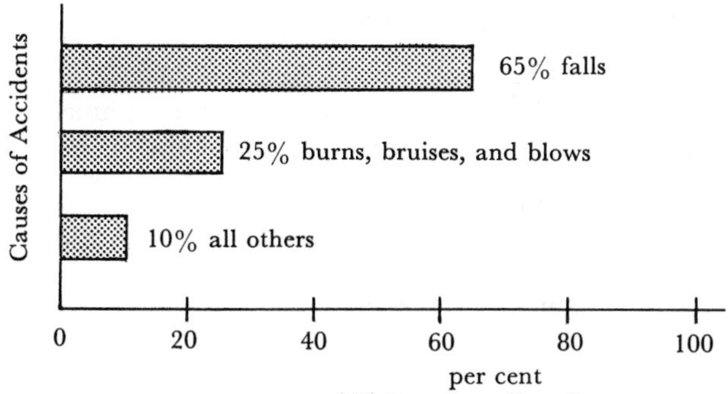

100 PER CENT BAR CHART

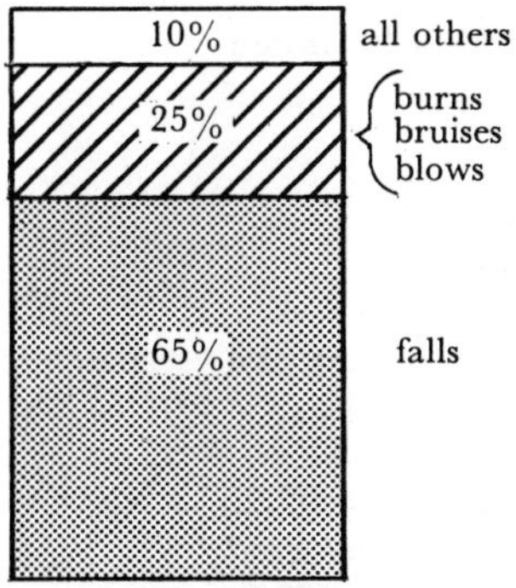

DIVIDED BAR CHART
(OR RECTANGLE GRAPH)

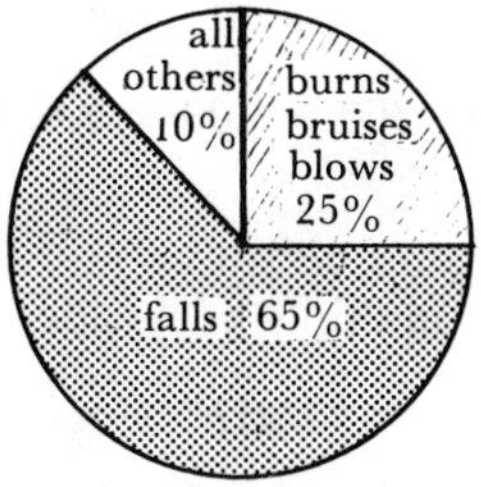

CIRCLE GRAPH

$$\begin{cases} .65 \times 360° = 234° \\ .25 \times 360° = 90° \\ .1 \times 360° = 36° \end{cases}$$

663. When is a circle graph or pie chart used, and how is it drawn?

It is used to show the relation of parts to the whole of something. It is used frequently in newspapers and magazines. You get the decimal fraction that each part represents with respect to the whole, and you multiply each fraction by 360° to get the central angle. With a protractor you lay out the central angle so found.

EXAMPLE: In a certain school the enrollment is as follows:

Freshmen	=	520
Sophomores	=	410
Juniors	=	380
Seniors	=	290
Total enrollment	=	1,600

Express the relationship between the enrollment in each class and the total enrollment, using a circle graph.

$$\text{Freshmen} = \frac{520}{1{,}600} = \frac{n^\circ}{360^\circ} \quad \text{or} \quad 1{,}600\,n = 520 \times 360 = 187{,}200$$
$$n = 117^\circ \text{ for freshmen}$$

If the enrollment is given or figured in per cents,

Freshmen	=	32.5%
Sophomores	=	25.625%
Juniors	=	23.75%
Seniors	=	18.125%
		100.000%

Then

$$\frac{32.5}{100} = \frac{n^\circ}{360^\circ} \quad \text{or} \quad 100n = 32.5 \times 360 = 11{,}700$$
$$n = 117^\circ = \text{freshmen}$$

But as $360 = 100 \times 3.6$, we have

$$\frac{32.5}{100} = \frac{n}{100 \times 3.6}$$

Now multiply each side by 100 and get

$$32.5 = \frac{n}{3.6} \quad \text{or} \quad 32.5 \times 3.6 = n = 117^\circ$$

So in each case multiply the % by 3.6.

$$\therefore \text{ Sophomores} = 25.625 \times 3.6 = 92.25^\circ$$
$$\text{Juniors} = 23.75 \times 3.6 = 85.5^\circ$$
$$\text{Seniors} = 18.125 \times 3.6 = 62.25^\circ$$

Draw the circle chart, using a protractor to lay off each angle in degrees.

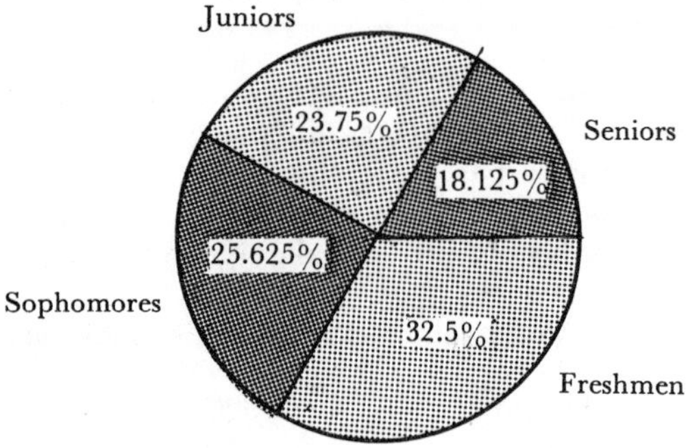

664. How is the same information shown in the form of a long bar chart?

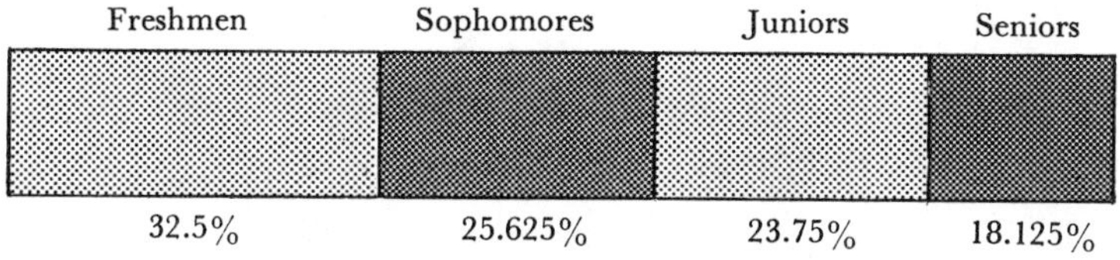

Divide the 100% length into the fractional parts represented by the %. This chart may be preferred to the circle chart for easier comparisons of lengths rather than the less easily comprehended sectors of a circle.

665. What is a block graph?

It is a rectangular block whose length indicates the quantity to be compared.

EXAMPLE: Compare, using a block graph, a school budget for the year 1950 of $286,000,000 with that for 1960 of $465,000,000.

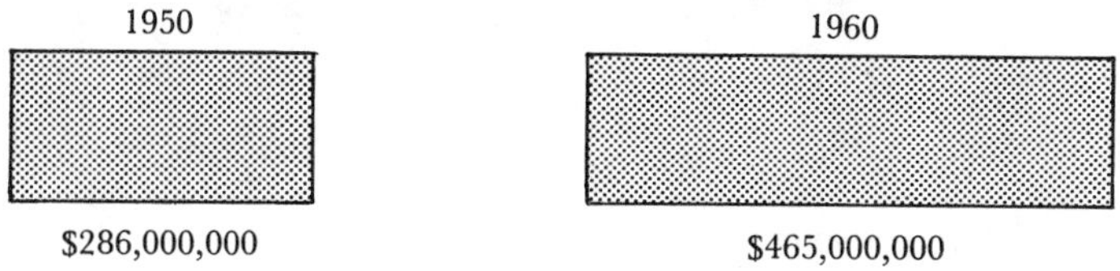

666. What is a broken-line graph, or line diagram, and when is it used?

When you select suitable scales, plot points in accordance with the given data, and join the points by straight line segments, you get a broken-line graph or line diagram.

The values between plotted points may or may not have significance, depending upon the nature of the quantities represented, and the implication is that successive values change uniformly and continuously. For example, on a graph of average monthly bank balances the in-between values have no meaning.

A line diagram is used when there is a long series of relatively continuous items. It is especially adapted to represent a time series.

Ex. (*a*) Show with a line graph the probable millions of dollars in auto sales for each month of 1960 in the United States.

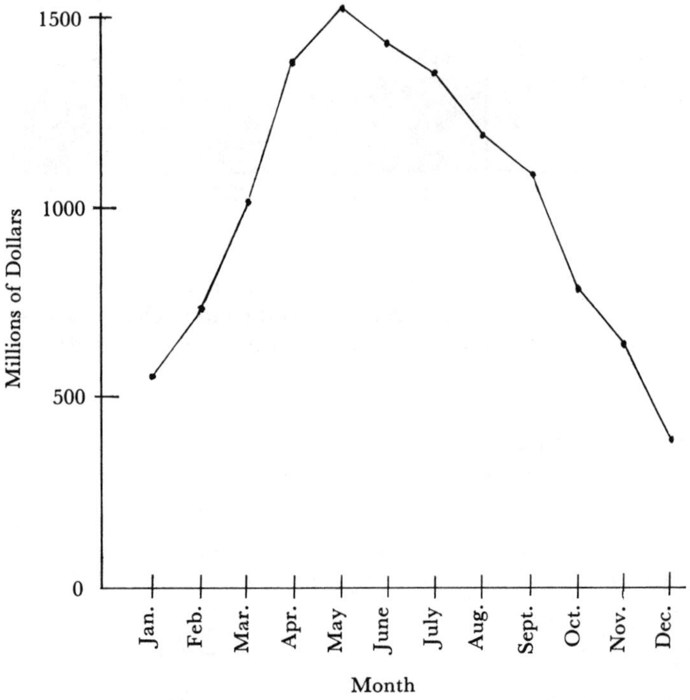

Ex. (*b*) Show a fever chart as a line diagram.

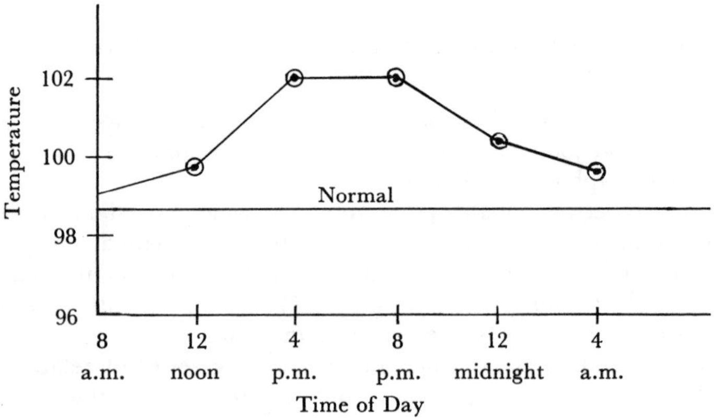

Here, rate of change is indicated because between the times the temperature is taken the patient's temperature is slowly going up or down. When the line is level or nearly so, the change is slow, and when the line goes up or down steeply, the change is rapid.

Line charts are useful in showing rate of change even with non-continuous data.

667. What is a curved graph (smooth-line graph) and when is it used?

It is very similar to a broken-line graph. When the "in-between" values vary continuously and uniformly (or nearly so) from one observed or measured value to the next, a smooth-curve line is drawn between the points either free hand or with a French curve.

Two or more graphs may be shown, one under the other, and these are known as comparative curve graphs. In comparing the relative amounts of collections and sales during each month of a year in a business, the upper curve may be sales and the lower collections.

EXAMPLE: Show with a curve graph the monthly normal temperatures in New York City.

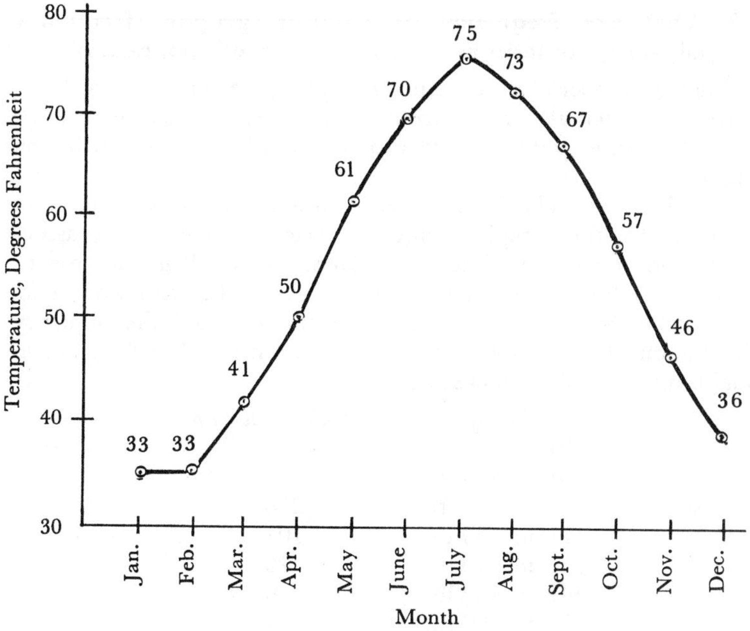

668. What are pictographs and when are they used?

They are graphs that use images or pictures to represent numbers. They portray kinds and quantities of things at a glance with a minimum of explanation. They are not commonly used, except for large distribution.

Example: Show with a pictograph the comparative apple production in the commercial counties of the United States for the years 1930 and 1956.

1930 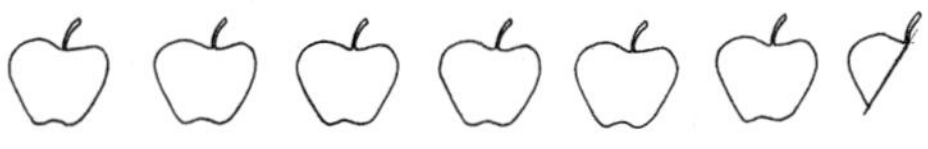 130,398,000 bushels

20 million bushels

1956  97,077,000 bushels

20 million bushels

669. What are frequency distribution graphs (frequency polygons, sometimes called "staircase" diagrams)?

When a number of measurements or phenomena are grouped into convenient intervals, the distribution of these frequencies can be shown by a time graph or histograph called a frequency distribution graph.

This shows at a glance the range of measurements (weights) most predominant, the complete range between the extreme measurements, the prevalence of extremely large and small measurements, symmetrical distribution on either side of a central tendency or mode.

Example: Show with a frequency distribution graph the frequency distribution of the weights of a class of women 5 feet 4 inches in height and 21 to 25 years of age.

Weight	*Number of women*
91 to 100 lb.	12
101 to 110 lb.	124
111 to 120 lb.	268
121 to 130 lb.	107
131 to 140 lb.	26
141 to 150 lb.	8
151 to 160 lb.	4

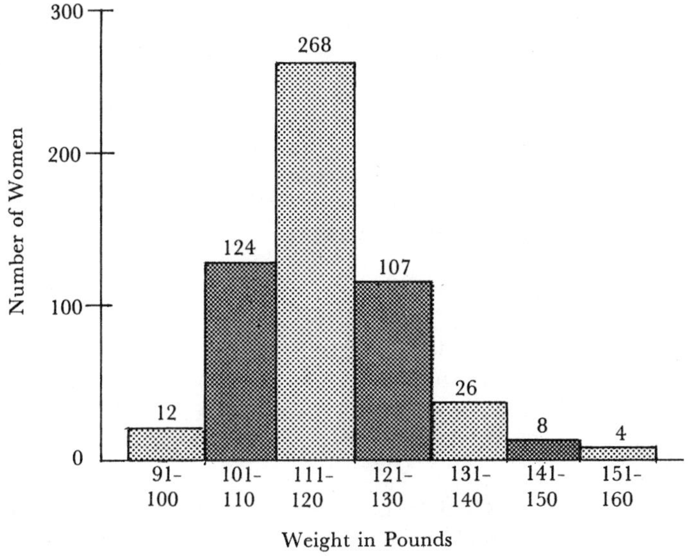

670. What is meant by an index number and how is it obtained?

An index number is a calculated or an assumed number used as a base for comparison with other values.

Instead of comparing the actual cost of living of a typical family for each year over a number of years, we can average the income for the period, and use the average figure as an index. The average figure (or index) is then considered to be 100%, and the figure for each single year can be expressed as a per cent of that index.

EXAMPLE: If the average cost of living for a family for the years 1955 to 1960 is \$6,000—which we call the index—and if we find that the cost is \$8,000 for 1961, then

$$\frac{\$8,000}{\$6,000} = \frac{n}{100} \quad \text{or} \quad n = \frac{800,000}{6,000} = 133\tfrac{1}{3}\%$$

This means that the cost of living in 1961 is $133\frac{1}{3}\%$ of the average for 1955–1960.

$133\frac{1}{3}\%$ = an index figure based on the 1955–1960 figure as the index.

671. What are the advantages of index numbers?

Changes are shown more vividly with index numbers.

To discover a trend, it is much easier to compare numbers in terms of 100% than to compare the numbers themselves.

EXAMPLE: 52% as compared with 100% is easier to understand than 346 compared with 665.

Using index numbers we can more readily compare present conditions with conditions in the past or with a more normal period.

We can use either a single year or an average of a period of years as an index.

672. What is meant by interpolation?

Interpolation is the reading between two points or values on a graph, of a missing point that is desired.

Ex. (*a*) If one book costs $3.25 and four books cost $13.00, it is reasonable to interpolate that two books cost $6.50 and seven books cost $22.75.

Ex. (*b*) If in 1958, 32,860 people were injured by falls from step ladders, and in 1960, 38,400 people were so injured, are we justified in saying that in 1959, the year in between, the number of people so injured must be 35,630, midway between 32,860 and 38,400? No, we cannot say so.

673. What is meant by extrapolation?

To extrapolate is to draw a conclusion (to predict) that a process will go on in the same direction as it seemed to be going when the data gave out and the graph ended.

EXAMPLE: If the number of juvenile crimes in 1958 were given as 282,346, and as 341,692 in 1959, we cannot extrapolate (predict) the figure for 1960. Too many factors may enter to change the picture.

674. When are interpolation and extrapolation advisable?

Only when the data are moving according to a predictable path or mathematical law.

675. When would we handle data in per cent form?

When the data are too large, per cents bring them down to a smaller, more comparable basis.

676. What is easier to compare, two areas or the lengths of two lines?

The length of two lines.

677. When and how are statistical maps used?

They are used to show geographic distribution. They combine figures with geographical areas. Sometimes various colors, shadings, or cross-hatching are used to indicate data.

EXAMPLE: To show graphically the distribution of telephones in the states of the United States, tabulate the phones for each state and choose a scale in which one dot represents a certain number of phones. The number of phones in any state is then indicated by the density of the dots in that state.

678. What is meant by Cartesian coordinates?

A system of coordinates, in a plane, that defines the position of a point, with reference to two mutually perpendicular lines called the axes of coordinates.

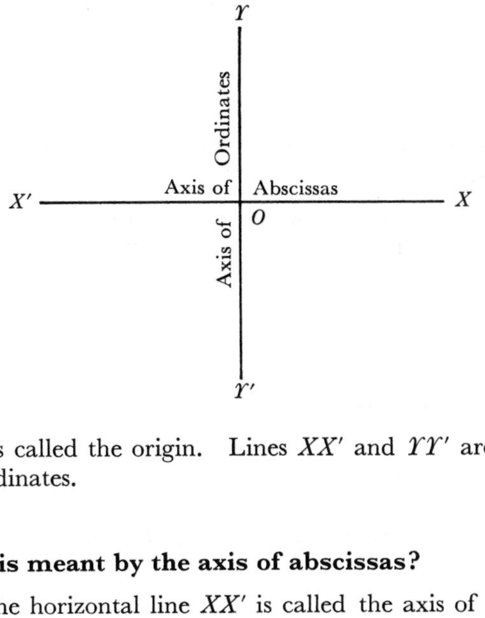

Point O is called the origin. Lines XX' and YY' are called the axes of coordinates.

679. What is meant by the axis of abscissas?

Usually the horizontal line XX' is called the axis of abscissas, or x axis.

680. What is meant by the axis of ordinates?

The line perpendicular to the x axis is called the axis of ordinates, or the y axis. YY' is the axis of ordinates.

681. In what order are the four quadrants formed by the axes of coordinates designated?

The four quadrants that are formed by the axes of coordinates are numbered from right to left, or counterclockwise, as shown in the figure.

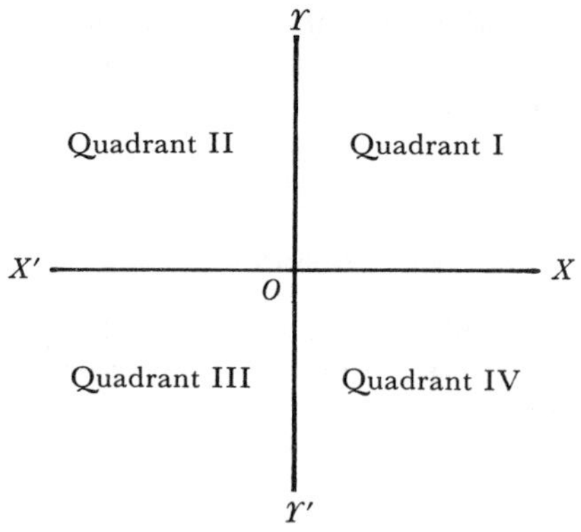

682. What directions are considered positive and what directions negative?

Distances measured to the right of the y axis are positive (+).
Distances measured to the left of the y axis are negative (−).
Distances measured above the x axis are positive (+).
Distances measured below the x axis are negative (−).

683. How are points located in Cartesian coordinates?

Each point is located by both its abscissa and ordinate. The abscissa is given first.

EXAMPLE: The coordinates of point P_1 are abscissa $x = 2$ and ordinate $y = 6$.

Point P_2 coordinates are $(-4, 5)$.
Point P_3 coordinates are $(-5, -4)$.
Point P_4 coordinates are $(7, -3)$.
These show a point in each quadrant.

Note that in each case the abscissa and the ordinate are taken from the axis to the point P.

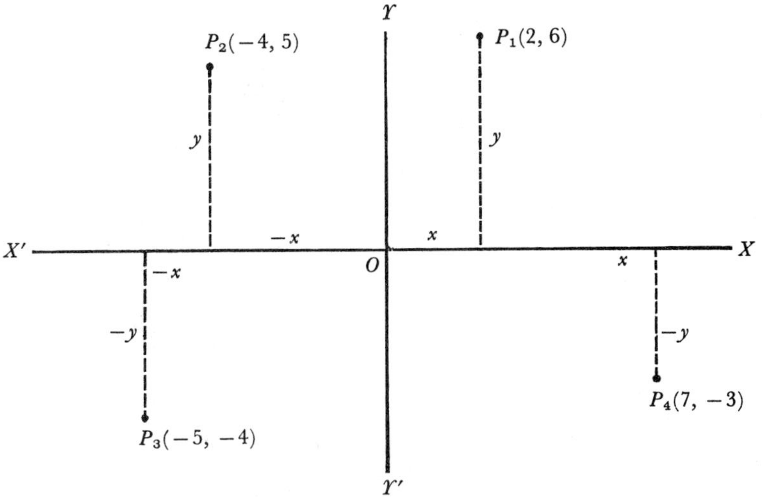

684. How do we plot a straight line relationship?

Whenever two quantities are directly proportional, the graph of their relationship is a straight line.

EXAMPLES: 1 cubic foot of water weighs	62.5 lb.
2 cubic feet of water weigh	125 lb.
4 cubic feet of water weigh	250 lb.
6 cubic feet of water weigh	375 lb.
10 cubic feet of water weigh	625 lb.

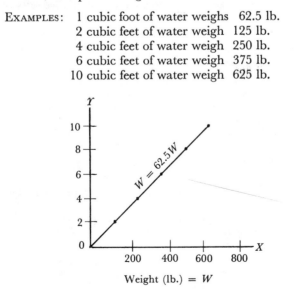

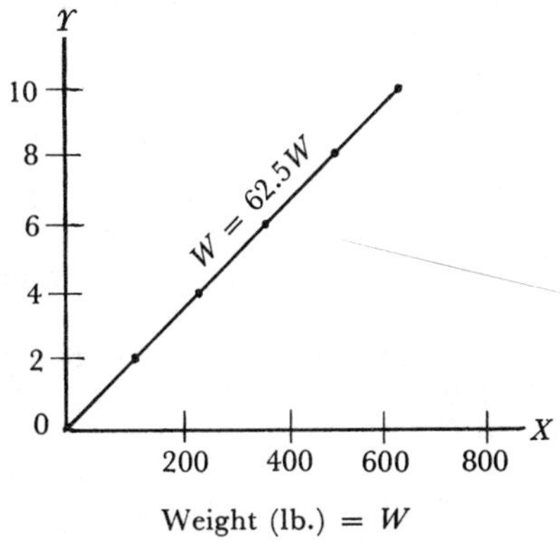

685. How do we plot the graph of a quadratic formula?

It is a curved line graph.

$S = 16t^2$ = a quadratic formula (parabola)
s = distance in feet (a body falls)
t = time in seconds (time of fall)

t	S
0	0
1	16
$1\frac{1}{2}$	36
2	64
$2\frac{1}{2}$	100
3	144
$3\frac{1}{2}$	196
4	256

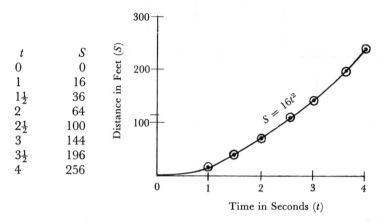

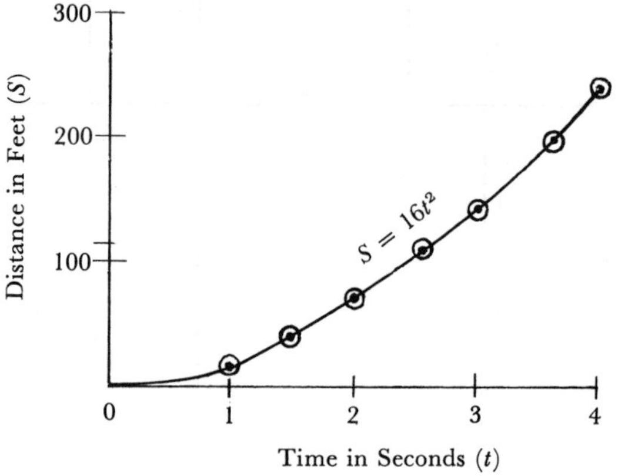

PROBLEMS

1. Draw a horizontal bar graph to show the comparative sales of an auto agency for the months of January and May, when the sales for January were $396,000 and for May $874,000.

2. Show with a horizontal bar graph the income of the employees of a firm, when the statistics are:

Income	Number of employees
$4,000–$4,999	8,400
$5,000–$5,999	3,200
$6,000–$6,999	2,100
$7,000–$7,999	1,800
$8,000–$8,999	760
$9,000–$9,999	139
$10,000 and over	68

3. Show with a vertical bar graph the comparison of income for the years 1950 to 1960, when the statistics are:

1950—$54,000,000	1956—$46,000,000
1951—$52,000,000	1957—$45,000,000
1952—$51,000,000	1958—$39,000,000
1953—$47,000,000	1959—$47,000,000
1954—$37,000,000	1960—$52,000,000
1955—$48,000,000	

4. Show with a 100% bar chart, divided bar chart, and circle graph where each dollar went in the following:

Materials and services purchased	\$620,000,000	53.40%
Wages and salaries	\$421,350,000	36.25%
Pensions, social security taxes, insurance, etc.	\$26,500,000	2.28%
Depreciation and patent amortization	\$21,100,000	1.82%
Interest on long term debt	\$6,200,000	.53%
Taxes on income and property	\$35,400,000	3.05%
Preferred and common stock dividend	\$18,300,000	1.57%
Reinvestment in the business	\$12,800,000	1.10%
Total =	\$1,161,650,000	100.00%

5. Express the relationship between the enrollment in each class, and the total enrollment, using a circle graph:

Freshmen	650
Sophomores	530
Juniors	480
Seniors	390

6. Show the information of (5) in the form of a long bar chart.

7. Compare, using a block graph, the budget of a town for the year 1959 of \$135,500, with that for 1960 of \$194,000.

8. Show with a line graph the average construction cost per new dwelling unit of one-family structures for the years 1950 to 1956.

Year	Cost
1950	\$8,675
1951	\$9,300
1952	\$9,475
1953	\$9,950
1954	\$10,625
1955	\$11,350
1956	\$12,225

9. Show a fever chart as a line diagram.

8 a.m.	99°F.	12 noon	99.8°F.	4 p.m.	101.8°F.
8 p.m.	102.7°F.	12 midnight	100.1°F.	4 a.m.	100°F.

10. Show with a curve graph the length of day for New York City, for each month.

Length of day at New York City for the first of each month given as:

January	9.2 hr.	February	10.6 hr.	March	11.2 hr.
April	12.2 hr.	May	13.6 hr.	June	14.6 hr.
July	15.4 hr.	August	14.2 hr.	September	13.6 hr.
October	11.4 hr.	November	10.2 hr.	December	9.4 hr.

11. Show with a pictograph the comparative peach production in the United States for 1955 and 1956.

1955—51,852 thousand bushels
1956—68,973 thousand bushels

12. Show with a frequency distribution graph the distribution of the heights of a class of men weighing 140 lb., and 20 to 24 years old.

Height	Number of men	Height	Number of men
5′ 4″	88	5′ 8″	438
5′ 5″	205	5′ 9″	94
5′ 6″	361	5′ 10″ and over	51
5′ 7″	782		

13. If the retail price index of dairy products in the United States for 1947–1949 is 100%, and the index figure for 1956 is 108.7%, what would be the cost of a quart of milk in 1956 if the cost in 1947 was 20¢?

14. How can we more readily compare 285 with 679?

15. If one gallon of paint cost $8.75 and four gallons cost $33, how much will seven gallons cost?

16. Locate the points (4, 3), (−2, 8), (−7, −3), (4, −8), (0, 4), and (−4, 0) in Cartesian coordinates.

17. Plot the relationship $P = 62.5h$ where P = pressure in lb. per sq. ft., and h = height in feet.

18. Plot the relationship $v = \sqrt{2gh}$ (the velocity acquired by a body falling a distance h feet through space), where $g = 32.2$ = constant.

CHAPTER XVIII

BUSINESS — FINANCE

686. What are the two types of cost?

(*a*) Net or prime cost = cost of goods alone.

(*b*) Gross cost = net cost + buying expenses, as handling or freight, storage, carrying charges, insurance, commissions, and additional charges connected with the cost of delivered goods.

687. Into what two groups is profit divided?

(*a*) Gross profit (margin of profit) = selling price − gross cost.

(*b*) Net profit = gross profit − total cost of doing business.

688. What constitutes cost of doing business?

Cost of doing business (overhead or operating expenses) includes advertising, taxes, selling expenses, employees' salaries, light, heat, delivery expenses, depreciation, and other expenses except those that constitute the gross cost of goods.

689. What is meant by (*a*) gross sales, (*b*) net sales, (*c*) gross purchases, (*d*) return purchases, (*e*) net purchases, (*f*) depreciation?

(*a*) Gross sales = total of sales over a period of time at invoice prices.

(*b*) Net sales = amount of sales after deducting returns and allowances.

(*c*) Gross purchases = total amount of goods bought for trading purposes.

(*d*) Return purchases = total amount of goods sent back to firms.

(*e*) Net purchases = gross purchases − return purchases.

(*f*) Depreciation = decrease in value of property because of use or changes resulting in disuse, recorded as a certain per cent of the cost value of the property, usually at the end of each business year.

690. What are (*a*) trade discounts, (*b*) cash discounts?

(*a*) Trade discounts = deductions from list price made to the trade.

(*b*) Cash discounts = deductions from invoice price when payment is made within a specified time, as 10 days, 30 days, etc. 2/10 means 2% discount if bill is paid within 10 days; 4/10, n/60 means 4% discount within 10 days, and full amount 60 days from date of invoice.

691. What is (*a*) a sales commission, (*b*) a buying commission?

(*a*) Sales commission = a percentage of a selling transaction charged by a salesman, agent, broker, or jobber for services in selling goods.

(*b*) Buying commission = a percentage of a buying transaction for services of buying goods.

692. When is there (*a*) a profit, (*b*) a loss?

(*a*) There is a profit when selling price is greater than cost of goods + all expenses (operating, shipping, selling, buying, etc.).

(*b*) There is a loss when selling price is less than that of goods + the other expenses.

When selling price = buying price + other expenses, there is no profit or loss.

Profits and losses are usually computed on the gross cost or on the net sales.

693. In figuring profit or loss, what is (*a*) the base, (*b*) the rate, (*c*) the percentage?

(*a*) Base = gross cost.
(*b*) Rate = per cent of gain or loss.
(*c*) Percentage = actual gain or loss.

694. How do we find the selling price when the net cost and the rate of profit are given?

Multiply the cost by the per cent of profit, and add this to the net cost.

EXAMPLE: What is the selling price if goods cost $20, and you want to make a profit of 60% of the cost?

$$\text{Selling price} = \text{cost} \times \%\ \text{profit} + \text{net cost}$$

$$\therefore\ (\$20 \times .6) + \$20 = 12 + 20 = \$32 = \text{selling price. Ans.}$$

695. How do we find the selling price when there is a loss, and you are given the net cost and the rate of loss?

Multiply the cost by the per cent of loss and subtract this from the cost.

EXAMPLE: What is the selling price if the cost is \$20, and the loss is 60% of the cost?

$$\text{Selling price} = \text{net cost} - (\text{cost} \times \%\ \text{loss})$$

$$\therefore\ \$20 - (\$20 \times .6) = \$20 - \$12 = \$8 = \text{selling price. Ans.}$$

696. How do we find the per cent of profit, given the cost and selling price?

Subtract the cost from the selling price to get the profit.

Divide the profit by the cost and multiply by 100 to get the per cent of profit.

EXAMPLE: What is the per cent of profit if the selling price is \$120 and the cost is \$80?

$$\text{Profit} = \$120 - \$80 = \$40$$

$$\therefore\ \frac{\$40}{\$80} \times 100 = 50\%\ \text{profit on cost. Ans.}$$

697. How do we find the per cent of loss, given the cost and the selling price?

Subtract the selling price from the cost to get the loss.

Divide the loss by the cost and multiply by 100 to get the per cent of loss.

EXAMPLE: What is the per cent of loss if the selling price is \$80, and the cost is \$120?

$$\text{Loss} = \$120 - \$80 = \$40$$

$$\therefore\ \frac{\$40}{\$120} \times 100 = 33\tfrac{1}{3}\%\ \text{loss on cost. Ans.}$$

698. How do we figure a discount or a commission?

Multiply the cost or the selling price of the item by the per cent of the trade discount.

Ex. (*a*) If the trade discount is 10%, and the cost of the item is \$2, then

$$\tfrac{1}{10} \times \$2 = \$\tfrac{2}{10} = \$.20 = 20¢ = \text{trade discount. Ans.}$$

Ex. (*b*) If the trade discount is .40, and the selling price is \$2, then

$$.4 \times \$2 = \$.8 = 80¢ = \text{trade discount. Ans.}$$

Note the difference between the forms in which the discount is given, percentages and decimals.

699. How do we find the cash discount when the amount of the bill and the rate of discount are given?

Multiply the rate of discount by the amount of the bill to get the discount.

EXAMPLE: If the terms are 4/10; n/60 and the bill is \$1,240, what are the cash discount and the net amount?

$$\$1{,}240 \times .04 = \$49.60 = \text{discount}$$
$$\$1{,}240 - \$49.60 = \$1{,}190.40 = \text{net amount}$$

700. What is meant by bank discount?

Bank discount is interest charged by a bank for advancing money on notes and time drafts.

The owner of the note endorses it to the bank, which holds it to maturity as security. Then the bank collects the face amount from the maker or from the one who signed the note. Should the maker not pay, then either the party who had the note discounted, or the endorser, has to pay it.

701. How is simple bank discount figured?

The same way that simple interest is figured.

Interest is figured for the actual number of days between the discount date and the due date.

EXAMPLE: Find the bank discount at 6%, and the net proceeds of a 92-day note for \$3,000, when the date of the note is August 1, 1960, and the due date is November 1, 1960.

92 days
August 1, 1960 (discount date) ⟷ November 1, 1960 (due date)

August = 30 days	\$3,000 note	
September = 30 days	\$30	= discount for 60 days at 6%
October = 31 days	\$15	= discount for 30 days at 6%
November = 1 day	\$1	= discount for 2 days at 6%
Total = 92 days	\$46	= discount for 92 days at 6%

$\therefore$ \$3,000 − \$46 = \$2,954 = net proceeds. Ans.

702. How do we figure the net price of an item when there is a series of discounts, as 40, 5, and 2 (meaning 40%, 5%, and 2%)?

(*a*) Multiply the cost of the item by the first discount %, and subtract this from the cost, getting result (I).

(*b*) Multiply result I by the second discount %, and subtract this from result I, getting result II.

(*c*) Multiply result (II) by the third discount %, and subtract this from result (II), getting the net price of item.

EXAMPLE: Given, cost \$300 and discounts 40, 5, and 2, find the net price.

(*a*) $\$300 \times 40\% = \$300 \times .4 = \$120$
$\$300 - \$120 = \$180 = \text{result I}$
(*b*) $\$180 \times 5\% = \$180 \times .05 = \$9$
$\$180 - \$9 = \$171 = \text{result II}$
(*c*) $\$171 \times 2\% = \$171 \times .02 = \$3.42$

$\therefore \$171 - \$3.42 = \$167.58 =$ net price of item. Ans.

Note: The discounts may be two or three in number or they may be a combination of trade and cash discounts.

In any case deduct the first discount in the series from the total amount and follow this by deducting the next discount from the remainder, etc.

703. How may the above process be shortened by obtaining a single equivalent of the *remainder* after deducting all the discounts?

Take 100 as the base regardless of the cost of the goods.

EXAMPLE: If the gross cost (or list price) = \$300, and the discounts are 40, 5, and 2, find the net cost.

If $100 =$ base, then $100 - 40 = 60 =$ remainder.
Now 5% of $60 = .05 \times 60 = 3$
Therefore $60 - 3 = 57 =$ remainder
Then 2% of $57 = .02 \times 57 = 1.14$
and $57 - 1.14 = 55.86 = 55.86\% =$ single equivalent remainder

$\therefore \$300 \times .5586 = \$167.58 =$ net cost. Ans.

704. What is the procedure for getting a single discount which is equal to two discounts, by mental calculation?

Subtract $\frac{1}{100}$ of their product from their sum.

EXAMPLE: Find a single discount equal to 30% and 4%.
Their sum is $30 + 4 = 34\%$
$\frac{1}{100}$ of their product is

$$\frac{30 \times 4}{100} = \frac{120}{100} = 1.2\%$$

The difference is

$\therefore 34 - 1.2 = 32.8\% =$ single equivalent discount. Ans.

705. Using this method, how can we get a single discount which is equal to a series of discounts?

(*a*) Find a single discount equal to the first two.
(*b*) Combine the result of the first two with the third.
(*c*) Combine the last result with the fourth, etc.

EXAMPLE: Find a single discount equal to 40%, 10%, and 5%.

(*a*) Combine 40% with 10%.

$$40 + 10 - \frac{40 \times 10}{100} = 50 - 4 = 46\%$$

(*b*) Combine the result 46% with 5%.

$$\therefore\ 46 + 5 - \frac{46 \times 5}{100} = 51 - 2.3 = 48.7\%$$

= single equivalent discount. Ans.

706. If after 8% and 4% discounts are deducted, the net cost of an invoice of goods is \$1,684.36, what is the list price?

$$8 + 4 - \frac{8 \times 4}{100} = 12 - .32 = 11.68\%$$

= single equivalent discount of 8% and 4%

$$100\% - 11.68\% = 88.32\% = \text{net cost}$$

$$\therefore\ \frac{\$1{,}684.36}{.8832} = \$1{,}900.71 = \text{list price. Ans.}$$

707. If the amount of discount is \$398.42, and the discounts are 40% and 2%, what is the net cost of the goods?

$$40 + 2 - \frac{40 \times 2}{100} = 41.2\% = \text{single equivalent discount}$$

Now

$$\$398.42 = 41.2\% = \text{discount}$$

Then

$$\frac{\$398.42}{.412} = \$967.04 = \text{list price}$$

$$\therefore\ \$967.04 - \$398.42 = \$568.62 = \text{net cost. Ans.}$$

708. If the terms on a \$2,680 invoice of goods are 4/10; n/60, how much do you gain if you borrow money from a bank at 6% for 60 days, and pay cash for the merchandise?

$.04 \times \$2{,}680 = \$107.20$ = discount
$\$2{,}680 - \$107.20 = \$2{,}572.80$ = net cost
\$2,680 at 6% for 60 days = \$26.80 = interest on loan
$\therefore\ \$107.20 - \$26.80 = \$80.40$ = gain. Ans.

709. If the gross cost of an article is \$6.72, and the article is sold at a profit of 30% on the selling price, how much is the net profit if 21% is charged to the cost of doing business?

$$100\% - 30\% = 70\% = .70$$

$$\therefore \$6.72 \text{ (gross cost)} = 70\% \text{ of the selling price}$$

and

$$\frac{\$6.72}{.7} = \$9.60 = \text{selling price}$$

$$30\% - 21\% = 9\% = \text{per cent of net profit}$$

$$\therefore \$9.60 \text{ (selling price)} \times .09 = \$.864 = \text{amount of net profit. Ans.}$$

710. What is the procedure for getting the selling price, given the net cost, percentage of profit, and cost of selling?

(*a*) Find the net profit. Net cost × % profit = net profit.
(*b*) Add net profit to net cost.
(*c*) This is what % of the selling price?
(*d*) Find the selling price by dividing by this %.

EXAMPLE: What is the selling price when the net cost of an article is \$12.60, and it is to be sold to make a net profit of 15% of the cost, and when the cost of doing business is 20% of the selling price?

(*a*) Net cost × % profit = \$12.60 × 15% = \$1.89 = net profit.
(*b*) Net profit + net cost = \$1.89 + \$12.60 = \$12.49.
(*c*) \$12.49 is 80% (= 100% − 20%) of the selling price.
(*d*) $\therefore \dfrac{\$12.49}{.80} = \$15.61 = \text{selling price. Ans.}$

711. How can we find the relation of net profit to selling price in percentage?

(*a*) Find net cost. Net cost = list buying price − discounts.
(*b*) Find selling price.

Selling price = list selling price − discounts.

(*c*) Find gross profit. Gross profit = selling price − net cost.
(*d*) Find % of gross profit on sales.

$$\% \text{ of gross profit on sales} = \frac{\text{gross profit}}{\text{selling price}} \times 100\%$$

(*e*) Find % of net profit on sales.

$$\% \text{ of net profit on sales} = \% \text{ of gross profit on sales} - \% \text{ cost of doing business}$$

EXAMPLE: An appliance dealer buys a color TV set for \$460, less 30% and 5%. He sells it for \$490, less 15%. If the cost of doing business is 14% of the sales, what % of the selling price is his net profit?

(*a*) Net cost. $(30 + 5 - \dfrac{30 \times 5}{100} = 35 - 1.5 = 33.5\%$

$= \text{single equivalent discount})$

$\$460 \times (100\% - 33.5\%) = \$460 \times .665 = \$305.90 = \text{net cost}$

(*b*) Selling price $= \$490 - (490 \times .15) = \$490 - \$73.50$
$= \$416.50 = \text{selling price.}$

(*c*) Gross profit $= \$416.50 - \$305.90 = \$110.60.$

(*d*) % of gross profit on sales $= \dfrac{\$110.60}{\$416.50} \times 100 = 26.55\%$

(*e*) % of net profit on sales $= 26.55\% - 14\% = 12.55\%$ Ans.

712. How can we find the relation of net profit to gross cost or to net cost, expressed as a percentage?

(*a*) Find net cost = list buying price − discounts.
(*b*) Find selling price = list selling price − discount.
(*c*) Find gross profit = selling price − net cost.
(*d*) Get cost of doing business = % × selling price.
(*e*) Find net profit on net cost = gross profit − cost of doing business.

(*f*) % of profit $= \dfrac{\text{net profit}}{\text{net cost}} = 100$

EXAMPLE: If basketballs cost \$12 a dozen, less 30% and 5%, and are sold for \$7 each, less 15%, and the cost of doing business is 20% of the sales, what is the % of profit on net cost?

(*a*) Net cost $30 + 5 - \dfrac{30 \times 5}{100} = 35 - 1.5 = 33.5\%$

$= \text{single equivalent discount}$

$\$72 \times (100\% - 33.5\%) = \$72 \times .665 = \$47.88 = \text{net cost}$

(*b*) Selling price:
$\$84\ (= 12 \times 7) - (\$84 \times .15) = \$84 - \$12.60 = \$71.40$

(*c*) Gross profit $= \$71.40 - \$47.88 = \$23.52$
(*d*) Cost of doing business $= .20 \times \$71.40 = \$14.28$
(*e*) Net profit on net cost $= \$23.52 - \$14.28 = \$9.24$

(*f*) % of profit $= \dfrac{\$9.24}{\$47.88} \times 100 = 19.3\%$ Ans.

713. If shirts are bought for \$5.60, less 14% and 8%, and are sold for \$7.40, less 10%, and the buying expenses are 4% of the net cost, and selling expenses are 5% of net sales, what % of the gross cost is the net profit?

$$(a)\ \text{Net cost } 14 + 8 - \frac{14 \times \overset{2}{\cancel{8}}}{\underset{25}{\cancel{100}}} = 22 - 1.12$$

$$= 20.88\% = \text{single equivalent discount}$$

\$5.60 × (100% − 20.88%) = \$5.60 × .7912 = \$4.43 = net cost

(*b*) Selling price = \$7.40 − \$7.40 × .10 = \$7.40 − \$.74 = \$6.66
(*c*) Cost of buying = \$4.43 × .04 = \$.18
(*d*) Gross cost = \$4.43 + \$.18 = \$4.61
(*e*) Net profit.
\$6.66 (selling price) − \$6.66 × .15 (cost of doing business)
= \$6.66 − \$1.00 = \$5.66

∴ \$5.66 − \$4.61 = \$1.05 = net profit

$$(f)\ \%\ \text{of profit on gross cost} = \frac{\$1.05}{\$4.61} \times 100 = 22.78\%\ \text{Ans.}$$

714. If we know the amount of profit, the per cent of profit on the gross cost, and the per cent of buying cost, how do we get the net cost and the cost of buying?

(*a*) Find the gross cost. Divide amount of profit by the % of profit on gross cost.

(*b*) Find gross cost %. Add % of buying cost to 100% (the net cost).

(*c*) Find net cost. Divide gross cost by gross cost %.

(*d*) Find cost of buying. Gross cost − net cost.

EXAMPLE: If 30% = % of profit on gross cost of an article, and the profit is \$12.93 and 7% = cost of buying, what are the net cost of the article and the cost of buying?

(*a*) Gross cost: \$12.93 = 30% = profit

$$\therefore \frac{\$12.93}{.30} = \$43.10 = \text{gross cost}$$

(*b*) Gross cost % = 7% (cost of buying) + 100% (net cost)
= 107%

$$(c)\ \text{Net cost} = \frac{\$43.10}{1.07} = \$40.28\ \text{Ans.}$$

(*d*) Cost of buying = \$1.07
\$43.10 (gross cost) − \$40.28 (net cost) = \$2.82 = cost of buying

715. If we know the net cost, per cent of buying expenses, and the amount of profit, how do we find the per cent of profit and the selling price?

(*a*) Get the cost of buying (% of buying expenses × net cost).
(*b*) Get the gross cost (net cost + buying expenses).
(*c*) Find % profit on gross cost (profit/gross cost × 100).
(*d*) Find selling price (profit + gross cost).

EXAMPLE: The net cost of an article is \$56. The buying expenses are 5% of net cost. What is the % of profit on the gross cost if the article is sold at a profit of \$18.60, and what is the selling price?

(*a*) Cost of buying = .05 × \$56 = \$2.80.

(*b*) Gross cost = \$56 + \$2.80 = \$58.80.

(*c*) % profit on gross cost $= \frac{\$18.60}{\$58.80} \times 100 = 26.53\%$ Ans.

(*d*) Selling price = \$58.80 + \$18.60 = \$77.40 Ans.

716. If you buy an article invoiced at \$34.60, less 3% discount, and sell it at 30% profit, what is the selling price?

Discount = \$34.60 × .03 = \$1.04
Net cost = \$34.60 − \$1.04 = \$33.56
Profit = .30 × \$33.56 = \$10.07
Selling price = \$33.56 + \$10.07 = \$43.63 Ans.

717. If a dealer buys a TV set for \$360, pays \$12 freight and cartage, and sells it at a profit of $33\frac{1}{3}\%$, what is the selling price?

Gross cost = \$360 + \$12 = \$372
Profit = $33\frac{1}{3}\%$ or $\frac{1}{3}$ × \$372 = \$124
Selling price = \$372 + \$124 = \$496 Ans.

718. If a merchant pays \$18.60 for an article, and sells it at a profit of 25% of the selling price, what is the selling price?

Selling price = 100%
Cost = 100% − 25% = 75% = \$18.60

$$\therefore \frac{\$18.60}{.75} = \$24.80 = \text{selling price. Ans.}$$

719. If the gross cost of an article is \$8.65, and it is sold at a profit of 25% on the selling price, what is the net profit if the cost of doing business is 12%?

As the profit is 25% on the selling price, then the gross cost \$8.65 = 75% of the selling price and selling price = \$8.65/.75 = \$11.53.

$$\text{Now \% of net profit} = 25\% - 12\% = 13\%$$

$$\therefore \text{ amount of net profit} = \$11.53 \times .13 = \$1.50 \text{ Ans.}$$

720. If a merchant sells apples at \$5.50 a bushel at $4\frac{1}{2}\%$ commission, and his commission amounts to \$148.50, while other charges are 35¢ a bushel, how many bushels does he sell and how much are the net proceeds?

$$\text{Amount of sales} = \frac{\$148.50}{.045} = \$3{,}300$$

$$\text{Number of bushels} = \frac{\$3{,}300}{\$5.50} = 600 \text{ Ans.}$$

Other charges = 600 × \$.35 = \$210
Total charges = \$148.50 + \$210 = \$358.50
Net proceeds = \$3,300 − \$358.50 = \$2,941.50 Ans.

721. The cost of a TV set to an appliance dealer is \$360, less 40% and 2%. What should he mark the set if he wants to make a profit of 25% on the net cost, and allow the customer a 15% discount on the marked price?

$$40 + 2 - \frac{40 \times 2}{100} = 42 - .8 = 41.2\% = \text{single equivalent discount}$$

$$\$360 \times (100\% - 41.2\%) = \$360 \times .588 = \$211.68 = \text{his net cost}$$

$$\$211.68 \times .25 = \$52.92 = \text{his 25\% profit on net cost}$$

$$\$211.68 + \$52.92 = \$264.60 = \text{net selling price}$$

$$100\% \text{ (marked price)} - 15\% \text{ (customers' discount)} = 85\% = \text{selling price}$$

or, selling price is 85% of the marked price.

$$\therefore \text{ The marked price} = \frac{\$264.60}{.85} = \$311.30 \text{ Ans.}$$

722. What is meant by the "future worth" or value of a sum of money?

We have seen that money at interest increases or accumulates as time passes.

Future value = amount in question (principal) × interest accumulation factor

Future value $S = P(1 + rt)$, at simple interest
P = principal, and $(1 + rt)$ = interest accumulation factor
r = rate of interest, t = time
At compound interest, $S = P(1 + r)^t$
$(1 + r)^t$ = accumulation factor

723. What is meant by the "present worth" or value of a sum of money?

It is the *principal* which, if put at interest at a given rate for a given time, will equal some assumed or desired amount in the future.

A sum of money is worth less today than in the future because you can invest the money today and allow it to accumulate.

For simple interest, $$P = \frac{S}{(1 + rt)} = \frac{\text{future value}}{\text{accumulation factor}}$$

For compound interest, $$P = \frac{S}{(1 + r)^t} = \frac{\text{future value}}{\text{accumulation factor}}$$

724. What is meant by the true discount?

True discount = the difference between future worth and present worth of a debt = interest on present worth of a debt for the time it has to run before maturity.

725. What are the present worth and the true discount of a debt for \$1,800, due in 8 months, if money is worth 6% interest?

$$\$1 \text{ for 8 months at } 6\% = \$1.04 \quad \text{or}$$
$$\$1 \text{ principal} + \underset{(t)}{\tfrac{8}{12}} \times \underset{(r)}{.06} = \$1 + .04$$
$$= \$1.04 = \text{accumulation factor}$$

$$\text{Then } P = \frac{S}{(1 + rt)} = \frac{\$1,800}{1.04} = \$1,730.77 = \text{present worth}$$

The \$1,800 debt which is due in 8 months is worth \$1,730.77 now.
And \$1,800 − \$1,730.77 = \$69.23 = the true discount

To prove this, we have:

$$\$1,730.77 \text{ for 8 months at } 6\%$$
$$= \$1,730.77 \times \tfrac{8}{12} \times .06 = \$69.23 \text{ interest}$$

and

$$\$1,730.77 + \$69.23 = \$1,800 = \text{the amount at maturity}$$

726. If A owes B \$1,000, which is not due until 3 years from now, and A offers to pay B today, what sum should A pay now at compound interest, assuming the money to be worth 4%?

$$P = \frac{S}{(1+.04)^3} \begin{array}{l} = \text{future worth} \\ = \text{accumulation factor} \end{array}$$

$$= \frac{\$1{,}000}{1.12486} = \$889.00 = \text{present worth}$$

This means that A should pay \$889.00 now.

Also it follows that

present value × accumulation factor = future worth

or

$$\$889.00 \times 1.12486 = \$1{,}000$$

Accumulation factors can be obtained from appropriate tables.

727. What is meant by the present value of 1, and how is it used?

The *present* value of 1 = the reciprocal of the accumulation factor.

It is much easier to multiply than to divide with numbers of many places, and that is why the present value of 1 is useful.

EXAMPLE: Find the present value of \$1,000 due in 3 years at 4% compound interest.

The accumulation factor is 1.12486.

So, instead of finding

$$\frac{\$1{,}000}{1.12486} = \$889.00$$

multiply \$1,000 by the reciprocal of the accumulation factor (or the present value of 1).

$$\therefore \$1{,}000 \times .88900 = \$889.00$$

Reciprocals of accumulation factors are given directly by a table of present values of 1. See Table 4, Appendix B, for a section of such a table.

728. In what two ways may consumer finance be considered?

(*a*) *Cash.* Loan and finance agencies give cash, and allow the borrower a certain time to repay the principal and interest.

(*b*) *Installment credit.* Businessmen offer installment credit and permit purchases to be paid for in installments at specified regular intervals.

729. What is meant by installment buying or buying goods "on time"?

Part of the purchase price is paid on possession, and the balance in fractional payments at stated intervals, until the entire sum is paid.

The merchant is considered to extend credit to the consumer. The purchaser is considered to borrow money indirectly.

EXAMPLE: If a TV set is priced at $150 cash, and the advertised payment plan is $25 down and $3 a week for 45 weeks, how much more does it cost on the installment plan?

$25 = down payment
$135 = 45 weeks at $3
$160 = total cost on installment plan
$150 = cash price
$10 = carrying charge. Ans.

730. If you buy a washing machine for $280, are given a $50 trade-in allowance for your old machine, and agree to pay the balance in 10 monthly installments plus a final installment of $35, how much would you save by buying for cash?

$280 − $50 = $230 = balance for cash
$230 = 10 × $23 = 10 equal monthly payments
$35 = final payment
$265 = total installment payments
$230 = cash
$35 = saved by buying for cash

You pay the equivalent of $280 + $35 = $315 for the machine instead of $280.

731. If you borrow $2,400 from a bank and pay it back in monthly payments of $38.05 over 6 years, how much do you pay the bank for the loan?

6 × 12 = 72 monthly payments to make
72 × $38.05 = $2,739.60 = total payment
$2,400.00 = amount borrowed
$339.60 = amount paid for loan

732. Why is buying goods on credit the same as borrowing money?

You actually keep for a time the money that belongs to the merchant and on this you must pay interest.

The additional money you pay on the installment plan represents a definite interest rate.

733. Why does credit or installment buying cost more?

It is more expensive to the merchant. He has to wait for what you owe him. You use the goods while you are still paying for them. The merchant has to keep a record of what you owe him. He takes extra risks because should you not be able to finish payment he can recover the goods but cannot sell them as new again.

734. Why do some merchants prefer the credit plan to cash despite all this?

They get more money for goods, even with all the risk, since the customer pays a comparatively high rate of "interest," carrying charge, or financing charge on credit purchases.

They can also sell more to those unable to afford cash buying.

735. What are some of the ranges of interest charged in consumer finance?

(*a*) Personal finance companies: $2\frac{1}{2}$ to $3\frac{1}{2}\%$ per month on unpaid balances.

(*b*) Contract interest rate: 6–12% per year.

Note that a charge of $2\frac{1}{2}\%$ per month = an annual effective rate of 34.5%. A charge of 3% per month = an annual effective rate of 42.6%.

(*c*) Credit unions: 12% per year or 1% per month.

(*d*) Industrial banks: 12–34% per year.

Note that to avoid an illegal rate of interest installment-buying contracts generally do not mention interest, but refer to a financing charge, or carrying charge, which includes interest, bookkeeping cost, and other expenses involved in installment buying.

736. What is the 6% method offered by some credit companies, and how do we find the monthly payment?

One-half per cent is added to the unpaid balance for each month, up to a limit of 12 months. You divide this result by the number of payments to find the monthly payment.

EXAMPLE: If you buy a refrigerator for \$480 and make a down payment of \$150, then pay the balance of \$330 in 1 year, what would be your monthly payment?

To \$330 you add 12 months × $\frac{1}{2}$% = 6%

$$\$330 \times .06 = \$19.80$$
$$\$19.80 + \$330 = \$349.80$$
$$\therefore \frac{\$349.80}{12} = \$29.15 = \text{monthly payment}$$

Note that this 6% plan is not the same as 6% interest, as will be shown later.

737. If you, as a merchant, decide to charge an additional 14% on the goods you sell "on time," what would be the price on a 10-equal-payment plan and the amount of each payment on a clock radio that sells for \$88.60 cash?

$$\$88.60 \times \$1.14 = \$101.00 = \text{price on installment plan}$$
$$\therefore \frac{\$101}{10} = \$10.10 = \text{each payment}$$

738. What is the key in figuring the annual rate of interest charge you pay, when you buy on the installment plan, or when you borrow money from a finance company to be repaid in monthly installments?

You must add up the number of months specified in the plan, divide it by 12 to convert to years, and substitute this in I = P*rt* (I = interest amount; P = principal; *t* = time in years; *r* = annual interest rate).

EXAMPLE: If the interest or carrying charge is \$8, and there are 6 monthly payments of \$10 on a purchase, what is the interest rate?

These terms mean you actually owe the merchant (or a loan company if it is a loan) \$60 cash, which you pay back in monthly installments.

You thus have kept or borrowed the

First payment of \$10 for 1 month
Second payment of \$10 for 2 months
Third payment of \$10 for 3 months
Fourth payment of \$10 for 4 months
Fifth payment of \$10 for 5 months
Sixth payment of \$10 for 6 months

Or, you kept \$10 for a total of 21 months = $\frac{21}{12}$ years = t

The sum of the months from 1 to 6 can be obtained directly from the sum of a series

$$S = (a + l)\frac{n}{2} \quad \text{or} \quad S = (1 + 6)\frac{6}{2} = 7 \times 3 = 21$$

P = principal = \$10 here.
I = interest, or carrying charge = \$8.

$$\therefore r = \frac{I}{Pt} = \frac{\$8}{\$10 \times \frac{21}{12}} = \frac{\overset{4}{\cancel{8}} \times \overset{4}{\cancel{12}}}{\underset{5}{\cancel{10}} \times \underset{7}{\cancel{21}}} = \frac{16}{35} = 45.7\% \text{ per year}$$

739. How much interest and financing charge do you pay when you buy a TV set for \$280, if you are allowed \$50 for your old set as trade-in allowance, and you agree to pay the balance in 10 monthly installments of \$23, plus a final installment of \$35?

\$280 − \$50 = \$230 = balance = 10 × \$23 in payments.

I = \$35 = final installment = interest and financing charge.

You keep or borrow the

First	\$23 payment for	1 month
Second	\$23 payment for	2 months
Third	\$23 payment for	3 months
Fourth	\$23 payment for	4 months
Fifth	\$23 payment for	5 months
Sixth	\$23 payment for	6 months
Seventh	\$23 payment for	7 months
Eighth	\$23 payment for	8 months
Ninth	\$23 payment for	9 months
Tenth	\$23 payment for	10 months
You keep	\$23 for a total of	55 months

Sum of months:

$$S = (1 + 10) \times \tfrac{10}{2} = 11 \times 5 = 55 \text{ months} = \tfrac{55}{12} \text{ years} = t$$

Here P = \$23

$$\therefore r = \frac{I}{Pt} = \frac{\$35}{\$23 \times \frac{55}{12}} = \frac{\overset{7}{\cancel{35}} \times 12}{23 \times \underset{11}{\cancel{55}}} = \frac{84}{253} = 33.2\% \text{ per year}$$

740. What precaution must you take in getting the sum of the number of months you keep or borrow the installment payment?

When the total of payments results in a sum greater than the cash price of the goods, find the payment number nearest to the cash price.

Then get the part of that payment that has gone toward the actual cost of the goods, and by proportion find the part of the time this payment has been kept by you.

EXAMPLE: If you buy a living room suite for $870 and pay $150 down and the balance in 10 monthly installments of $84, what is the rate of financing charge?

$870 − $150 = $720 = cash balance you owe

10 × $84 = $840 = amount paid in 10 installments

$840 − $720 = $120 = amount of financing or carrying charge

You keep or borrow the

First	$84 payment for 1	month
Second	$84 payment for 2	months
Third	$84 payment for 3	months
Fourth	$84 payment for 4	months
Fifth	$84 payment for 5	months
Sixth	$84 payment for 6	months
Seventh	$84 payment for 7	months
Eighth	$84 payment for 8	months
Ninth	$84 payment for $5\frac{1}{7}$	months

At the end of the eighth payment you have paid back 8 × $84 = $672.

The cash balance you owe is $720.

$720 − $672 = $48, which goes toward meeting the actual cash balance.

Since during the ninth month only $48 goes toward the actual cost of the suite, you must consider the ninth payment as having been kept only

$$\frac{n \text{ months}}{9 \text{ months}} = \frac{\$48}{\$84} \quad \text{or} \quad n = \frac{48}{84} \times 9 = 5\frac{1}{7} \text{ months}$$

Thus, the $84 payment is kept only $5\frac{1}{7}$ months.

Sum of months from 1 to 8 is S = $(1 + 8) \times \frac{8}{2} = 36$

$$36 + 5\frac{1}{7} = 41\frac{1}{7} \text{ months} \quad \text{or} \quad \frac{41\frac{1}{7}}{12} \text{ years} = t$$

$$\therefore r = \frac{I}{Pt} = \frac{\$120}{\$84 \times \dfrac{41\frac{1}{7}}{12}} = \frac{120}{7 \times \dfrac{288}{7}} = \frac{120}{288} \times 100$$

$$= 41.67\% \text{ per year financing charge}$$

741. How can we solve for the rate of interest by getting the total amount of the installment money you keep or borrow for onè month, in the example of Question 740?

You keep or borrow:

$84 for 1 month.
$168 for 1 month. This is the same as borrowing $84 for 2 months
$252 for 1 month. This is the same as borrowing $84 for 3 months
$336 for 1 month. This is the same as borrowing $84 for 4 months
$420 for 1 month. This is the same as borrowing $84 for 5 months
$504 for 1 month. This is the same as borrowing $84 for 6 months
$588 for 1 month. This is the same as borrowing $84 for 7 months
$672 for 1 month. This is the same as borrowing $84 for 8 months

Now, you do not keep the entire amount, $756 (= 9 × $84), of the next installment because you need only $720 − $672 = $48 to reach the cash balance of $120 you owe.

Then, by proportion:

$$\frac{x}{\$756} = \frac{\$48}{\$84}$$

$$x = \overset{108}{\cancel{\$756}} \times \frac{\overset{4}{\cancel{\$48}}}{\underset{\underset{1}{\cancel{7}}}{\cancel{\$84}}} = \$432$$

Thus, you finally keep $432 for 1 month.
Get the sum of amounts from $84 to $672.

$$S = (a + l) \times \frac{n}{2} = (\$84 + \$672)\,\frac{8}{2} = \$756 \times 4 = \$3{,}024$$

$n = 8$ terms in the progression.

To this add the last amount = $ 432

Total amount of money kept for 1 month = $3,456 = P

$$\therefore\ r = \frac{I}{Pt} = \frac{\$120}{\$3{,}456 \times \frac{1}{12}} = \frac{\overset{5}{\cancel{120}} \times \overset{1}{\cancel{12}}}{\underset{\underset{12}{\cancel{288}}}{\cancel{3{,}456}}} = \frac{5}{12} \times 100$$

$= 41.67\%$ per year

Question 740, in which you get the total number of months you keep the $84 payment, is somewhat simpler.

742. If you borrow $300 from a finance company to pay a surgical bill, and you are charged 3% per month interest on the unpaid balance of the loan, while you are required to repay the loan in 12 monthly installments of $25 each, how much do you pay back for the $300 loan, and what is the annual interest rate, using the installment plan method?

Payment number	Payment on principal	Interest 3% on balance	Total amount paid	New balance
1	$25	$9.00	$34.00	$275
2	$25	$8.25	$33.25	$250
3	$25	$7.50	$32.50	$225
4	$25	$6.75	$31.75	$200
5	$25	$6.00	$31.00	$175
6	$25	$5.25	$30.25	$150
7	$25	$4.50	$29.50	$125
8	$25	$3.75	$28.75	$100
9	$25	$3.00	$28.00	$75
10	$25	$2.25	$27.25	$50
11	$25	$1.50	$26.50	$25
12	$25	$.75	$25.75	0
Total 78 mo.		$58.50	$358.50	

Total months:

$$S = (1 + 12) \times \frac{12}{2} = 13 \times 6 = 78 \text{ months} = \frac{78}{12} \text{ years} = t$$

Total interest:

$$S = (9 + .75) \times \frac{12}{2} = 9.75 \times 6 = \$58.50 = I$$

P = principal = $25

Total amount paid on loan:

$$S = (\$34 + \$25.75) \times \frac{12}{2} = \$59.75 \times 6 = \$358.50$$

$$\therefore r = \frac{I}{Pt} = \frac{\$58.50}{25 \times \frac{78}{12}} = \frac{58.50 \times \overset{2}{\cancel{12}}}{25 \times \underset{13}{\cancel{78}}} = \frac{117}{325} \times 100 = 36\%$$

We see that 3% a month = 36% a year.

743. If you borrow $300 from a credit union, where the interest charge is 1% a month on the unpaid balance, and you pay back the loan in 12 monthly payments of $25, plus interest charge, how much do you pay back, and what is the annual interest rate? How does this compare with a secured bank loan of $300 for 1 year at 6%?

Payment number	Payment on principal	Interest 1% on balance	Principal, plus interest	New balance
1	$25	$3.00	$28.00	$275
2	$25	$2.75	$27.75	$250
3	$25	$2.50	$27.50	$225
4	$25	$2.25	$27.25	$200
5	$25	$2.00	$27.00	$175
6	$25	$1.75	$26.75	$150
7	$25	$1.50	$26.50	$125
8	$25	$1.25	$26.25	$100
9	$25	$1.00	$26.00	$75
10	$25	$.75	$25.75	$50
11	$25	$.50	$25.50	$25
12	$25	$.25	$25.25	$0
Total 78 mo.		$19.50	$319.50	

Total months:

$$S = (1 + 12) \times \tfrac{12}{6} = 13 \times 6 = 78 \text{ months} = \tfrac{78}{12} \text{ years} = t$$

Total interest:

$$S = (\$3 + .25) \times \tfrac{12}{6} = \$3.25 \times 6 = \$19.50 = I$$

Total amount paid on loan:

$$(\$28 + \$28.25) \times \tfrac{12}{2} = \$53.25 \times 6 = \$319.50$$

$$\therefore r = \frac{I}{Pt} = \frac{\$19.50}{\$25 \times \frac{78}{12}} = \frac{\overset{3.9}{\cancel{19.50}} \times \overset{2}{\cancel{12}}}{\underset{5}{\cancel{25}} \times \underset{13}{\cancel{78}}} = \frac{7.8}{65} \times 100 = 12\%$$

We see that 1% a month = 12% a year, but the $300 is not kept one full year, but is being paid back every month.

A second loan from a bank would be:

$$\$300 \times .06 = \$18.00 = \text{interest paid}$$

Here you keep the $300 the entire year.

This is almost as cheap as a credit union loan where you pay back every month.

744. If you get a loan of \$2,500 at 5% interest per year, and you agree to pay it back in 20 years at \$16.50 per month, how much is the total amount of repayment, and how much does it cost you?

20 × 12 = 240 months = number of payments
240 × \$16.50 = \$3,960 = total repayment on loan

∴ \$3,960 − \$2,500 = \$1,460 = cost to you Ans.

745. How does the above cost compare with a bank loan of \$2,500 for 20 years at 5%?

\$2,500 × .05 = \$125 per year
20 × \$125 = \$2,500 = cost of loan

You pay less when you pay back the money each month. Ans.

746. If you get a loan of \$7,000 at 5% a year on the unpaid balance from a mortgage company, to finance your home, and you agree to pay it back in 8 years at \$88.62 per month, what is the total repayment on the loan and how much does it cost you?

Note: The \$88.62 per month is obtained by multiplying \$7,000 by an annuity factor, .01265992, obtained from a table based on an annuity formula used by the mortgage company.

8 × 12 = 96 months = number of payments
96 × \$88.62 = \$8,507.52 = total repayment
\$8,507.52 − \$7,000 = \$1,507.52 = cost to you for loan

Repayment schedule:

Month	Amount of payment	Interest 5% on balance	Amount paid on principal	New balance
First	\$88.62	\$29.17	\$59.45	\$6,940.55
Second	\$88.62	\$28.91	\$59.71	\$6,880.84
Third	\$88.62	\$28.67	\$59.95	\$6,820.89
Fourth	\$88.62	\$28.42	\$60.20	\$6,760.69
Fifth	\$88.62	\$28.17	\$60.45	\$6,700.24

747. What is a commonly used method of determining the annual rate of interest when you buy or borrow on the installment plan?

This is a method based on the assumption that each installment payment contains principal and interest in the ratio of the starting unpaid balance to the carrying charge.

EXAMPLE: If a loan is for \$180 to be paid in 10 months, at \$20 a month, and there is a carrying charge of \$20, then the \$180 principal is $\frac{9}{10} = .9$ of the total debt of \$200 and the interest is $\frac{1}{10}$ of \$200, or \$20. Thus

$$\tfrac{9}{10} \text{ of each monthly payment} = \tfrac{9}{10} \times \$20 = \$18 = \text{principal}$$

and

$$\tfrac{1}{10} \text{ of each monthly payment} = \tfrac{1}{10} \times \$20 = \$2 = \text{interest}$$

Here all the installments are equal and the procedure gives a reasonable approximation to a true interest rate, as you will see.

This is known as the equal installment, constant-ratio method of determining annual interest rate in installment plans.

748. What is the formula for the equal installment constant-ratio method of finding annual interest rate in installment plans?

$$r = \frac{2mI}{P(n+1)}$$

r = annual interest rate (as a decimal fraction)
m = payment periods per year
$\begin{cases} m = 12 \text{ for monthly payments} \\ m = 52 \text{ for weekly payments, etc.} \end{cases}$
I = total interest or carrying charge, in dollars
P = unpaid balance at beginning of credit period or cash price, less any down payment
n = number of payments called for, excluding down payment

EXAMPLE: What is the per cent interest per year on a loan of \$180, plus \$20 carrying charge, to be paid in 10 equal monthly installments?

m = 12 (payments are monthly)
I = \$20 = carrying charge
P = \$180 = balance due (no down payment)
n = 10 = number of installments

$$\therefore\ r = \frac{2 \times 12 \times \$20}{\$180(10+1)} = \frac{2 \times \overset{4}{\cancel{12}} \times \cancel{20}}{\underset{\underset{3}{\cancel{9}}}{\cancel{180}} \times 11} = \frac{8}{33} = .2424 = 24.24\%$$

749. How is the constant-ratio formula obtained?

We know that $I = Prt$ = simple interest formula. From this we get

$$r = \frac{I}{Pt}$$

Now find the average length of time the installments are in the hands of the borrower.

If m = number of payment periods in a year, then

$$1 \text{ payment is with the borrower } \frac{1}{m} \text{ th year}$$

$$1 \text{ payment is with the borrower } \frac{2}{m} \text{ th year}$$

$$1 \text{ payment is with the borrower } \frac{3}{m} \text{ th year, etc.}$$

$$\text{up to } \frac{n}{m} \text{ th year.}$$

The sum of the time progression or series is obtained from

$$S = \frac{n}{2}(a + l)$$

where S = sum, n = number of terms (payments), a = first term = $1/m$, and l = last term = n/m. Then

$$S = \frac{n}{2}\left(\frac{1}{m} + \frac{n}{m}\right) = \frac{n}{2m}(1 + n) = \text{time in years}$$

Now, divide this by n to get the average time the payments are held or borrowed, or

$$\frac{\frac{n}{2m}(1 + n)}{n} = \frac{1 + n}{2m} = t = \text{time in years}$$

$$\therefore r = \frac{I}{P \times \left(1 + \frac{n}{2m}\right)} = \frac{2mI}{P(n + 1)}$$

750. If a TV set is priced at \$150 cash, and the advertised payment plan is \$25 down and \$3 a week for 45 weeks, what is the interest rate?

Here

m = 52, since payments are weekly
I = \$25 + 45 × \$3 − \$150 = \$160 − \$150 = \$10
= total interest = carrying charge
P = \$150 − \$25 down payment = \$125 = unpaid balance

$$\therefore r = \frac{2mI}{P(n+1)} = \frac{2 \times 52 \times \$10}{\$125(45+1)} = \frac{2 \times 52 \times \overset{\overset{1}{2}}{\cancel{10}}}{\underset{25}{\cancel{125}} \times \underset{23}{\cancel{46}}} = \frac{104}{575}$$

$$= .181 = 18.1\% \text{ Ans.}$$

751. A clock radio is offered for \$45 cash or on time payments for 10% more, with a down payment of \$9.50, and the balance in 13 weekly payments. What is the annual rate of interest?

Here

m = 52, since payments are weekly
I = 10% of \$45 = .1 × \$45 = \$4.50 = carrying charge
P = \$45 − \$9.50 down payment = \$35.50
n = 13 payments

$$r = \frac{2mI}{P(n+1)} = \frac{2 \times 52 \times \$4.50}{\$35.50(13+1)} = \frac{\cancel{2} \times 52 \times \overset{.9}{\cancel{4.5}}}{\underset{7.1}{\cancel{35.5}} \times \underset{7}{\cancel{14}}}$$

$$\therefore r = \frac{46.8}{49.7} = .942 = 94.2\% \text{ Ans.}$$

752. A hi-fi set can be bought for \$380 cash with a discount of \$19, or in 12 equal monthly installments by paying \$130 and adding a \$30 carrying charge. What is the annual rate of interest?

Here

m = 12, since payments are monthly
I = (\$380 + \$30) − (\$380 − \$19) = \$410 − \$361 = \$49
= total carrying charge
P = \$361 cash − \$130 down payment = \$231
n = 12 payments

$$\therefore r = \frac{2mI}{P(n+1)} = \frac{2 \times 12 \times \$49}{\$231(12+1)} = \frac{2 \times \overset{4}{\cancel{12}} \times \overset{7}{\cancel{49}}}{\underset{\underset{11}{\cancel{33}}}{\cancel{231}} \times 13} = \frac{56}{143}$$

$$= .392 = 39.2\% \text{ Ans.}$$

753. If you borrow \$150 from a loan company for 10 months, and repay it in 10 equal installments of \$17.34, what rate of interest do you pay?

Here

$m = 12$
$P = \$150$
$I = 10 \times \$17.34 - \$150 = \$23.40 = \text{carrying charge}$
$n = 10$

$$\therefore r = \frac{2mI}{P(n+1)} = \frac{\overset{1}{\cancel{2}} \times \overset{4}{\cancel{12}} \times \$23.40}{\underset{\underset{\$25}{\cancel{\$75}}}{\cancel{\$150}} \times 11} = \frac{93.6}{275} = .34 = 34\% \text{ Ans.}$$

754. How can we get the annual rate paid in Question 753, by finding the amount of money the borrower had the use of for 1 month?

The borrower had the use of \$15 for 1 month, \$30 for another month, \$45 for 1 month, etc., to \$150 for 1 month.

Sum of the series from \$15 to \$150:

$$S = (\$15 + \$150) \times \tfrac{10}{2} = \$165 \times 5 = \$825$$

The borrower had the use of \$825 for 1 month.

$$\therefore r = \frac{I}{Pt} = \frac{\$23.40}{\$825 \times \frac{1}{12}} = \frac{\$23.40 \times \overset{4}{\cancel{12}}}{\underset{\$275}{\cancel{\$825}}} = \frac{93.6}{275} = .34$$

$$= 34\% \text{ Ans.} \quad \text{(as in Question 753)}$$

755. How can we get the annual rate paid in Question 753 by finding the total time the borrower had the amount of the installment available for use?

The borrower had \$15 available for 1 month
\$15 available for 2 months
\$15 available for 3 months
\$15 available for 4 months, etc., to
\$15 available for 10 months.

Sum of the series from 1 to 10 months:

$$S = \tfrac{10}{2}(1 + 10) = 55 \text{ months}$$

The borrower had \$15 available for use for a total of 55 months or $\frac{55}{12}$ years $= t$.

$$\therefore\ r = \frac{\text{I}}{\text{P}t} = \frac{\$23.40}{\$15 \times \frac{55}{12}} = \frac{23.4 \times \overset{4}{\cancel{12}}}{\underset{5}{\cancel{15}} \times 55} = \frac{93.6}{275} = .34$$

$$= 34\% \text{ Ans.} \quad \text{(as in Question 753.)}$$

756. If you borrow \$300 from a bank for 15 months and pay back \$21.57 per month, what annual rate are you paying, as figured by the three methods shown?

(*a*) Constant-ratio method:

$$\frac{2m\text{I}}{\text{P}(n + 1)}$$

Each \$21.57 consists of \$20 payment on principal and \$1.57 carrying charge.

Here $m = 12$, $\text{I} = 15 \times \$1.57 = \$23.55$, $\text{P} = \$300$, and $n = 15$.

$$\therefore\ r = \frac{\overset{1}{\cancel{2}} \times \overset{1}{\cancel{12}} \times \$23.55}{\underset{\$25}{\cancel{\$300}} \times \underset{8}{\cancel{16}}} = \frac{23.55}{200} = .1178 = 11.78\% \text{ Ans.}$$

(*b*) Total-amount-used-for-1-month method.

Sum of series of \$20 for 1 month to \$300 for 1 month:

$$\text{S} = (\$20 + \$300) \times \tfrac{15}{2} = \$320 \times \tfrac{15}{2} = \$160 \times 15$$
$$= \$2{,}400 \text{ used for 1 month}$$

$$\therefore\ r = \frac{\text{I}}{\text{P}t} = \frac{\$23.55}{\$2{,}400 \times \frac{1}{12}} = \frac{23.55 \times \overset{1}{\cancel{12}}}{\underset{200}{\cancel{2{,}400}}} = \frac{23.55}{200}$$

$$= .1178 = 11.78\% \text{ Ans.}$$

(*c*) Total time amount of installment was available for use method.

Sum of series of 1 month to 15 months the \$20 was kept:

$$\text{S} = (1 + 15) \times \frac{15}{2} = \frac{\overset{8}{\cancel{16}} \times 15}{\underset{1}{\cancel{2}}} = 120 \text{ months borrower kept } \$20$$

$$\therefore\ r = \frac{\text{I}}{\text{P}t} = \frac{\$23.55}{\$20 \times \frac{120}{12}} = \frac{23.55}{200} = .1178 = 11.78\% \text{ Ans.}$$

757. If you buy, on time, a set of dishes that costs \$86 cash, and \$12 is added for carrying charges on a payment plan of \$14 down and \$14 a month for 6 months, what is the rate of interest you pay?

P = principal = \$86 − \$14 down = \$72 = the amount of money of which the borrower actually has the use
I = \$12 = total carrying charge
$m = 12$
$n = 6$

$$\therefore r = \frac{2m\mathrm{I}}{\mathrm{P}(n+1)} = \frac{2 \times \overset{2}{\cancel{12}} \times \overset{\$1}{\cancel{\$12}}}{\underset{\underset{\$1}{\cancel{\$6}}}{\cancel{\$72}}(6+1)} = \frac{4}{7} = .5714 = 57.14\% \text{ Ans.}$$

758. What is the interest on the time plan, if a clothes dryer sells for \$189 cash, or \$20 down and \$21 per month for 10 months?

Here

P = \$189 cash − \$20 down = \$169
I = (\$20 + 10 × \$21) − \$789 = \$230 − \$189 = \$41 carrying charge

$$\therefore r = \frac{2m\mathrm{I}}{\mathrm{P}(n+1)} = \frac{2 \times 12 \times \$41}{\$169(10+1)} = \frac{984}{1{,}859} = .529 = 52.9\% \text{ Ans.}$$

759. What is the constant-ratio formula for finding the interest rate when all payments are equal except the last one?

The last payment may be different from the regular one to take care of any remaining balance.

$$r = \frac{2m\mathrm{I}(\mathrm{P}+\mathrm{I})}{\mathrm{P}n(\mathrm{P}+\mathrm{I}+l)}$$

where l = last payment, in dollars

EXAMPLE: What is the interest rate per year on the time plan of a set of cooking utensils that is advertised at \$28 cash, or \$5 down and \$5 per week for 5 weeks, with a last payment of \$2 in the sixth week?

\$5 down + 5 × \$5 + \$2	= \$32
Cash cost	= \$28
Carrying charge	= \$4 = I

$m = 52$, since payments are weekly
P = principal = \$28 cash − \$5 down = \$23
l = \$2 = last payment
n = 6 payments

$$\therefore r = \frac{\overset{1}{\not{2}} \times 52 \times \$4(\$23 + \$4)}{23 \times \underset{3}{\not{6}}(\$23 + \$4 + \$2)} = \frac{208 \times \overset{9}{\not{27}}}{\underset{23}{\not{69}} \times 29} = \frac{1{,}872}{667} = 2.81$$

$= 281\%$ a year Ans.

760. What is the interest rate per year if a clock costs \$25 cash, or \$5 down and \$5 per month for 4 months, with a \$3.75 payment the fifth month?

Here

$m = 12$, since payments are monthly
I = (\$5 + 4 × \$5 + \$3.75) − \$25 = \$3.75
P = \$25 cash − \$5 down = \$20
l = \$3.75

$$\therefore r = \frac{2m\text{I}(\text{P} + \text{I})}{\text{P}n(\text{P} + \text{I} + l)} = \frac{\overset{1}{\not{2}} \times \overset{6}{\not{12}} \times \$3.75(\$20 + \$3.75)}{\underset{\underset{\$5}{\$\not{10}}}{\$\not{20}} \times 5(\$20 + \$3.75 + \$3.75)}$$

$$= \frac{22.5 \times 23.75}{25 \times 27.50} = \frac{534.375}{687.5}$$

$= .777 = 77.7\%$ Ans.

761. What is meant by partial payments?

They are payments on an obligation or a note in which a part of the indebtedness is paid each time.

Example: A promissory note for \$5,000 given for 6 months should normally be paid in full when due. However, substantial payments may be made on it, and the date and the amount should be entered on the back of the note.

762. What two rules are used to solve partial payment problems and upon what does the method used depend?

(*a*) The United States rule.

This rule was first used by the United States government when payments and interest were involved. Many states adopted the method when it was approved by the Supreme Court of the United States so that compound interest would not be charged.

It is used when partial payments are made on an interest bearing note of over one year maturity.

(*b*) The merchants' rule.

The method used depends upon agreement or the law in the state in which the maker of the note lives.

763. How do banks, accepting partial payments of notes submitted for discount, collect compound interest and yet avoid the Supreme Court ruling?

They have the old note canceled and a new one drawn for the amount still unpaid. In this way they are able to collect compound interest because they collect the interest in advance.

764. For how long do notes and accounts on which no payments have been made remain in full force?

Under the Statute of Limitations the time is 6 years from the due date. During this time the creditor may take court action to recover.

765. Must mortgages made for a definite time be paid on maturity?

Yes, but very often they are permitted to continue indefinitely as long as the interest payments are made when due. Generally, banks holding mortgages accept partial payments on any interest date.

766. What is the procedure for solving partial payment problems by the merchants' rule?

(*a*) Get the interest on the face of the note from its date to the date it is paid in full.

(*b*) Get the interest on each payment from its date to the date of payment in full.

(*c*) Subtract the sum of the payments, plus their interest from the face of the note, plus its interest.

EXAMPLE: A note for $1,000 dated April 16, 1961, has the following payments endorsed on the back: July 14, 1961, $250; September 30, 1961, $200; November 24, 1961, $100. If the maker wishes to pay in full on December 31, 1961, what is the amount due at that time when the interest is 6%?

The payment periods are found by compound subtraction unless more readily determined otherwise.

December 31, 1961	12 — 31 — 1961	
April 16, 1961	4 — 16 — 1961	
	8 — 15	= 8 mo. 15 days

Interest on $1,000 for 8 mo. at 6%	= $40.00
Interest on $1,000 for 15 days at 6%	= $2.50
Total	= $42.50

Amount due on note = $1,000 + $42.50 = $1,042.50

Interest on $250 (July 14 to December 31) for 5 mo. 17 days at 6%	= $6.96
Interest on $200 (September 30 to December 31) for 3 mo. 1 day at 6%	= $3.03
Interest on $100 (November 24 to December 31) for 1 mo. 7 days at 6%	= $.62
Total = $550	Total = $10.61

∴ $1,042.50 − $560.61 (= $550 + $10.61)

= $481.89 = balance due on December 31, 1961. Ans.

767. What is the procedure for solving partial payment problems by the United States rule?

(*a*) Get the interest on the original principal from date of note to date of first payment.

(*b*) Subtract first payment from sum of principal and interest, if the first payment is greater than the interest then due. The result becomes the new principal on which interest is figured until the second payment is made.

(*c*) The partial payment for any period should be greater than the interest for that period; otherwise, you must add this payment to the next payment or payments, until their sum is equal to or greater than the interest for the combined periods.

(*d*) The same procedure is continued until the time when the amount due on the note is desired.

EXAMPLE: Find the balance due on December 31, 1961, on the note of Question 766 for $1,000 dated April 16, 1961, where the partial payments endorsed on the back of the note are: July 14,

$250; September 30, $200; and November 24, $100, and interest is 6%.

Face of note	$1,000.00
Add interest on $1,000 (April 16 to July 14 = 2 mo. 28 days)	+$14.67
Amount due on July 14	$1,014.67
Subtract payment of July 14	−$250.00
New principal on July 14	$764.67
Add interest on $764.67 (July 14 to September 30 = 2 mo. 16 days)	+$9.69
Amount due on September 30	$774.36
Subtract payment of September 30	−$200.00
New principal on September 30	$574.36
Add interest on $574.36 (September 30 to November 24 = 1 mo. 24 days)	+$5.17
Amount due on November 24	$579.53
Subtract payment of November 24	−$100.00
New principal on November 24	$479.53
Add interest on $479.53 (November 24 to December 31 = 1 mo. 7 days)	+$2.96
Balance due on December 31, 1961, by United States rule	$482.49

We see that

$482.49 = balance by United States rule
$481.89 = balance by merchants' rule
$0.60 = difference. Ans.

768. By the United States rule, how much is required to settle on August 1, 1961, a demand note for $10,000 dated February 1, 1960, with interest at 6% and with the following payments endorsed upon it: April 10, 1960, $2,000; August 4, 1960, $100; February 1, 1961, $4,000; June 1, 1961, $1,000?

Face of note	$10,000.00
Add interest on $10,000 (February 1 to April 10, 1960 = 2 mo. 9 days)	+$115.00
Amount due on April 10, 1960	$10,115.00
Subtract payment of April 10, 1960	−$2,000.00
New principal April 10, 1960	$8,115.00
Add interest on $8,115 (April 10 to August 4, 1960 = 3 mo. 24 days)	$154.19

We see that the payment of $100 on August 4, 1960, is less than the interest $154.19 of August 4.

We must then find and add the interest for two interest periods, and subtract the sum of the two payments from this amount due on February 1, 1961.

New principal April 10, 1960	$8,115.00
Add interest on $8,115	
(April 10, 1960 to February 1, 1961 = 9 mo. 21 days)	+$393.59
Amount due on February 1, 1961	$8,508.59
Subtract two payments of $100 and $4,000	−$4,100.00
New principal February 1, 1961	$4,408.59
Add interest on $4,408.59	
(February 1, 1961 to June 1, 1961 = 4 mo.)	+$88.18
Amount due on June 1, 1961	$4,496.77
Subtract payment of June 1, 1961	−$1,000.00
New principal on June 1, 1961	$3,496.77
Add interest on $3,496.77	
(June 1, 1961 to August 1, 1961 = 3 mo.)	+$52.46
Balance due on August 1, 1961	$3,549.23

769. What are the two general kinds of taxes?

(*a*) Direct taxes levied on personal income, profits, value of property or business.

(*b*) Indirect taxes, levied on imported goods, tobacco, sales tax on goods, war tax, etc. These ultimately are passed on to the consumer in the prices of the things he buys.

770. What is (*a*) a poll tax, (*b*) a property tax, (*c*) an income tax, (*d*) a surtax?

(*a*) Poll tax = tax as a requirement for voting in certain communities.

(*b*) Property tax = tax levied on property.

(*c*) Income tax = tax levied on income.

(*d*) Surtax = an additional tax added to regular tax rate.

771. What is (*a*) a licence, (*b*) an assessment?

(*a*) A permit to do something you desire or to enjoy some specific privilege.

(*b*) Assessment = tax levied by appointed or elected assessors against an individual or a company on real property or use of some property.

772. In what form are assessments usually stated?

In terms of per cent, in terms of mills per $1.00, so much per $100, or so much per $1,000.

773. What are the three items that are usually involved in taxation?

(*a*) Base = amount to be taxed = assessed valuation.

(*b*) Rate = tax rate.

(*c*) Tax amount = tax expressed in dollars.

Ex. (*a*) What is the tax on a property valued at \$8,000 (base) at $2\frac{1}{4}\%$ (rate)?

$$\$8,000 \times .0225 = \$180 = \text{tax amount. Ans.}$$

Ex. (*b*) What is the tax on a \$9,000 property when the rate is 30 mills per \$1.00?

$$1 \text{ mill} = \tfrac{1}{10}¢,\ 30 \text{ mills} = 30 \times \tfrac{1}{10}¢ = 3¢ = \$.03 \text{ per } \$1.00 = \text{rate}$$
$$\text{Base} = \$9,000$$
$$\therefore\ \$9,000 \times .03 = \$270 = \text{tax amount. Ans.}$$

774. What is the tax on a property assessed for \$7,500, if the rate is \$2.885 per \$100, and the collectors' fee is 2%?

\$2.885 per \$100 = 2.885% = .02885

$$\$7,500 \times .02885 = \$216.38$$
$$\$216.38 \times .02 = \$4.33$$
$$\therefore\ \text{Tax} + \text{fee} = \$220.71 \text{ Ans.}$$

775. How do we find the tax rate when given the base (assessed valuation) and the tax amount?

Divide the tax amount by the base.

Ex. (*a*) What is the tax rate on a \$4,000 property when the tax is \$80?

$$\frac{\$80}{\$4,000} = \frac{1}{50} = \frac{2}{100} = 2\% \text{ Ans.}$$

Ex. (*b*) If the assessed valuation of taxable property in a town is \$2,383,015, and the tax to be raised is \$68,750, what should be the tax rate expressed as a per cent, the rate per \$100 valuation, and the rate per \$1,000 valuation?

$$\$68,750 \div \$2,383,015 = .02885 = 2.885\%$$
$$2.885\% = \$2.885 \text{ per } \$100$$
$$2.885\% = \$28.85 \text{ per } \$1,000$$

776. How do we find the assessed valuation when given the tax rate and the tax?

Divide the tax by the tax rate.

Ex. (*a*) What is the base (assessed valuation) when the tax is \$300 and the rate is 3%?

$$\frac{\$300}{\frac{3}{100}} = \overset{100}{\cancel{300}} \times \frac{100}{\underset{1}{\cancel{3}}} = \$10,000 \text{ Ans.}$$

Ex. (*b*) What is the value of the assessable property of a town, if the tax roll is $68,750 and the tax rate is $2.885 per $100?

$$\$2.885 \text{ per } \$100 = 2.885\% = .02885$$

$$\therefore \$68{,}750 \div .02855 = \$2{,}383{,}015 \text{ Ans.}$$

777. How do we calculate (*a*) surtax, (*b*) total tax?

(*a*) Multiply the base by the surtax rate.

(*b*) Multiply the base by the regular tax rate.

Add (*a*) and (*b*).

Ex. (*a*) What is the total tax on $16,000 if the regular tax is 5% and the surtax is 3%?

$$\$16{,}000 \times .05 = \$800$$
$$\$16{,}000 \times .03 = \$480$$
$$\therefore \overline{\$1{,}280} = \text{total tax. Ans.}$$

The surtax may not start at the same point as the regular tax.

Ex. (*b*) What is the total income tax on $8,000, if the regular tax is 5% and the surtax is 2% after the first $3,000 of income?

$$\$8{,}000 \times .05 = \$400 = \text{regular tax}$$
$$\$8{,}000 - \$3{,}000 = \$5{,}000$$
$$\$5{,}000 \times .02 = \$100 = \text{surtax}$$
$$\therefore \$400 + \$100 = \$500 = \text{total tax. Ans.}$$

PROBLEMS

1. What is the selling price if goods cost you $30, and you want to make a profit of 40% of the cost?

2. What is the selling price if the cost is $30, and the loss is 40% of the cost?

3. What is the per cent of profit if the selling price is $180 and the cost is $130?

4. What is the per cent of loss if the selling price is $130, and the cost is $180?

5. If the trade discount is 20% and the cost of the item is $15, what is the amount of the discount?

6. If the trade discount is .30 and the selling price is $15, what is the amount of the discount?

7. If the terms are 6/10; n/60 and the bill is $1,800, what are the cash discount and the net amount?

8. Find the bank discount at 6% and the net proceeds of a 92-day note for $1,000, when the date of the note is July 1, 1961, and the due date is October 1, 1961.

9. Given cost \$500 and discounts 40%, 6%, and 3%, find the net price of the goods.

10. If the gross cost (or list price) is \$425 and the discounts are 40%, 6%, and 2%, find the net cost by first getting a single equivalent remainder considering 100 as the base.

11. By mental calculation, find a single discount equal to 35% and 5%.

12. Find a single discount equal to 40%, 5%, and 3%.

13. If after 10% and 3% discounts are deducted, the net cost of an invoice of goods is \$1,232.86, what is the list price?

14. If the amount of discount is \$285.15 and the discounts are $33\frac{1}{3}$% and 3%, what is the net cost of the goods?

15. If the terms on a \$1,800 invoice of goods are 4/10; n/60, how much would you gain if you borrow money at a bank at 6% for 60 days, and pay cash for the goods?

16. If the gross cost of an article is \$12 and it is sold at a profit of 35%, how much is the net profit if 18% is charged to the cost of doing business?

17. If the net cost of an article is \$18.40, what is the selling price if it is to be sold to make a net profit of 20% of the cost, and the cost of doing business is 18% of the selling price?

18. A dealer buys a TV set for \$380, less 40% and 2%. He sells it for \$425, less 10%. If the cost of doing business is 18% of the sales, what per cent of the selling price is his net profit?

19. If shirts cost \$66 a dozen, less 40% and 2%, and are sold for \$6.25 each, less 10%, and the cost of doing business is 18% of the sales, what is the per cent of profit on net cost?

20. If trousers are bought for \$8.40, less 20% and 5%, and are sold for \$10.20, less 10%, and the buying expenses are 3% of the net cost and selling expenses are 16% of net sales, what per cent of the gross cost is the net profit?

21. If 35% = % of profit on gross cost of an article, and the profit is \$16.40, and 6% = cost of buying, what are the net cost of the article and the cost of buying?

22. The net cost of an article is \$60. The buying expenses are 4% of net cost. What is the per cent of profit on the gross cost if the article is sold at a profit of \$14.30, and what is the selling price?

23. If you buy an article invoiced at \$42.80, less 10% discount, and sell it at 25% profit, what is the selling price?

24. If a dealer buys a refrigerator for \$380, pays \$15 freight and cartage, and sells it at a profit of 30%, what is the selling price?

25. If a merchant pays \$26.70 for an article and sells it at a profit of 28% of the selling price, what is the selling price?

26. If the gross cost of an article is \$12.35, and it is sold at a profit of 30% on the selling price, what is the net profit if the cost of doing business is 15%?

27. What is the income tax on $7,500 if the regular tax is 4% and the surtax is 2% after the first $2,500 of income?

28. The cost of a washer-dryer to an appliance dealer is $340, less 40% and 2%. What should he mark the set if he wants to make a profit of 28% on the net cost, and allow the customer a 12% discount on the marked price?

29. What are the present worth and the true discount of a debt for $2,400 due in 9 months, if money is worth 6% interest?

30. If *A* owes *B* $2,400 which is not due until 2 years from now, and *A* offers to pay *B* today, what sum should *A* pay now at compound interest, assuming money to be worth 6%?

31. Find the present value of $2,400 due in 3 years at 4% compound interest.

32. If a TV set is priced at $195 cash, and the advertised payment plan is $35 down and $4.50 a week for 40 weeks, how much more does it cost on the installment plan?

33. If you buy a washing machine for $240 with a $35 trade-in allowance on your old one, and agree to pay the balance in 10 monthly installments, plus a final installment of $30, how much would you save by buying for cash?

34. If you borrow $1,800 from a bank and pay it back in monthly payments of $42.29 over 4 years, how much would you pay the bank for the loan?

35. On the basis of the 6% method offered by some credit companies, if you buy a refrigerator for $450, make a down payment of $150, and then pay the balance of $300 in 1 year, what would be your monthly payment?

36. If a merchant wishes to charge an additional 16% on the goods he sells on time, what would be the price on a 10-equal-payment plan, and the amount of each payment on a radio that sells for $98 cash?

37. If the interest or carrying charge is $12, and there are 8 monthly payments of $12 each, what is the interest rate per year by the "sum of the time" method?

38. How much interest and financing charge do you pay when you buy a TV set for $250 with a $40 trade-in allowance on your old set, and you agree to pay the balance in 10 monthly installments of $21, plus a final installment of $30, using the "sum of the time" method?

39. If you buy some furniture for $760 and pay $140 down and the balance in 10 monthly installments of $73 each, what is the rate of financing charge by the "sum of the time" method?

40. Solve Problem 39 by the "total installment money kept for one month" method.

41. If you borrow $200 from a finance company with a 3% per month charge on the unpaid balance of the loan, and you are required to repay the loan in 10 monthly installments of $20 each, how much do you pay back for the \$200 loan, including interest, and what is the annual interest rate, using the "sum of the time" method?

42. If you borrow \$200 from a credit union, and are charged 1% a month on the unpaid balance, and you pay back the loan in 10 monthly installments of \$20 plus interest charge, how much do you pay back, and what is the annual interest rate by the "sum of the time" method?

43. If you get a loan of \$2,000 at 5% interest per year, and you agree to pay it back in 20 years at \$12.50 per month, how much is the total amount of repayment, and how much does it cost you?

44. How does the cost in Problem 43 compare with a bank loan of \$2,000 for 20 years at 5%?

45. If you get a loan of \$6,000 at 5% a year on the unpaid balance from a mortgage company to finance your home, and you agree to pay it back in 12 years at \$55.49 per month, what is the total repayment on the loan, and how much does it cost you?

46. What is the per cent interest per year on a loan of \$200 plus \$25 carrying charge, to be paid in 10 equal monthly installments, using the "constant-ratio" method?

47. If a TV set is priced at \$200 cash, and advertised on a payment plan of \$30 down, and \$5 a week for 37 weeks, what is the interest rate, using the "constant-ratio" method?

48. A radio is offered for \$65 cash, or on time payments for 10% more with a down payment of \$12, and the balance in 12 weekly payments. What is the annual rate of interest, using the "constant-ratio" method?

49. A "hi-fi" set can be bought for \$640 cash with a discount of \$20, or in 12 equal monthly installments by first paying \$150 and adding a \$32 carrying charge. What is the annual rate of interest, using the "constant-ratio" method?

50. If you borrow \$250 from a loan company for 10 months, and repay it in 10 equal installments of \$28.80, what rate of interest do you pay? Solve this by the (*a*) "constant-ratio" method, (*b*) "sum of the time" method, (*c*) "total installment money" method.

51. If you borrow \$500 from a bank for 16 months and pay back \$33.65 per month, what annual rate are you paying, as figured by the three methods studied?

52. If you buy, on time, a typewriter that costs \$98 cash, and \$14 is added for carrying charges on a payment plan of \$14 down and \$12 a month for 7 months, what is the rate of interest you pay, using the "constant-ratio" method?

53. What is the interest on the time plan, if a clothes dryer sells for \$215 cash or \$25 down and \$22.80 per month for 10 months? Use the "constant-ratio" method.

54. What is the interest rate per year, on a time plan, on a set of cooking utensils that is advertised at \$34 cash, or \$5 down and \$6 a week for 5 weeks, with a last payment of \$3 in the sixth week, using the special "constant-ratio" method?

55. What is the interest rate per year if a clock costs \$30 cash, or \$6 down and \$6 per month for 4 months, with a \$2.50 payment the fifth month. Use the special "constant-ratio" method?

56. A note for \$2,000 dated May 15, 1961, has the following payments endorsed on the back: August 12, 1961, \$500; October 28, 1961, \$400; November 29, 1961, \$200. If the maker desires to pay in full on December 31, 1961, what is the amount due at that time, by the merchants' rule with interest at 6%?

57. Find the balance due on December 31, 1961 on the note of Problem 56, using the United States rule.

58. By the United States rule, how much is required to settle on September 1, 1961, a demand note for \$8,000 dated March 1, 1960, with interest at 6%, and with the following payments endorsed upon it: May 12, 1960, \$1,600; September 3, 1960, \$80; March 4, 1961, \$3,200; July 5, 1961, \$800?

59. What is the tax on a property valued at \$10,000 at $2\frac{1}{4}$% rate?

60. What is the tax on a \$12,000 property when the rate is given as 35 mills per \$1.00?

61. What is the total tax on a property assessed for \$9,500, if the rate is \$2.963 per \$100, and the collector's fee is 2%?

62. What is the tax rate on a \$6,000 property when the tax is \$120?

63. If the assessed valuation of taxable property in a town is \$3,875,680 and the tax to be raised is \$89,430, what would be the tax rate expressed as a per cent, the rate per \$100 valuation, and the rate per \$1,000 valuation?

64. What is the assessed valuation of a property when the tax amount is \$340, and the rate is 2.9%?

65. What is the value of the assessable property of a town if the tax roll is \$89,430 and the tax rate is \$2.910 per \$100?

66. What is the total tax on \$12,000 if the regular tax is 5% and the surtax is 3%?

CHAPTER XIX

VARIOUS TOPICS

A. Working rates of speed

778. What factors are involved in any problem relating to men working?

(*a*) The number of men that are working.
(*b*) The amount of work to be done.
(*c*) The time involved.

779. How can we find the time it will take one man to do the amount of work done by a number of men who work at equal rates of speed?

Multiply the number of men by the given time.

EXAMPLE: Seven men working at equal rates of speed take 10 days to finish a job. How long will it take one man to do the job?

$$7 \text{ men} \times 10 \text{ days} = 70 \text{ man-days}$$

$$\therefore \ 1 \text{ man takes } 70 \text{ days. Ans.}$$

780. How can we find the time it will take a number of men (working at equal rates of speed) to do a job when we know the time it takes one man to do it?

Divide the given time by the number of men.

EXAMPLE: One man works 8 days to finish a job. How long will it take four men to do the same job (all working at equal rates of speed)?

$$1 \text{ man} \times 8 \text{ days} = 8 \text{ man-days}$$

$$\therefore \ \frac{8 \text{ man-days}}{4 \text{ men}} = 2 \text{ days (for 4 men). Ans.}$$

781. How can we find the time it will take a number of men to do a job when given the time for a different number of men (working at equal rates of speed) to do the job?

Multiply the given number of men by the given time to get the man-days equal to the time it takes one man to do the work. Then divide this by the required number of men.

EXAMPLE: How long will it take 5 men to do a job that is done by 8 men in 50 days?

$$8 \text{ men} \times 50 \text{ days} = 400 \text{ man-days} = \text{time for one man}$$

$$\therefore \frac{400}{5 \text{ men}} = 80 \text{ days for 5 men. Ans.}$$

782. If the rates of speed of the men are unequal, how can we find the time it will take one of the men to do a job when given the time and the ratio of the speeds with which a number of men do the job?

(*a*) Assume the slowest man as a base of 1 and set up a ratio to get the "equal" number of man-days based on the work of the slowest man.

(*b*) Multiply the given time by the "equal" number of man-days to get the time of the slowest man to do the job himself.

(*c*) Divide this product by the number of "equal" man-days required.

EXAMPLE: If three men do a job in 10 days and two of the men are twice as fast as the third, how long will it take one of the faster men to do the job?

The slow man = 1 = base
Ratio is 2:1:2
Therefore 2 + 1 + 2 = 5 = number of "equal" man-days, based on the work of the slowest man
Now, given time, 10 days × 5 ("equal" man-days) = 50 days = time for slowest man to do the job himself
Since one of the faster men is twice as fast,

$\therefore \frac{50}{2} = 25$ days = time of a fast man to do the job himself. Ans.

783. How do we find the amount of work a man will do in part of the time when we know the time it takes him to do the entire job?

Express the times as a fraction.

EXAMPLE: If it takes a man 9 days to do a job, how much of the work will he do in 3 days?

Express as a fraction $\frac{3 \text{ days}}{9 \text{ days}} = \frac{1}{3}$

$\therefore$ He will do in 3 days $\frac{1}{3}$ of the work that he would do in 9 days. Ans.

784. Knowing the time necessary to complete a fraction of a job, how can you find the time necessary to do the entire job?

Divide the given time by the fraction.

EXAMPLE: If $\frac{3}{4}$ of the job is done in 6 days, how long will it take to complete the job?

$$\frac{6 \text{ days}}{\frac{3}{4}\text{ (fraction)}} = \overset{2}{\cancel{6}} \times \frac{4}{\cancel{3}} = 8 \text{ days to complete the job. Ans.}$$

785. How can we find the time it will take a number of men working together to do a job, when we know their respective rates of work?

Find the part of the job each would do in 1 day.

Add these fractions to get the combined part of the job done in 1 day.

Divide 1 by this combined fraction.

EXAMPLE: If it takes A 3 days to paint a house, B 4 days for the same job, and C 8 days, how long will it take them to do the job working together?

In 1 day A will do $\frac{1}{3}$ of the job
In 1 day B will do $\frac{1}{4}$ of the job
In 1 day C will do $\frac{1}{8}$ of the job

Then $\frac{1}{3} + \frac{1}{4} + \frac{1}{8} = \frac{8}{24} + \frac{6}{24} + \frac{3}{24} = \frac{17}{24}$ of the work will be done in 1 day, all working together.

$$\therefore \frac{1}{\frac{17}{24}} = \frac{24}{17} = 1\frac{7}{17} \text{ days to do the job together. Ans.}$$

786. Knowing the time it takes a number of men to complete a job and the individual rates of work except one, how can we find the time it would take the man with the unknown rate to do the job by himself?

(*a*) From the given time, get the fraction of the work done in 1 day when all work together.

(*b*) Get the fraction of the work done by each whose rate is known, and add these fractions.

(*c*) Subtract sum of (*b*) from (*a*) to get the fraction or part of the job done in 1 day by the man with the unknown rate.

(*d*) Divide 1 by fraction resulting in (*c*) to get the time it would take him to do the job by himself.

Example: A can do a job in 6 days and B in 8 days. A, B, and C working together can do it in 3 days. How long will it take C to do the job by himself?

In 1 day A can do $\frac{1}{6}$ of the job
In 1 day B can do $\frac{1}{8}$ of the job
In 1 day $A + B = \frac{1}{6} + \frac{1}{8} = \frac{4}{24} + \frac{3}{24} = \frac{7}{24}$
In 1 day $A + B + C = \frac{1}{3}$ of the job
In 1 day $\frac{1}{3} - \frac{7}{24} = \frac{8}{24} - \frac{7}{24} = \frac{1}{24}$ of the job for C alone

$$\therefore \frac{1}{\frac{1}{24}} = 24 \text{ days for } C \text{ to do the job by himself. Ans.}$$

B. Mixtures — Solutions

787. What is the procedure for solving an ordinary mixture problem?

(*a*) Consider the element of the mixture that does not change (the constant ingredient) and find its amount in the original mixture.

(*b*) Find the per cent this amount is of the final mixture.

(*c*) From this get the amount of the final mixture.

(*d*) Subtract the original mixture from the final mixture to get the quantity or amount to be added.

Example: How much alcohol would you add to a 20% alcohol mixture of 180 gallons of alcohol and ammonia to make a 25% alcohol mixture?

(*a*) Ammonia is the constant ingredient, which is 80% of the original mixture, or

$$.8 \times 180 \text{ gal.} = 144 \text{ gal. ammonia}$$

(*b*) 144 gal. ammonia = 75% of the final mixture.

(*c*) 144 = $\frac{3}{4}$ of the final mixture. Therefore

$$\frac{144}{\frac{3}{4}} = 144 \times \frac{4}{3} = 192 \text{ gal.} = \text{the final mixture}$$

(*d*) (Final) 192 gal. − 180 gal. (original) = 12 gal. of alcohol to be added to make a 25% alcohol mixture.

Proof: The alcohol originally was 20% of 180 gal. = 36 gal.
Add 12 gal. +12 gal.
48 gal.

$$\therefore \frac{48 \text{ gal.}}{192 \text{ gal.}} = \tfrac{1}{4} = 25\% \text{ alcohol. Ans.}$$

788. When given two different grades of an article in a mixture, how can we find the amount of each?

EXAMPLE: How many pounds of groats that sell for 16¢ per lb. should be mixed with groats that sell for 24¢ per lb., to get a total mixture of 100 lb. to sell for 18¢ per lb.?

(*a*) Find the value of the total mixture at the given price

100 lb. × 18¢ = $18.00

(*b*) Find the value of the total mixture at the lower price

100 lb. × 16¢ = $16.00

(*c*) Subtract the lower from the higher value

$18.00 − $16.00 = $2.00

(*d*) Subtract the price of the lower item from the price of the higher item

24¢ − 16¢ = 8¢

(*e*) Now the difference between the values, $2.00, divided by the difference between the prices, 8¢, is

$$\frac{\$2.00}{\$.08} = 25$$

or 25 lb.—the number of pounds of the higher-grade ingredient.

(*f*) 100 lb. − 25 lb. = 75 lb. = amount of the lower grade in the mixture.

∴ You need 75 lb. of the 16¢ groats and 25 lb. of the 24¢ groats to make a 100 lb. mixture of the 18¢ groats. Ans.

789. If we know the percentage concentrations of several simple ingredients of a mixture, how can we find the percentage strength of the mixture?

EXAMPLE: What is the percentage strength of alcohol in a mixture of 6 gal. of 12% alcohol, 8 gal. of 14% alcohol, and 12 gal. of 35% alcohol?

If we have 1 gallon of 12% alcohol, .12 of the gallon is pure alcohol and .88 of the gallon is water.

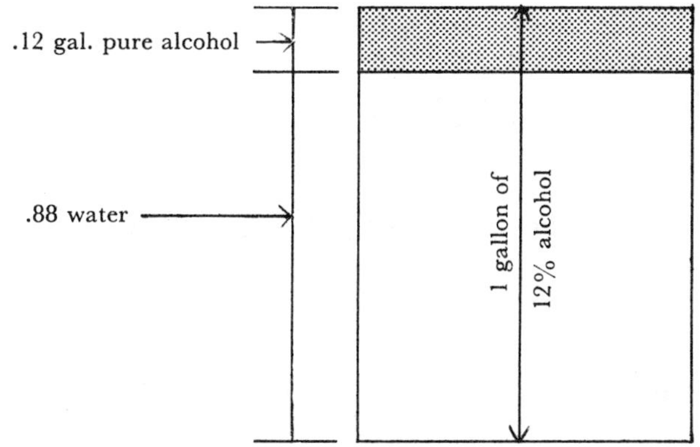

Now if we add 11 gallons of water to make a total volume of 12 gallons of solution, the concentration or proportion of alcohol is reduced to

$$\frac{.12 \text{ gal. alcohol}}{12 \text{ gal. solution}} = \frac{1}{100} = 1\% \text{ concentration}$$

Thus 1 gal. of 12% alcohol = 12 gal. of 1% alcohol and

6 gal. of 12% alcohol =	6 × 12 =	72 gal. of 1% alcohol
8 gal. of 14% alcohol =	8 × 14 =	112 gal. of 1% alcohol
12 gal. of 35% alcohol =	12 × 35 =	420 gal. of 1% alcohol
Total = 26 gal. of x% alcohol =	Total =	604 gal. of 1% alcohol

Thus 26 gal. of mixture contains as much pure alcohol as 604 gal. of 1% alcohol.

$$\therefore x = \frac{604 \text{ gal.}}{26 \text{ gal.}}$$

$= 23.23\%$ = percentage concentration of the mixture Ans.

790. How many quarts of water must be added to 5 quarts of a 35% solution of hydrochloric acid to reduce it to a 25% solution?

As above:

1 qt. of a 35% solution of hydrochloric acid = 35 qt. of a 1% solution of hydrochloric acid

Then 5 qt. of a 35% solution = 5 × 35 = 175 qt. of a 1% solution of hydrochloric acid

And x qt. of 25% solution = 175 qt. of a 1% solution, or

$$x = \frac{175}{25} = 7 \text{ qt.} = \text{total mixture}$$

∴ 7 qt. − 5 qt. = 2 qt. to be added to make it a 25% solution. Ans.

791. How is the above solved by the procedure of Question 787?

The hydrochloric acid does not change (is the constant ingredient), and is 35% of the original mixture:

.35 × 5 qt. = 1.75 qt. hydrochloric acid

Now 1.75 qt. = 25% of the final mixture

Then $\frac{1.75}{.25} = 7$ qt. = the final mixture

∴ 7 qt. − 5 qt. = 2 qt. water to be added to make a 25% solution of hydrochloric acid and water. Ans.

792. How much alcohol must we add to 3 quarts of a 25% solution of alcohol and water to make a 40% solution?

$$\begin{aligned} 3 \text{ qt. of } 25\% \text{ solution} &= 3 \times 25 = 75 \text{ qt. of } 1\% \text{ alcohol solution} \\ x \text{ qt. of } 100\% \text{ solution} &= 100x = (100x) \text{ qt. of } 1\% \text{ alcohol solution} \\ \text{Total} &= (3 + x) \text{ qt. of } 40\% \text{ solution} \\ &= (75 + 100x) \text{ qt. of } 1\% \text{ alcohol solution} \\ (3 + x) \times 40 &= 75 + 100x \\ 120 + 40x &= 75 + 100x \\ 45 &= 60x \end{aligned}$$

$$\therefore\ x = \tfrac{45}{60} = \tfrac{3}{4} \text{ qt. alcohol to be added. Ans.}$$

793. How is the above solved by the procedure of Question 787?

Water is the constant ingredient, which is 75% of the original mixture, or

$$\begin{aligned} .75 \times 3 &= 2.25 \text{ qt. of water} \\ \text{Now } 2.25 \text{ qt. of water} &= .6 \text{ of the final mixture} \end{aligned}$$

$$\text{Therefore } \frac{2.25}{.6} = 3.75 \text{ qt.} = \text{final mixture}$$

And $3.75 - 3.00 = .75$ qt. of alcohol to be added to make a 40% alcohol solution

Proof: Alcohol originally $= .25 \times 3 = .75$ qt.
Add alcohol $= .75$ qt.
Total $= 1.50$ qt. alcohol

$$\therefore \frac{1.5 \text{ qt.}}{3.75 \text{ qt.}} = \frac{1}{2.5} = \frac{2}{5} = 40\% \text{ concentration of final mixture. Ans.}$$

794. How can we use the procedure of Question 789 to determine the amount of each of several simple ingredients whose percentage of concentration is known to produce a mixture of a desired concentration?

EXAMPLE: In what proportion should 45% and 85% alcohol mixture be mixed to give an alcohol mixture of 68% strength? Percentages are by volumes.

x volumes of 45% alcohol $= x \times 45 = 45\,x$ volumes of 1% alcohol
y volumes of 85% alcohol $= y \times 85 = 85\,y$ volumes of 1% alcohol
Total $x + y$ volumes $= 45\,x + 85\,y$ volumes of 1% alcohol

Or $(x + y)$ volumes of mixture contain as much pure alcohol as $45x + 85y$ volumes of 1% alcohol. Therefore the strength of the mixture is as many per cent as the number of $(x + y)$ volumes contained in $45x + 85y$, or

$$\frac{45x + 85y}{x + y} = 68 \text{ (given \%)}$$

Then $45x + 85y = 68(x + y) = 68x + 68y$
and $85y - 68y = 68x - 45x$

$$17y = 23x$$

$$\frac{17}{23} = \frac{x\ (45\%)}{y\ (85\%)}$$

$\therefore$ Mix 17 volumes of 45% alcohol with 23 volumes of 85% alcohol to get a 68% alcohol. Ans.

795. How may the above be shown diagrammatically?

Place the desired (new) percentage concentration at the intersection of two diagonal lines. Place the percentage concentrations to be mixed at the left-hand corners. Merely take the difference between the center figure and each left-hand figure and place it at the corresponding end of the diagonal. This gives at once the part or volume to be mixed of the given solution concentration.

45% — (85 − 68) = 17 volumes of 45% alcohol

68%

85% — (68 − 45) = 23 volumes of 85% alcohol

This method is the result of the calculation in Question 794, and gives the same answer.

796. How may the above method be applied to mixtures of different quantities of liquids of known specific gravities?

EXAMPLE: How many gallons of water should be mixed with 12 gallons of glycerine of specific gravity 1.24 to get a desired 1.07 specific gravity?

water = 1.00 — (1.24 − 1.07) = .17 volumes of water

1.07

glycerine = 1.24 — (1.07 − 1.00) = .07 volumes of glycerine

Thus .17 volumes of water must be mixed with each .07 volumes of glycerine of sp. gr. 1.24 to produce a mixture of 1.07 sp. gr., or

$$\frac{.17}{.07} = \frac{17 \text{ water}}{7 \text{ glycerine}}$$

Then by proportion:

$$17 \text{ water} : 7 \text{ glycerine} = x \text{ gal. water} : 12 \text{ gal. glycerine}$$

$$\frac{17}{7} = \frac{x}{12} \quad \text{or} \quad 7x = 204$$

$$\therefore\ x = 29\tfrac{1}{7} \text{ gal. of water. Ans.}$$

Note: The above calculations apply only when the mixed liquids do not contract in volume when mixed. When alcohol and water are mixed in equal volumes there is a shrinkage of over 5.5% in volume. The solution of sugar in water also results in a contraction of volume.

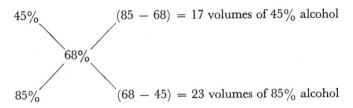

797. What types of percentage solutions occur in practice?

(*a*) Weight in weight, designated *w/w*.

This means that a definite weight of a substance is to be dissolved to produce 100 weights of solution.

(*b*) Weight in volume, designated *w/v*.

This means that a definite weight of substance is to be dissolved in enough solvent to produce 100 volumes of solution.

(*c*) Volume in volume, designated *v/v*.

This means that a definite volume of liquid is to be mixed with enough solvent to produce 100 volumes of solution. In the United States *v/v* concentration is designated for liquids and *w/v* for solids dissolved in liquids.

798. How may we convert (*a*) fluid ounces (United States) into avoirdupois ounces? (*b*) avoirdupois ounces into fluid ounces?

(*a*) Avoirdupois ounces = 1.04 fluid ounces (United States).
(*b*) Fluid ounces (United States) = avoirdupois ounces/1.04.

799. How many ounces of aluminum chloride should be dissolved to make a gallon of 25% *w/v* aqueous solution?

One United States gallon = 128 fluid ounces

$$25\% = .25$$
$$.25 \times 128 = 32$$
$$\therefore\ 32 \times 1.04 = 33.28 \text{ avoirdupois ounces of aluminum chloride. Ans.}$$

800. How much of 4.45% potassium sulfite and 6.7% of morpholine of specific gravity 1.0016 should be used to make a gallon of solution?

One United States gallon = 128 fluid ounces

$$4.45\% = .0445$$
$$.0445 \times 128 = 5.696 \text{ fluid ounces}$$
$$5.696 \times 1.04 = 5.92 \text{ avoirdupois ounces of potassium sulfite}$$
$$6.7\% = .067$$
$$.067 \times 128 = 8.576 = 8.58$$

Since morpholine is a fluid, it is more convenient to measure than to weigh; so to find the equivalent volume divide 8.58 by the sp. gr. 1.0016 and by 1.04 to convert to fluid ounces

$$\therefore\ \frac{8.58}{1.002 \times 1.04} = 8.25 \text{ fluid ounces of morpholine. Ans.}$$

801. How much pure lysol (100%) is needed to make 1,000 cc. of 3% lysol solution?

Write this in the form of a proportion:

$$\frac{\text{small }\%}{\text{large }\%} = \frac{\text{small quantity before dilution }(q)}{\text{large quantity after dilution }(Q)}$$

or

$$\frac{3\%}{100\%} = \frac{q}{1{,}000\text{ cc.}} \quad \text{and} \quad 100q = 3 \times 1000$$

$$\therefore\ q = \frac{3{,}000}{100} = 30 \text{ cc. of lysol to be diluted. Ans.}$$

802. A mixture of 54 pints of acid and water contains 24 pints of pure acid and 30 pints of water. How much water must be added to make a mixture that is 25% pure acid?

The constant ingredient is the acid = 24 pints. 24 pints = 25% of final solution, or

$$\frac{24}{\frac{1}{4}} = 24 \times 4 = 96 \text{ pints} = \text{final solution}$$

$$\therefore\ 96 - 54 = 42 \text{ pints of water to be added. Ans.}$$

C. Tanks and Receptacles (Filling, Emptying)

803. When we are given the time it takes to fill a tank, how can we express the part of the tank filled in a unit of time?

Expressed by a fraction, 1 divided by the time.

EXAMPLE: If it takes 10 minutes to fill a tank, how much of the tank is filled in 1 minute?

$\frac{1}{10}$ of the tank is filled in 1 minute. Ans.

804. When we are given the fraction of the tank filled in a unit of time, how can we find the time it takes to fill the whole tank?

Divide 1 by the fraction of the tank.

EXAMPLE: If in 1 minute a pipe can fill $\frac{5}{12}$ of a tank, then

$$\frac{1}{\frac{5}{12}} = 1 \times \frac{12}{5} = 2\frac{2}{5} \text{ minutes to fill the tank. Ans.}$$

805. How do we find the time it takes to fill a tank when we have several pipes acting at the same time, and we are given the time each takes to fill it when acting alone?

(*a*) Find the part of the tank filled in 1 minute by each pipe, in fraction form.

(*b*) Add the fractions.

(*c*) Invert the sum to get the time needed when all act together.

EXAMPLE: A 2-inch pipe fills a tank in 8 minutes, a 3-inch pipe fills it in 5 minutes. How long will it take to fill the tank with both pipes acting together?

The 2-in. pipe fills $\frac{1}{8}$ of the tank in 1 min.
The 3-in. pipe fills $\frac{1}{5}$ of the tank in 1 min.

Together, $\frac{1}{8} + \frac{1}{5} = \frac{5 + 8}{40} = \frac{13}{40}$ of the tank filled in 1 min.

$\therefore\ \frac{40}{13} = 3\frac{1}{3}$ min. for both pipes to fill the tank, acting together. Ans.

806. What is the procedure for solving a tank problem when filling and emptying take place at the same time?

(*a*) For each pipe acting alone, find the fractional part of the tank being filled or emptied in a unit of time.

(*b*) Add the fractions for filling.

(*c*) Add the fractions for emptying.

(*d*) Compare the sums by finding the lowest common denominator of both fractions. The one with the greater numerator will be the larger quantity and the faster process.

EXAMPLE: Will a tank eventually remain filled or be emptied if it has a pipe (#1) which can fill it in 10 hours, a pipe (#2) which can fill it in 6 hours, a pipe (#3) which can empty it in 7 hours, and a pipe (#4) which can empty it in 5 hours, and all pipes are in simultaneous operation?

Pipe #1 can fill $\frac{1}{10}$ of the tank in 1 hr. = rate of filling
Pipe #2 can fill $\frac{1}{6}$ of the tank in 1 hr. = rate of filling
Pipe #3 can empty $\frac{1}{7}$ of the tank in 1 hr. = rate of emptying
Pipe #4 can empty $\frac{1}{5}$ of the tank in 1 hr. = rate of emptying

Sum of filling rates:

$$\frac{1}{10} + \frac{1}{6} = \frac{3 + 5}{30} = \frac{8}{30} = \frac{4}{15} \text{ of the tank filled in 1 hour}$$

Sum of emptying rates:

$$\frac{1}{7} + \frac{1}{5} = \frac{5+7}{35} = \frac{12}{35} \text{ of the tank emptied in 1 hour}$$

$15 = 3 \times 5 \qquad 35 = 7 \times 5 \qquad \therefore \text{ L.C.D.} = 3 \times 5 \times 7 = 105$

Thus $\frac{4}{15} = \frac{28}{105}$ and $\frac{12}{35} = \frac{36}{105}$

$\frac{28}{105}$ of the tank is filled in 1 hour

$\frac{36}{105}$ of the tank is emptied in 1 hour

The tank will eventually be emptied when all the pipes are open.

$\frac{36}{105} - \frac{28}{105} = \frac{8}{105}$ of the tank will be emptied in 1 hour

$$\therefore \frac{1}{\frac{8}{105}} = \frac{105}{8} = 13\tfrac{1}{8} \text{ hours to empty the tank. Ans.}$$

Note that here, where the emptying fraction is greater than the filling fraction, the tank must be filled at the beginning of the operation.

807. How can we find the number of gallons a container can hold?

Multiply its contents (expressed in cubic feet) by $7\frac{1}{2}$.

1 cu. ft. = 12 in. × 12 in. × 12 in. = 1,728 cu. in.

1 standard United States gallon contains 231 cu. in.

$\therefore \frac{1728}{231} = 7.48$ gallons in a cu. ft. $= 7\frac{1}{2}$ gallons (approx.). Ans.

Example: How many gallons in a container 6′ × 10′ × 4′?

$$6' \times 10' \times 4' = 240 \text{ cu. ft.}$$

$$\therefore 240 \times \tfrac{15}{2} = 1{,}800 \text{ gallons. Ans.}$$

D. Scales for Models and Maps

808. When do we have a true scale model of any structure?

When the ratio of the length of any part of a model to the length of the same part in the actual structure is the same for all parts, then we have a true scale model of the structure.

Example: What is the scale of a model of a tower on a suspension bridge if the actual height is 200 ft. and the height on the model is 10 inches?

$$10 \text{ in.} = 200 \text{ ft.} \quad \text{or} \quad 1 \text{ in.} = 20 \text{ ft.}$$

This means that 1 in. anywhere on the model represents 20 ft. or

$$12 \times 20 = 240 \text{ in. on the structure}$$

$$\therefore 1:240 \text{ or } \tfrac{1}{240} \text{ is the scale of the model Ans.}$$

809. If the uniform recommendation for airplane models is 1:72, what is the wing span of a model if the wing span of the actual plane is 80 ft.?

Scale is 1 : 72 or $\frac{1}{72}$

This means 1 in. on the model represents 72 in. on the structure, then

$$\frac{\text{wing span of model}}{12 \times 80 \text{ ft.} = 960 \text{ in.}} = \frac{1}{72}$$

$$\therefore \text{ Wing span of model} = \frac{\overset{40}{\cancel{960}}}{\underset{3}{\cancel{72}}} = \frac{40}{3} = 13\tfrac{1}{3} \text{ in. Ans.}$$

810. If the scale of the model of an airplane is 1:72, how far away from the model would you have to stand so that it would appear the same as if you were 900 yards from the actual plane?

Scale is 1:72 or 1 yd.: 72 yd. Then

$$\frac{\text{distance}}{900 \text{ yd.}} = \frac{1}{72}$$

$$\therefore \text{ Distance} = \frac{\overset{25}{\cancel{900}}}{\underset{2}{\cancel{72}}} = \frac{25}{2} = 12\tfrac{1}{2} \text{ yd. Ans.}$$

811. What is a map and how is its scale expressed?

A map is a scale diagram showing geographic features on the earth, located with reference to one another.

The scale is sometimes given diagrammatically, as,

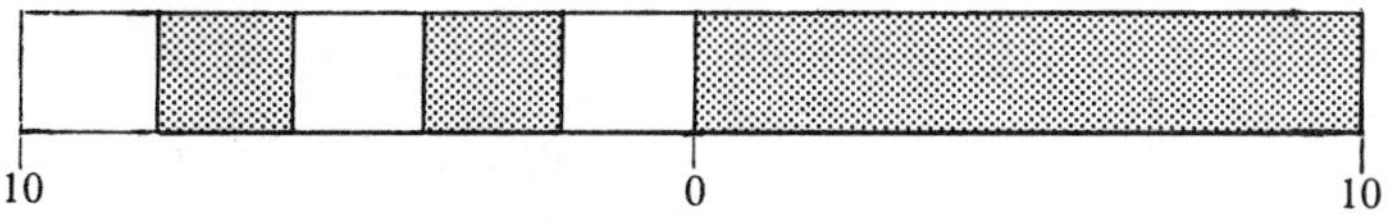

and is sometimes expressed as a ratio.

In sectional charts of the United States, the scale is 1:500,000.

In regional charts of the United States, the scale is 1:1,000,000.

Ex. (*a*) What is the ratio of a map that is drawn to the scale of 1 inch to the mile?

$$\frac{1 \text{ inch}}{1 \text{ mile}} = \frac{1 \text{ inch}}{(12 \times 5{,}280) \text{ inches}} = \frac{1}{63{,}360}$$

The scale or ratio is thus 1 : 63,360.

Ex. (*b*) How many miles does 1 inch represent on a sectional chart? Scale is 1 : 500,000 or 1 in. represents 500,000 in. on the earth.

$$\therefore\ 500{,}000 \text{ in.} = \frac{\overset{3{,}125}{\cancel{500{,}000}}}{\underset{33}{\cancel{5{,}280}} \times 12} = \frac{3{,}125}{396} = 7.9 \text{ miles. Ans.}$$

812. How many miles will $2\frac{3}{4}$ inches represent on a map drawn to a scale of 1:5,000,000?

Scale is 1 :5,000,000 or 1 in. represents 5,000,000 in.

$$\frac{\overset{31{,}250}{\cancel{5{,}000{,}000}}}{\underset{33}{\cancel{5{,}280}} \times 12} = \frac{31{,}250}{396} = 78.91 \text{ miles per inch}$$

$$\therefore\ 78.91 \times 2.75 = 217 \text{ miles. Ans.}$$

813. If the scale of a map is 1: 21,120, what would be the distance between two towns which are 24 in. apart on the map?

(*a*) By the ratio method:

$$\frac{1}{21{,}120} = \frac{24 \text{ in.}}{x \text{ in.}}$$

$$\therefore\ x = 24 \times 21{,}120 \text{ in.} = \frac{\overset{2}{\cancel{24}} \times \overset{4}{\cancel{21{,}120}}}{\underset{1}{\cancel{12}} \times \underset{1}{\cancel{5{,}280}}} = 8 \text{ miles. Ans.}$$

(*b*) By the method of getting the value of 1 inch on the map first, and then multiplying by the number of inches on the map:

1 in. on map represents 21,120 in. on the earth

24 in. on map represents 24 × 21,120 in. on the earth

$$\therefore\ \frac{\overset{2}{\cancel{24}} \times \overset{4}{\cancel{21{,}120}}}{\underset{1}{\cancel{12}} \times \cancel{5{,}280}} = 8 \text{ miles. Ans.}$$

814. If the scale of a map is $4\frac{1}{2}$ inches to the mile, what would be the actual area in acres of a lake on this map that has been found by a planimeter to have an area of 56 square inches?

Scale is $4\frac{1}{2}$ inches : 1 mile
Therefore $4\frac{1}{2}$ in. × $4\frac{1}{2}$ in. (= 20.25 sq. in.) = 1 sq. mi.
Now 1 sq. mi. = 640 acres

(*a*) By ratio method:

$$\frac{20.25 \text{ sq. in.}}{640 \text{ acres}} = \frac{56 \text{ sq. in.}}{x \text{ acres}} \quad \text{or} \quad 20.25x = 56 \times 640$$

$$\therefore\ x = \frac{56 \times \overset{128}{\cancel{640}}}{\underset{4.05}{\cancel{20.25}}} = \frac{7{,}168}{4.05} = 1{,}769.88 \text{ acres}$$

(*b*) By getting the value of 1 square inch on the map first and then multiplying by the given square inches.

Rule: If you want to get the value of one unit of any element in the problem, you should divide by that element.

We want the number of acres in 1 square inch, so we divide by square inches. Thus

$$\frac{640 \text{ acres}}{20.25 \text{ sq. in.}} = \text{number of acres in 1 sq. in.}$$

$$\therefore\ 56 \text{ sq. in.} = 56 \times \frac{640}{20.25} = \frac{35{,}840}{20.25} = 1{,}769.88 \text{ acres Ans.}$$

E. Angle measurement

815. What is an angle?

The opening between two lines intersecting at a point is called an angle. The gable of a roof and the intersection of two streets are practical examples of angles.

Also, angle = amount of turning required to rotate BA to position BC.

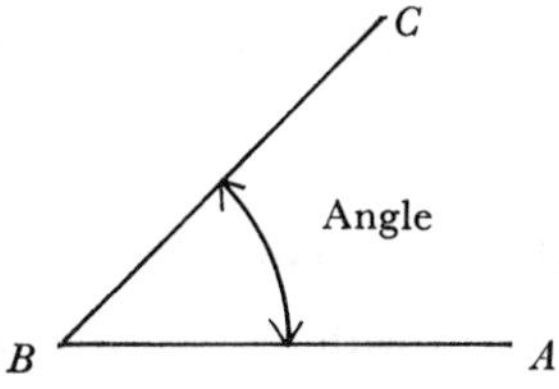

816. What are the parts of an angle?

An initial line, a terminal line, and a vertex constitute an angle.

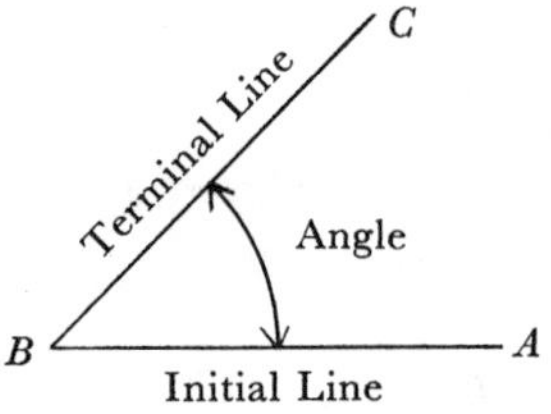

BA = initial line   BC = terminal line   B = vertex

817. What is meant by (*a*) an angle of 1 degree, (*b*) an arc of 1 degree?

(*a*) Divide the circumference of a circle into 360 equal parts and draw lines from the center of the circle to the points of division. 360 small angles will be formed, each of which is called an angle of 1 degree, or 1 deg. = $\frac{1}{360}$ of circumference.

(*b*) Each of the 360 equal parts of the circumference is called an arc of 1 degree.

A quarter of a circle = a right angle = 90 deg. = ninety 1-degree angles side by side. Half a circle = 180 degrees.

The symbol for a degree is [°]. Thus, 90° = 90 degrees.

818. What is meant by an angle of 1 minute?

Divide an angle of 1 deg. into 60 equal angles. Each of these is called an angle of 1 minute. The symbol for a minute is [′]. Thus $1' = \frac{1}{60}^\circ$. Each corresponding arc division is called an arc of 1 minute.

819. What is meant by an angle of 1 second?

Divide an angle of 1 minute into 60 equal angles. Each of these is called an angle of 1 second. The symbol for a second is [″]. Thus $1'' = \frac{1}{60}' = (\frac{1}{60} \times \frac{1}{60}) = \frac{1}{3600}^\circ$. Each corresponding arc division is called an arc of 1 second.

820. How can an angle be measured?

An angle can be measured with an instrument called a protractor. Place the protractor on the angle with 00 on one side and point 0 on the vertex. Read the scale where the other side crosses it. This gives degrees of angular measurement.

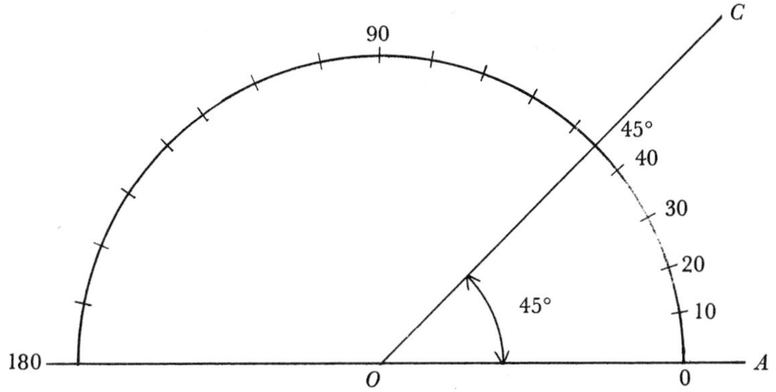

Vertex Protractor

PROBLEMS

1. Working at equal rates of speed, 8 men take 12 days to finish a job. How long will it take one man to do the job?

2. One man works 10 days to finish a job. How long will it take five men to do the same job, all working at equal rates of speed?

3. How long will it take 8 men to do a job that is done by 12 men in 40 days, working at equal rates of speed?

4. If 3 men do a job in 12 days and two of the men are three times as fast as the third, how long will it take one of the faster men to do the job?

5. If it takes a man 12 days to do a job, how much of the work will he do in 3 days?

6. If $\frac{5}{8}$ of a job is done in 15 days, how long will it take to complete the job?

7. If it takes *A* 4 days to build a boat, *B* 6 days for the same job, and *C* 10 days, how long will it take them to do the job working together?

8. *A* can do a job in 5 days and *B* in 7 days. *A*, *B*, and *C* working together can do it in 3 days. How long will it take *C* to do it by himself?

9. How much alcohol would you add to a 25% alcohol mixture of 160 gallons of alcohol and ammonia to make a 40% alcohol mixture?

10. How many pounds of rice that sells for 25¢ per lb. should be mixed with rice that sells for 35¢ per lb. to get a total mixture of 120 lb. to sell for 28¢ per lb.?

11. What is the percentage strength of alcohol in a mixture of 8 gal. of 14% alcohol, 10 gal. of 22% alcohol, and 16 gal. of 40% alcohol?

12. How many quarts of water must be added to 8 quarts of a 40% solution of hydrochloric acid to reduce it to a 16% solution?

13. How much alcohol must we add to 5 quarts of a 30% solution of alcohol and water to make a 60% solution?

14. In what proportion should 35% and 65% mixtures of alcohol be mixed to give an alcohol mixture of 54% strength? Percentages are by volumes.

15. Show how Problem 14 can be solved by a diagrammatic method.

16. How many gallons of water should be mixed with 14 gal. of glycerine of specific gravity 1.22 to get a desired 1.05 specific gravity? Use diagrammatic method to get the ratio of volumes of water to glycerine.

17. How many ounces avoirdupois of aluminum chloride should be dissolved to make a gallon of 30% *w/v* aqueous solution?

18. How much 3.35% potassium sulfite and 8.2% morpholine of specific gravity 1.002 should be used to make a gallon of solution?

19. How much pure lysol (100%) is required to make 2,500 cc. of 5% lysol solution?

20. A mixture of 98 pints of acid and water contains 42 pints of pure acid and 56 pints of water. How much water must be added to make a mixture that is 30% pure acid?

21. If it takes 12 minutes to fill a tank, how much of the tank is filled in 1 minute?

22. If in 1 minute a pipe can fill $\frac{2}{5}$ of a tank, how long will it take to fill the entire tank?

23. A 2-in. pipe can fill a tank in 12 min., a 3-in. pipe can fill it in 4 min. How long will it take to fill the tank with both pipes acting together?

24. Will a tank eventually be filled or emptied, if it has a pipe *A* which can fill it in 8 hours, a pipe *B* which can fill it in 6 hours, a pipe *C* which can empty it in 5 hours, a pipe *D* which can empty it in 6 hours, and all pipes are in operation simultaneously?

25. How many gallons are there in a container 8 ft. × 12 ft. × 6 ft.?

26. What is the scale of the model of a radio tower if the actual height is 450 ft. and the height of the model is 15 in.?

27. If the scale is 1:72, what is the wing span of a model when the wing span of the plane is 105 ft.?

28. If the scale of the model of a plane is 1:72, how far from the model should you be so that it will appear the same size as the real plane at a distance of 1,500 yd.?

29. What is the ratio of a map that is drawn to the scale of 1 in. to 4 miles?

30. How many miles does $2\frac{1}{2}$ in. represent on a sectional United States chart?

31. How many miles will $3\frac{1}{2}$ in. represent on a map drawn to a scale of 1:5,000,000?

32. If the scale on a map is 1:31,680, what would be the distance between two towns which are 30 in. apart?

33. If the scale of a map is 5 in. to the mile, what would be the actual area in acres of a lake on this map that has been planimetered to be 38 sq. in.?

34. How many minutes are there in an angle of 34 degrees?

35. How many degrees are there in 2 revolutions of the terminal line?

36. How many seconds are there in an angle of 34 minutes?

37. How many seconds are there in $\frac{1}{900}°$?

CHAPTER XX

INTRODUCTION TO ALGEBRA

821. What is algebra?

The Arabic word *al-jabr* is said to mean the reunion of broken parts. Algebra thus unifies arithmetic, completes it, and shortens mathematical solutions. It is the science treating the correct use of mathematics. By its use unknown quantities may become known.

822. Why is algebra said to be a shorthand extension of arithmetic?

In arithmetic, we are concerned with the *numbers of things*, as, 15 molecules, 20 apples, and 80 dollars. In each case we have a number representing the *quantity* of this and the particular thing itself with its name written out.

In algebra, we still have the number representing the quantity, but we select a *symbol* to represent the thing as, x, molecules; y, apples; and z, dollars. Then $15x$ represents 15 molecules, $20y$ represents 20 apples, and $80z$ represents 80 dollars.

The symbols provide us with a shorthand method of expressing facts.

When a letter is used to represent a number it is known as a *literal* number.

EXAMPLE: What is meant by x pounds or y dollars?

The x or y may represent any amount depending upon the circumstances in the problem that is being considered.

823. How are the letter symbols in algebra selected?

A symbol may be used to represent anything we please. The same letter may be used to represent a certain thing in one problem, and a different thing in a different problem, but in any one problem one symbol is always kept for one thing, and a different symbol for a different thing.

A letter from the beginning of the alphabet, such as a, b, c, d, etc., is chosen for a quantity that is *constant* in any one problem.

A letter from the end of the alphabet as v, w, x, y, or z is chosen for a quantity that is a *variable* in any one problem.

However, the symbols are frequently arbitrary as a years, b dollars, p pounds, and x feet.

Some symbols are frequently conventional and are self-suggestive of what they represent, such as R = rate, P = principal, t = time, A = area, r = radius, w = weight, V = volume, v = velocity, a = acceleration, etc.

Small numbers known as subscripts are often used to distinguish one symbol from another representing the same kind of quantity. For example, v_1 and v_2 are used to represent two different velocities in the same problem, t_1 and t_2 may represent two different temperatures, and A_1 and A_2 may indicate two different areas.

824. What is meant when two letters, or a number and a letter, are placed alongside each other?

It means that they are to be multiplied together.

EXAMPLE: $ab = a \times b$, $xy = x \times y$, $3m = 3 \times m$, and $20p = 20 \times p$.

If $p = \frac{1}{4}$, then $20p = 20 \times \frac{1}{4} = 5$.

825. What is meant by a coefficient?

The number or arithmetical part in front of the symbol is called a coefficient.

EXAMPLE: In $20p$, 20 is the coefficient of p.

826. What is meant by a term?

The number and symbol taken together are called a term. One term is called a monomial.

EXAMPLE: $20p$ = a term.

Note that when we deal with one article we usually omit the coefficient 1.

EXAMPLE: If we want to represent one dollar we write simply x, instead of $1x$.

827. What is a binomial?

An expression that contains two terms (from Latin *bi-*, meaning two).

EXAMPLE: $(a + b)$, $(3x - 2y)$, and $(6 - 4x)$ are binomials.

828. What is meant by (*a*) a factor of a product, (*b*) literal factors or numbers, (*c*) specific numbers?

(*a*) Each of several numbers or letters that are multiplied is a factor of the product.

EXAMPLE: In ab, a and b are factors of the product ab. In $3ab$, 3, a, and b are factors of the product $3ab$. In $5 \times 6 = 30$, 5 and 6 are factors of 30.

(*b*) Letters used to express numbers are called literal factors or literal numbers.

EXAMPLE: In $3ab$, a and b are literal factors.

(*c*) Signed numbers are often called directed or specific numbers.

EXAMPLE: -3, -7, and -9 are specific numbers.

829. What is meant by (*a*) an algebraic quantity, (*b*) an algebraic expression?

(*a*) An algebraic quantity is one that has all literal factors or a combination of literal and specific numbers.

EXAMPLE: ab^2c^3 is an algebraic quantity with all literal factors. $-3a^2b^2$ is an algebraic quantity with a combination of literal and specific numbers.

(*b*) An algebraic expression contains two or more factors or quantities, or a combination of both connected by signs of operation.

EXAMPLE: $2ab + x^2 + 5d$, $9y - 5$, and $x^2 - 2yx + y^2$ are algebraic expressions.

Thus, an algebraic expression is made up of terms.

830. What is meant by the coefficient of a product?

In any product each factor is the coefficient of every other factor or group of factors.

Ex. (*a*) In the product $3x$, 3 is the coefficient of x, and x is the coefficient of 3.

Ex. (*b*) In ay^2, a is the coefficient of y^2, and y^2 is the coefficient of a.

Ex. (*c*) In $(a - 1)b$, $(a - 1)$ is the coefficient of b, and b is the coefficient of $(a - 1)$.

Ex. (*d*) In $12xy$, 12 is the coefficient of xy, $12x$ is the coefficient of y, and $12y$ is the coefficient of x.

831. What is a polynomial?

A quantity of two or more terms connected by plus or minus signs is a polynomial.

EXAMPLE: $3x + 5y$, $4ab^2 - 3bc^2 + bcd^2$ are polynomials.

832. What symbols are used in algebra to indicate addition and subtraction?

The same symbols used in arithmetic.

Let x denote a thing.
(*a*) Then $4x + 7x = 11x$ = addition.
(*b*) And $7x - 4x = 3x$ = subtraction.

833. What symbols are used to indicate multiplication and division?

(*a*) $5x \times 3 = 15x$ (multiplication with a multiplication sign between the factors); or $5x \cdot 3 = 15x$ (using a dot for the multiplication sign).

Two or more letters written together with no sign between them means that they are to be multiplied together, as:
$a \times b = a \cdot b = ab = a$ multiplied by b.
$x \times y \times z = x \cdot y \cdot z = xyz = x$ multiplied by y multiplied by z.

(*b*) $\frac{10x}{2} = 5x$ = division.

834. What are the four elements of every algebraic term?

(*a*) A sign, (*b*) a coefficient, (*c*) a symbol, and (*d*) an index.

EXAMPLE: In $-4x^3$, the sign is $-$, the coefficient is 4, the symbol is x, and the index is 3.
The term is read: "minus $4x$ cubed."

835. On what occasions are some of the elements omitted?

(*a*) When the coefficient is 1, it is omitted.
Thus $-x^2$ is actually $-1x^2$ = "minus one x squared."

(*b*) When the index is 1, it is omitted.
Thus $-5x$ is actually $-5x^1$ = "minus five x to the first power."

(*c*) A plus sign is omitted when the term stands alone or at the beginning of an expression.
Thus $5x^2$ is actually $+5x^2$ = "plus five x squared."

(*d*) According to (*a*), (*b*), and (*c*):
x is actually $+1x^1$ = "plus one x to the first power."
The sign, coefficient, and index are omitted.
$-x$ is actually $-1x^1$ = "minus one x to the first power."
Here we omit the coefficient and index, but not the sign.

836. How is $+x^1 - 5x^2 + 1x^4 - 3y^3$ written in practice?

$x - 5x^2 + x^4 - 3y^3$ Ans.

837. What laws of addition, subtraction, multiplication, and division of numbers are also applicable to algebraic processes?

(*a*) Cumulative law for addition.
In arithmetic, $5 + 9 = 9 + 5 = 14$.
In algebra, $a + b = b + a$.
The sum is the same regardless of the order in which the terms are added.

(*b*) Associative law for addition.
In arithmetic, $(5 + 9) + 12 = 5 + (9 + 12) = 26$.
In algebra, $(a + b) + c = a + (b + c) = a + b + c$.
The sum is the same regardless of the groups that are formed.

(*c*) Cumulative law for multiplication.
In arithmetic, $5 \times 9 = 9 \times 5 = 45$.
In algebra, $ab = ba$.
The product is the same regardless of the order of the factors.

(*d*) Associative law for multiplication.
In arithmetic, $(5 \times 9) \times 12 = 5 \times (9 \times 12) = 540$.
In algebra, $(ab)c = a(bc) = abc$.
The product is the same regardless of the grouping of the factors.

(*e*) When you multiply a factor by the sum of several terms it is the same as taking the sum of the products of the factor multiplied by each term.
In arithmetic, $5(9 + 12) = 5 \times 9 + 5 \times 12$.
In algebra, $a(b + c) = ab + ac$.

(*f*) When you multiply a factor by the difference between two terms it is the same as taking the difference between the products of the factor multiplied by each term.
In arithmetic, $5(9 - 12) = 5 \times 9 - 5 \times 12$.
In algebra, $a(b - c) = ab - ac$.

Cases (*e*) and (*f*) are known as the *distributive* laws for multiplication with respect to addition and subtraction.

838. How may we regard two or more letters or numbers enclosed in parentheses?

We may regard them all as one quantity.

Ex. (*a*) In $3(a + b)$, we first add $a + b$ and then multiply by 3.

Ex. (*b*) In $5(a - 3)$, we first subtract 3 from a and then multiply by 5.

Ex. (*c*) In $8(m + n + p)$, we first add m, n, and p and then multiply by 8.

Ex. (*d*) In $\frac{P + R}{4}$, we first add P and R and then divide by 4.

839. In algebraic fractions, why may the fraction be considered to act as a set of parentheses?

Because the entire numerator is to be divided by the denominator.

Ex. (*a*) In $\frac{a + 3}{4}$, you first add 3 to a and then divide by 4.

Ex. (*b*) In $\frac{2a + 3b}{5}$, $2a + 3b$ is considered one quantity which is to be divided by 5.

Ex. (*c*) In $\frac{20x - 5}{5}$, first subtract 5 from $20x$ and then divide by 5. It is not $\frac{20x}{5} - 5$.

However, if you break up the numerator you must divide each part by the denominator, or

$$\frac{20x}{5} - \frac{5}{5} = 4x - 1$$

If $x = 2$, then

$$\frac{20x - 5}{5} = \frac{20 \times 2 - 5}{5} = \frac{40 - 5}{5} = \frac{35}{5} = 7$$

or

$$4x - 1 = 4 \times 2 - 1 = 7$$

840. In what ways may $\frac{3}{4}x$ be written?

(*a*) $\frac{3}{4}x$ (*b*) $\frac{3x}{4}$ (*c*) $.75x$

841. How are verbal expressions translated to algebraic symbols and terms?

By substituting coefficients, symbols, and signs for words.

(*a*) Three times a number $= 3a$.
(*b*) One-sixth the base B $= \frac{1}{6} \times B$.
(*c*) Three times a number increased by 5 $= 3a + 5$.
(*d*) A number less one-third itself $= a - \frac{a}{3}$.
(*e*) Cost plus 8 $= c + 8$.
(*f*) The sum of any three numbers $= a + b + c$.

(*g*) Height h, less 15 $= h - 15$.
(*h*) Twice the sum of any two numbers $= 2(a + b)$.
(*i*) One-third the difference of any two numbers $= \frac{1}{3}(a - b)$.
(*j*) Five times a number less twice another number $= 5a - 2b$.
(*k*) The product of any three numbers $= a \cdot b \cdot c$.
(*l*) Any even number $= 2a$.
(*m*) Any odd number $= 2a + 1$.
(*n*) Four times the product of any two numbers divided by a third number $= \frac{4ab}{c}$.

842. How are algebraic symbols converted to verbal expressions?

(*a*) $a - 5$ = five less than a.
(*b*) $a + 5$ = five more than a.
(*c*) $5mn$ = five times the product of m and n.
(*d*) $5x + 4y$ = five times x increased by four times y.
(*e*) $3p - 7$ = three times p diminished by seven.
(*f*) $\frac{ab}{5}$ = one-fifth of the product of a and b.
(*g*) $\frac{3}{8}k$ = three-eighths of k, or one-eighth of three times k.
(*h*) $2a + 3b - 5c$ = five times a number subtracted from the sum of twice another number and three times a third number.
(*i*) $6(a + 3)$ = six times the sum of a and 3.
(*j*) $\frac{h}{6}(m + n)$ = one sixth of h multiplied by the sum of m and n.
(*k*) $\frac{a + b}{3}$ = one third the sum of a and b.
(*l*) $\frac{1}{3}Bh$ = one third the product of B and h.
(*m*) $\sqrt{2gh}$ = the square root of the product of 2, g, and h.

843. What is the general procedure for expressing thoughts algebraically?

Do not set up a complete problem "in one step." Take care of each phrase or sentence that expresses a condition individually. Then combine the separate parts into one or more expressions.

Ex. (*a*) What is the total cost of golf balls to a dealer if he buys 10 dozen at \$6 a dozen, and 30 dozen at \$8 a dozen?

10 doz. × \$6 = \$60 = cost of first lot
30 doz. × \$8 = \$240 = cost of second lot
∴ \$300 = total cost. Ans.

Since all factors are specific numbers, we get a specific answer.

Ex. (*b*) What is the total value of sales when a merchant sells a shirts at \$12.50 per shirt, and b shirts at \$10.50 per shirt?

$$\$12.50a = \text{value of first lot}$$
$$\$10.50b = \text{value of second lot}$$

$$\therefore \$12.50a + \$10.50b = \text{total value of shirts}$$

The answer is not a specific number, because some of the terms are literal.

The answer cannot be simplified, but if we let $a = 48$ and $b = 72$,

$$\$12.50 \times 48 + \$10.50 \times 72 = \$600 + \$756 = \$1{,}356 \text{ Ans.}$$

844. How do we indicate a letter multiplied by itself a number of times?

$$a \times a = aa = a^2$$
$$a \times a \times a = a^3$$
$$a \times a \times a \times a \times a = a^5, \text{etc.}$$

Small figures called exponents are placed to the right above the letter and indicate how many times the factor is multiplied by itself.

Therefore a^5 does not mean 5 times a but a multiplied by itself five times over.

$$5 \times a = 5a \quad \text{but} \quad a^5 = a \times a \times a \times a \times a$$

$$\therefore 5 \times 2 = 10 \quad \text{but} \quad 2^5 = 2 \times 2 \times 2 \times 2 \times 2 = 32$$

The product of a factor times itself is called the power of the factor.

845. Why is a^2 called "*a* squared"?

When all four sides of a rectangle are of equal length, it is called a square.

The area is then $a \cdot a = a^2$ sq. units.

$$\therefore a^2 \text{ is called ``}a\text{ squared''}$$

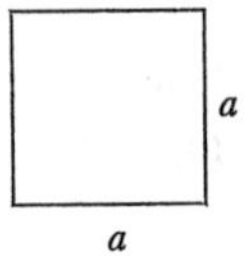

846. Why is a^3 called "*a* cubed"?

A rectangular solid with equal sides of length, breadth, and height is called a cube.

The volume of such a cube is $a \cdot a \cdot a = a^3$

$$\therefore a^3 \text{ is called ``}a\text{ cubed''}$$

By the same process, we can obtain expressions with higher exponents such as $a^4 = a$ to the fourth power.

$$a^5 = a \text{ to the fifth power}$$
$$a^{20} = a \text{ to the twentieth power}$$
$$a^m = a \text{ to the } m\text{th power}$$

We have seen that raising quantities or terms to given powers is called involution.

847. How do we raise an algebraic term to any power?

An algebraic term consists of a number and a symbol.

(*a*) Raise the number to the power indicated.
(*b*) Raise the symbol to the same power.
(*c*) Multiply the results.

Ex. (*a*) $3x$ squared means $3^2 \times x^2$.

$$3 \times 3 = 3^2 = 9$$
$$x \times x = x^2$$
$$\therefore\ 3^2 \times x^2 = 9x^2$$

Ex. (*b*) $3x$ cubed means $3^3 \times x^3$.

$$3 \times 3 \times 3 = 3^3 = 27$$
$$x \times x \times x = x^3$$
$$\therefore\ 3^3 \times x^3 = 27x^3$$

Ex. (*c*) $3x$ raised to the fourth power $= 3^4 \times x^4 = 81x^4$.

848. What is the rule for multiplying the same kind of letters or expressions together?

Add the exponents in

$$(a + b)^3 \times (a + b)^4 = (a + b)^{3+4} = (a + b)^7$$

Now $\qquad x^3 = x \cdot x \cdot x \quad \text{and} \quad x^4 = x \cdot x \cdot x \cdot x$

Therefore

$$x^3 \times x^4 = x \cdot x \cdot x \cdot x \cdot x \cdot x \cdot x = x^7 \quad \text{or} \quad x^3 \times x^4 = x^{3+4} = x^7$$

849. How do we multiply letters that have coefficients affixed?

First multiply the coefficients, then multiply the letters.

Ex. (*a*) $3x \times 4x = 3 \times 4 \times x \times x = 12x^2$. Note that $x = x^1$.

Ex. (*b*) $6x^3b^2 \times 3xb^5 = 6 \times 3 \times x^{3+1} \times b^{2+5} = 18x^4b^7$.

Ex. (*c*) $6a^{2b+5} \times 3a^{b-3} = 6 \times 3 \times a^{2b+5+b-3} = 18a^{3b+2}$.

850. What is the meaning of square root?

The area of a square is derived from the length of any one of its sides. We may thus consider the side as the root from which the square has evolved. We thus call the length of the side of a square the square root of the area of that square.

851. What is the rule for getting the square root of any power of a letter?

Take one half the exponent under the square root sign to get the exponent of the square root.

EXAMPLES: $\sqrt{x} = \sqrt{x^1} = x^{\frac{1}{2}} =$ square root

$\sqrt{x^2} = x^{\frac{2}{2}} = x^1 = x =$ square root

$\sqrt{x^3} = x^{\frac{3}{2}} =$ square root

$\sqrt{x^4} = x^{\frac{4}{2}} = x^2 =$ square root

$\sqrt{x^7} = x^{\frac{7}{2}} =$ square root

$\sqrt{3ab^2y^4} = \sqrt{3^1a^1b^2y^4} = 3^{\frac{1}{2}} \cdot a^{\frac{1}{2}} \cdot b^{\frac{2}{2}} \cdot y^{\frac{4}{2}}$

$= 3^{\frac{1}{2}}a^{\frac{1}{2}}by^2 = \sqrt{3a} \cdot b \cdot y^2$

852. What is meant by the root of a given number or term?

Each of the equal numbers or terms used to produce a power of a quantity or term is said to be a root of the power quantity or term.

Ex. (*a*) If $x^3 = x \cdot x \cdot x$, then x is a root of x^3, or $\sqrt[3]{x^3} = x =$ cube root of x^3.

Ex. (*b*) If $27x^3 = 3 \cdot 3 \cdot 3 \times x \cdot x \cdot x = 3^3 \times x^3$, then $3x$ is a root of $27x^3$, or $\sqrt[3]{27x^3} = 3x =$ cube root of $27x^3$.

Ex. (*c*) $\sqrt{9x^2} = 3x =$ square root of $9x^2$.

853. What is the rule for division of the same kind of symbols?

Subtract the exponent of the denominator from that of the numerator.

Ex. (*a*) Divide x^5 by x^3.

$$\frac{x^5}{x^3} = \frac{x \cdot x \cdot x \cdot x \cdot x}{x \cdot x \cdot x} = x^2 \quad \text{or} \quad \frac{x^5}{x^3} = x^{5-3} = x^2$$

Ex. (*b*) $\dfrac{x^{15}}{x^8} = x^{15-8} = x^7$

Ex. (*c*) $\dfrac{x^8}{x^4} = x^{8-4} = x^4$ (not x^2)

854. How can we show that a quantity to the zero power = 1?

$$\frac{x^4}{x^4} = x^{4-4} = x^0$$

But we know that $\frac{x^4}{x^4} = 1$ (anything divided by itself = 1)

$$\therefore\ x^0 = 1 \quad \text{or} \quad \text{any quantity to the zero power} = 1$$

EXAMPLE: $(a^2 \cdot y^3\sqrt{x})^0 = 1$

855. What is the result of (*a*) $\frac{a^6b^3c^5}{a^2bc^3}$, (*b*) $\frac{x^3y^4z^6}{z^2x^2}$, (*c*) $\frac{x^3yz}{xz}$, (*d*) $\frac{xy^2}{y^3}$, (*e*) $\frac{15x^5y^4z^2}{3x^4y^2z^4}$?

(*a*) $\frac{a^6b^3c^5}{a^2bc^3} = a^{6-2} \cdot b^{3-1} \cdot c^{5-3} = a^4b^2c^2$

(*b*) $\frac{y^3x^4z^6}{z^2x^2} = y^{3-0} \cdot x^{4-2} \cdot z^{6-2} = x^2y^3z^4$

(*c*) $\frac{x^3yz}{xz} = x^{3-1} \cdot y^1 \cdot z^{1-1} = x^2 \cdot y \cdot 1 = x^2y$

(*d*) $\frac{xy^2}{y^3} = x \cdot y^{2-3} = x \cdot y^{-1} = \frac{x}{y}$ $\left[\text{Note: } y^{-1} = \frac{1}{y},\ y^{-2} = \frac{1}{y^2}\right]$

(*e*) $\frac{15x^5y^4z^2}{3x^4y^2z^4} = 5x^{5-4} \cdot y^{4-2} \cdot z^{2-4} = \frac{5x \cdot y^2}{z^2}$

The numerical coefficients are divided by themselves.

856. What does $\frac{1}{y^2}$ mean?

$$\frac{y^4}{y^6} = y^{4-6} = y^{-2}$$

But $\frac{y^4}{y^6} = \frac{1}{y^2}$ (dividing numerator and denominator by y^4)

$$\therefore\ y^{-2} \text{ and } \frac{1}{y^2} \text{ mean the same thing.}$$

Similarly,

$y^{-3} = \frac{1}{y^3}$ $\qquad$ $10^{-1} = \frac{1}{10} = .1$

$y^{-4} = \frac{1}{y^4}$ $\qquad$ $10^{-2} = \frac{1}{10^2} = \frac{1}{100} = .01$

$y^{-1} = \frac{1}{y}$, etc. $\qquad$ $10^{-3} = \frac{1}{10^3} = \frac{1}{1{,}000} = .001$, etc.

857. When may we regard two terms as like terms?

When they contain like *symbols* with like *indices*, and are thus of equal value.

Ex. (*a*) x and x^2 are *not* like terms. The indices are different; x is simply x while $x^2 = x \cdot x$. If $x = 3$ then one term is 3 and the other is 9, and are thus not alike in value.

Ex. (*b*) b^2 and b^3, xy^2 and x^2y, a^2b^3 and a^3b^2 are also not alike.

Ex. (*c*) $7x$ and $12x$, $9y$ and $17y$, $3a^2$ and $5a^2$ are alike.

858. Does the order in which the symbols occur matter at all?

No.

EXAMPLE: xyz has the same value as xzy or as yxz. If $x = 3, y = 4$, and $z = 5$, then

$$xyz = 3 \times 4 \times 5 = 60 \quad \text{or} \quad 3 \times 5 \times 4 = 60 \quad \text{or} \quad 4 \times 3 \times 5 = 60$$

859. What is a simple test as to whether two terms are or are not alike in value?

Write out each term without indices and compare.

Ex. (*a*) Is a^2b^3 like a^3b^2?

$$a^2b^3 = a \cdot a \cdot b \cdot b \cdot b$$
$$a^3b^2 = a \cdot a \cdot a \cdot b \cdot b$$

$\therefore$ They are not alike

Ex. (*b*) Is $a^2b^3c^2$ like $a^2c^2b^3$?

$$a^2b^3c^2 = a \cdot a \cdot b \cdot b \cdot b \cdot c \cdot c$$
$$a^2c^2b^3 = a \cdot a \cdot c \cdot c \cdot b \cdot b \cdot b$$

$\therefore$ They are alike

860. What do [+] and [−] signs mean in algebra?

The sign [+] means a movement in a certain direction.
The sign [−] means a movement in the opposite direction.

Ex. (*a*) If you move 300 ft. toward the right from A to B in the following diagram and then move back 100 ft. to C, you are now only 200 ft. from A.

If movement to the right is [+] and movement to the left is [−], then

$$+300 \text{ ft.} - 100 \text{ ft.} = 200 \text{ ft. relative to A}$$

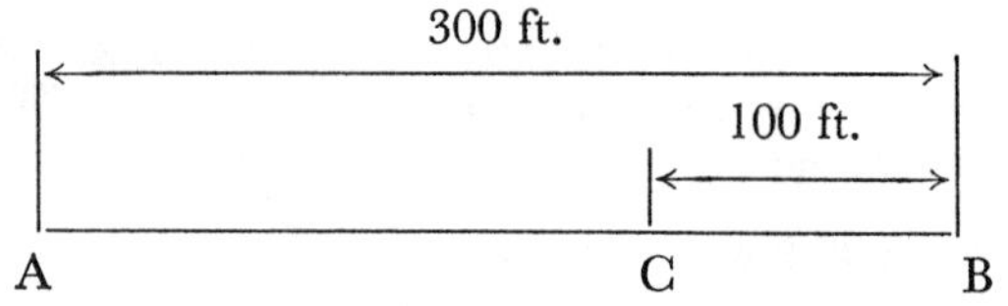

Ex. (*b*) If you move 300 ft. to the right to B and then move back 400 ft. to C, then

$$+300 \text{ ft.} - 400 \text{ ft.} = -100 \text{ ft. relative to A}$$

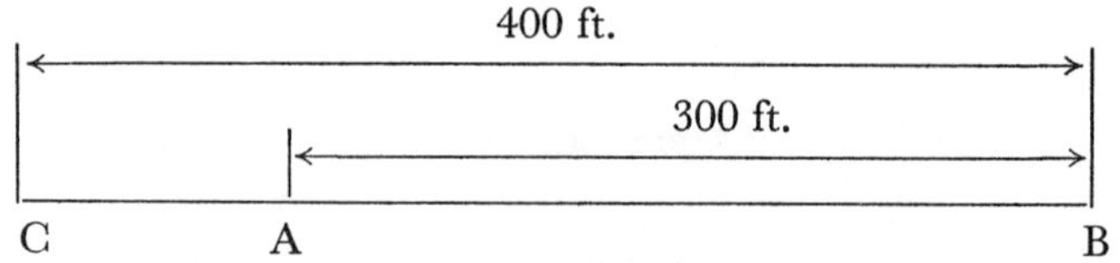

Ex. (*c*) If you rose 5,000 ft. in the air, then came down 1,000 ft., and you assume that up is [+] and down is [−], then, algebraically,

$$+5{,}000 \text{ ft.} - 1{,}000 \text{ ft.} = +4{,}000 \text{ ft.}$$

Now you are only 4,000 ft. above ground.

Ex. (*d*) If you went down in a mine 1,500 ft., then came up 800 ft., and you assume that up is [+] and down is [−], then, algebraically,

$$-1{,}500 \text{ ft.} + 800 \text{ ft.} = -700 \text{ ft.}$$

You are only 700 ft. down.

861. How are [+] and [−] quantities applied to debt and income?

Let [+] = income
Let [−] = debt (or what you have spent)

Ex. (*a*) What would be your financial position if you spend $25, then $10, and then get your salary of $150?

Algebraically

$$-\$25 - \$10 + \$150 = \$115 = \text{what you have left over}$$

Ex. (*b*) If you have $25x$ dollars, and you owe $35x$ dollars, what is the algebraic sum?

$$+25x - 35x = -10x \text{ dollars. Ans.}$$

This means that you pay as much of the debt as you can and you still owe $10x$ dollars.

862. What is the rule for subtraction of one plus quantity, from another plus quantity?

Change the sign of the plus quantity to be subtracted and add as usual.

Ex. (*a*) Subtract +8 from +15.

$15 - (+8) = 15 - 8 = 7$ (Change sign of +8 to −8)

Ex. (*b*) Subtract 25 from 60.

$60 - (+25) = 60 - 25 = 35$ (Change sign of +25 to −25)

Ex. (*c*) $12ab - (+8ab) = 12ab - 8ab = 4ab$

863. How can we show that two minuses mean a plus?

Ex. (*a*) If we subtract −8 from 15, we get

$15 - (-8) = 15 + 8 = 23$ (Minus a $-8 = +8$)

As [+] means a movement in a certain direction, and [−] means a movement in the opposite direction, then −(−8) means a step in the opposite direction to (−8), which must mean a step in the [+] direction.

$$\therefore\ -(-8) \text{ means } +8 \quad \text{and} \quad 15 + 8 = 23$$

Ex. (*b*) $12ab - (-8ab) = 12ab + 8ab = 20ab$

864. What is the rule for signs?

Like signs give plus.
Unlike signs give minus.

EXAMPLES:

$+(+8) = +8 \qquad +(+a) = +a = a$
$-(-8) = +8 \qquad -(-a) = +a = a$
$+(-8) = -8 \qquad +(-a) = -a$
$-(+8) = -8 \qquad -(+a) = -a$

865. What is the rule for numbers (or letters) that are multiplied together or are divided?

The rule for signs must be applied.

Ex. (*a*)

$+3(+8) = +24 \qquad +a(+b) = +ab = ab$
$-3(-8) = +24 \qquad -a(-b) = +ab = ab$
$+3(-8) = -24 \qquad +a(-b) = -ab$
$-3(+8) = -24 \qquad -a(+b) = -ab$

+3(−8) means 3 steps each of 8 units in the same direction as the minus direction.

−3(−8) means 3 steps of units in the opposite direction to the minus direction; that is, in the plus direction.

Ex. (*b*) Since division is the reverse of multiplication the rule of signs also applies.

$$\frac{+24}{+8} = +3 \qquad \frac{+a}{+b} = +\frac{a}{b} = \frac{a}{b}$$

$$\frac{-24}{-8} = +3 \qquad \frac{-a}{-b} = +\frac{a}{b} = \frac{a}{b}$$

$$\frac{+24}{-8} = -3 \qquad \frac{+a}{-b} = -\frac{a}{b}$$

$$\frac{-24}{+8} = -3 \qquad \frac{-a}{+b} = -\frac{a}{b}$$

866. How do we distinguish between $+3(-8)$ and $+3 - 8$?

$+3(-8) = 3$ steps each of 8 units to the left $= -24$
$+3 - 8 = 3$ steps to the right and then 8 steps to the left $= -5$

867. What is the result of $8(a - b) - 12(3a - 4b)$?

Remove parentheses by multiplication and rule of signs

$$8a - 8b - 36a + 48b$$

Combine *a*'s and *b*'s. Note that no sign in front of a letter or number means [+]

$$\therefore\ -28a + 40b \text{ Ans.}$$

868. What is the result of $7[3a - 4(5b - 6a) - 2b]$?

First remove the inner parentheses.

$$7[3a - 20b + 24a - 2b]$$

$$\therefore\ 7[27a - 22b] = 189a - 154b \text{ Ans.}$$

869. What is the result of $3[4x - \{(2x + y) + 5(3x + y)\} - 6y]$?

Remove inner parentheses first.

$$3[4x - \{2x + y + 15x + 5y\} - 6y]$$

Remove inner brackets.

$$3[4x - 2x - y - 15x - 5y - 6y] = 3[-13x - 12y]$$

$$\therefore\ 39x - 36y \text{ Ans.}$$

870. How can you check yourself to know whether your solution is correct?

Substitute small values for the different letters in the problem and in the answer.

EXAMPLE: Thus in Question 867, $8(a - b) - 12(3a - 4b)$, assume $a = 1$ and $b = 2$:

$$8(1 - 2) - 12(3 - 8) = -8 + 60 = 52$$

Now, in the answer $-28a + 40b$

$$-28 + 80 = 52 \text{ Check}$$

871. What is the procedure for evaluating algebraic terms?

Substitute the appropriate numbers for the letters.

Ex. (*a*) If $a = 3$, $b = 4$, $c = -6$, and $x = 5$, then the value of $a^2 - 2ax + x^2$ is

$$3^2 - 2 \times 3 \times 5 + 5^2 = 9 - 30 + 25 = 4$$

Ex. (*b*) $\frac{x^4}{b^3} = \frac{5^4}{4^3} = \frac{625}{64} = 9.766$

Ex. (*c*) $\frac{3xb\sqrt{b}}{a^3} = \frac{3 \times 5 \times 4\sqrt{4}}{3^3} = \frac{5 \times 4 \times 2}{3^2} = \frac{40}{9}$

872. What is the first important fact to remember in adding or subtracting algebraic terms?

Only those terms which are alike may be added or subtracted.

EXAMPLE: We *may* combine

$$7xy + 4xy - 3xy \quad \text{into} \quad (7 + 4 - 3)xy = 8xy$$

We *may not* combine

$$12x^2 - 9y^3 + 6z^4 \quad \text{beyond} \quad 3(4x^2 - 3y^3 + 2z^4)$$

873. What is the procedure for getting the algebraic sum of a number of terms?

(*a*) Arrange the signed terms with like symbols in separate columns.

(*b*) In each column get the sum of the minus terms and the sum of the plus terms separately.

(*c*) Subtract the smaller sum from the greater and affix the sign of the greater absolute number.

Ex. (*a*) Find the algebraic sum of $15x$, $-16y$, $8z$, $-17x$, $15y$, $-12z$, $16y$, $-20x$, $14z$, $11x$, $-6z$, and $-5y$.

$+15x$	$-16y$	$+8z$	
$-17x$	$+15y$	$-12z$	
$-20x$	$+16y$	$+14z$	
$+11x$	$-5y$	$-6z$	
$-11x$	$+10y$	$+4z$	= algebraic sums

$$\begin{array}{lll} +15x + 11x = +26x & +15y + 16y = +31y & +8z + 14z = +22z \\ -17x - 20x = -37x & -16y - 5y = -21y & -12z - 6z = -18z \\ \hline -37x + 26x = -11x & +31y - 21y = +10y & +22z - 18z = +4z \end{array}$$

Subtract smaller from greater and affix sign of greater:

$$\therefore \; -11x + 10y + 4z \text{ Ans.}$$

Ex. (*b*) Add $6a$, $3bc$, $4a^2d$, $-3a$, $-4bc$, $7a^2d$, $7a$, and $-6bcd$.

$$\begin{array}{l} 6a + 3bc + 4a^2d \\ -3a - 4bc + 7a^2d \\ 7a \qquad\qquad\qquad - 6bcd \\ \hline 10a - bc + 11a^2d - 6bcd \end{array}$$

874. Why is it that to any term you may add only other like terms if you want to give the result as a single term?

If x is apples, then the sum of 8 apples, 15 apples, and 6 apples is $8x + 15x + 6x = 29x$. But we may not represent the sum of $8x$, $15y$, and $6z$ as a single total (term) any more than we can represent the sum of 8 apples, 15 pears, and 6 peaches as a single total.

875. What is the procedure for subtraction of algebraic quantities?

Change the signs of the subtrahend and proceed as in addition.

EXAMPLE: From $10a - 4b + 5c$ subtract $5a + 7b + 3d$.

$$\begin{array}{l} 10a - 4b + 5c \\ -(5a + 7b \qquad + 3d) \end{array}$$

This becomes:

$$\begin{array}{l} 10a - 4b + 5c \\ -5a - 7b \qquad - 3d \quad \text{(Signs are changed)} \\ \hline 5a - 11b + 5c - 3d \text{ Ans.} \end{array}$$

876. What is the procedure for removing parentheses or brackets enclosing a number of algebraic terms?

On removing parentheses preceded by a [−] sign change the signs of all terms within the parentheses.

Ex. (*a*) $6 + (10 - 6) - (5 + 3)$ becomes

$$6 + 10 - 6 - 5 - 3 = 16 - 14 = 2$$

Ex. (*b*) $5a + (7 - [3a - 8])$. First remove the inner brackets, then remove the parentheses.

$$5a + 7 - 3a + 8 = 2a + 15$$

Ex. (*c*) $5a - (7 - [3a - 8])$. Remove the inner brackets first.
$5a - (7 - 3a + 8)$. Now remove the parentheses.

$$5a - 7 + 3a - 8 = 8a - 15$$

877. How may we illustrate the multiplication of a polynomial, algebraically?

Ex. (*a*) To multiply 7 by 14 we have:

$$\begin{array}{rrl} & 7 \times 14 & = 7(10 + 4) \\ \text{or} & 10 + 4 & = \text{multiplicand} \\ & \times 7 & = \text{multiplier} \\ \hline & 70 + 28 & = 98 = \text{product} \end{array}$$

Now substitute letters:

$$\begin{array}{rl} 3a + 7b & = \text{multiplicand} \\ \times\ 4a & = \text{multiplier} \\ \hline 12a^2 + 28ab & = \text{product} \end{array}$$

Multiply each term of the polynomial by the multiplier.

Ex. (*b*) To multiply algebraically 26×12:

$$\begin{array}{rl} 26 \times 12 & = (20 + 6)(10 + 2) \\ 20 + 6 & = \text{multiplicand} \\ 10 + 2 & = \text{multiplier} \\ \hline 200 + 60 & = 10(20 + 6) \\ 40 + 12 & = 2(20 + 6) \\ \hline 200 + 100 + 12 & = 312 = \text{product} \end{array}$$

Now substitute letters:

$$\begin{array}{rl} 3a + 7b & = \text{multiplicand} \\ \times 4a - 2b & = \text{multiplier} \\ \hline 12a^2 + 28ab & = 4a(3a + 7b) \quad (4a \text{ is the multiplier}) \\ - 6ab - 14b^2 & = -2b(3a + 7b) \quad (-2b \text{ is the multiplier}) \\ \hline 12a^2 + 22ab - 14b^2 & = \text{product} \end{array}$$

878. What is the product of:

(*a*) $-3a^2b^4$ by $5a^3c^3$,
(*b*) $4a^2 + 6ab - 8c^2$ by $7a^3$,
(*c*) $2a^2b^3 - 3b^2c^3 + 5c^2d^3 - 4a^2bc^2d^2$ by $-5a^2b^3c^2$,
(*d*) $6a^2 + 3b$ by $3a + 4b^2$?

(*a*) $-3a^2b^4 \times 5a^3c^3 = -3 \times 5 \times a^{2+3} \times b^4 \times c^3$

$$= -15a^5b^4c^3 \text{ Ans.}$$

(*b*) $4a^2 + 6ab - 8c^2 \times 7a^3 = 28a^5 + 42a^4b - 56a^3c^2$ Ans.

(*c*) $2a^2b^3 - 3b^2c^3 + 5c^2d^3 - 4a^2bc^2d^2 \times -5a^2b^3c^2$

$= -10a^4b^6c^2 + 15a^2b^5c^5 - 25a^2b^3c^4d^3 + 20a^4b^4c^4d^2$ Ans.

(*d*)
$$\begin{array}{l} 6a^2 + 3b \\ 3a \;\; + 4b^2 \\ \hline 18a^3 + 9ab \\ \qquad\qquad + 24a^2b^2 + 12b^3 \\ \hline 18a^3 + 9ab + 24a^2b^2 + 12b^3 \text{ Ans.} \end{array}$$

879. How can we show that the square of the sum of two terms is equal to the square of the first term plus twice the product of the two terms plus the square of the second term?

$$\begin{array}{rrr} a & + & b \\ \times a & + & b \\ \hline a^2 + & ab & \\ + & ab & + b^2 \\ \hline a^2 + & 2ab & + b^2 \end{array} \qquad (a + b)(a + b) = (a + b)^2$$

Ex. (*a*) $(a + 5)^2 = a^2 + 2 \times a \times 5 + 5^2 = a^2 + 10a + 25$ Ans.

Ex. (*b*) $(3x + 5y)^2 = \overline{3x}^2 + 2 \times 3x \times 5y + \overline{5y}^2$

$= 9x^2 + 30xy + 25y^2$ Ans.

Note: A line over a term with an index over it to the right means that the entire term is raised to the power of the index.

880. How can we show that the square of the difference of two terms is equal to the square of the first term minus twice the product of the two terms plus the square of the second term?

$$\begin{array}{rrr} a & - & b \\ \times a & - & b \\ \hline a^2 - & ab & \\ - & ab & + b^2 \\ \hline a^2 - & 2ab & + b^2 \end{array} \qquad (a - b)(a - b) = (a - b)^2$$

Ex. (*a*) $(a - 5)^2 = a^2 - 2 \times a \times 5 + 5^2 = a^2 - 10a + 25$ Ans.

Ex. (*b*) $(8x - 3y)^2 = \overline{8x}^2 - 2 \times 8 \times 3 \times xy + \overline{3y}^2$

$= 64x^2 - 48xy + 9y^2$ Ans.

881. How can we show that the product of the sum and difference of two terms is equal to the difference of their squares?

$$\begin{array}{r}
a + b \\
\times\, a - b \\
\hline
a^2 + ab \\
- ab - b^2 \\
\hline
a^2 - b^2
\end{array}$$

Ex. (*a*) $(a + 4)(a - 4) = a^2 - \overline{4}^2 = a^2 - 16$

Ex. (*b*) $26 \times 14 = (20 + 6)(20 - 6) = \overline{20}^2 - \overline{6}^2$
$= 400 - 36 = 364$

Ex. (*c*) $26 \times 24 = (25 + 1)(25 - 1) = \overline{25}^2 - \overline{1}^2$
$= 625 - 1 = 624$

Ex. (*d*) $(3x + 2)(3x - 2) = \overline{3x}^2 - \overline{2}^2 = 9x^2 - 4$

Ex. (*e*) $(5x - 2y)(5x + 2y) = \overline{5x}^2 - \overline{2y}^2 = 25x^2 - 4y^2$

882. What is the procedure for getting the direct answer to the multiplication of any binomial by another binomial?

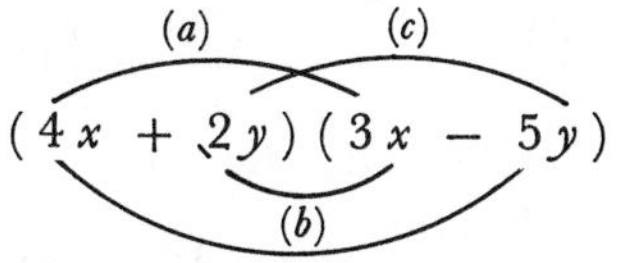

(*a*) Multiply the left terms for the *first* product: $4x \cdot 3x = 12x^2$.

(*b*) Multiply the outer terms and add the product to the product of the inner terms for the *second* product.

$$4x \times (-5y) + 2y \times 3x = -20xy + 6xy = -14xy$$

(*c*) Multiply the right terms for the *third* product

$$2y \cdot (-5y) = -10y^2$$

$$\text{Ans.} = 12x^2 - 14xy - 10y^2$$

883. What is the result of simplifying $2x(x + 5y) + 3y(x + 4y)$?

$$2x^2 + 10xy + 3xy + 12y^2 \quad \text{or} \quad 2x^2 + 13xy + 12y^2$$

884. What is the result of simplifying

$$(a + 2)(a + 4) + (a + 3)(a + 4) + (a + 2)(a + 5)?$$

$$a^2 + 6a + 8 + a^2 + 7a + 12 + a^2 + 7a + 10$$

or $\quad 3a^2 + 20a + 30$ Ans.

Use method of Question 882.

Check by assuming that $a = 1$ and substituting in original expression and in answer:

In original expression

$$(1 + 2)(1 + 4) + (1 + 3)(1 + 4) + (1 + 2)(1 + 5)$$
$$= 3 \times 5 + 4 \times 5 + 3 \times 6$$
$$= 15 + 20 + 18 = 53$$

In answer $3 + 20 + 30 = 53$ Check

885. What is the result of simplifying

$$\mathbf{(2a - 2b)(2a + 4b) - (2a + 3x)(2a - 5x) - 2b(2a - 4b)?}$$

$$4a^2 + 4ab - 8b^2 - (4a^2 - 4ax - 15x^2) - 4ab + 8b^2$$

Change signs on removing parentheses.

$$4a^2 + 4ab - 8b^2 - 4a^2 + 4ax + 15x^2 - 4ab + 8b^2$$
$$\therefore\ 15x^2 + 4ax \text{ Ans.}$$

886. What is the procedure for dividing a polynomial by a single term?

Divide each term in the polynomial by the single term.

Ex. (*a*) Divide $24x^3 - 12x^2 + 6x$ by $3x$.

$$\frac{24x^3 - 12x^2 + 6x}{3x} = \frac{24}{3}x^{3-1} - \frac{12}{3}x^{2-1} + \frac{6}{3}x^{1-1} = 8x^2 - 4x + 2$$

Ex. (*b*) Divide $96x - 56y - 88z$ by -8.

$$\frac{96}{-8}x \frac{-56}{-8}y \frac{-88}{-8}z = -12x + 7y + 11z$$

Use rule of signs.

Ex. (*c*) Divide $18a^4b^5 - 13ab + 7ab^4$ by $3a^2b^2$.

$$\frac{18a^4b^5 - 13ab + 7ab^4}{3a^2b^2} = \frac{18}{3}a^{4-2}.b^{5-2} - \frac{13}{3}a^{1-2}.b^{1-2} + \frac{7}{3}a^{1-2}.b^{4-2}$$
$$= 6a^2b^3 - \frac{13}{3ab} + \frac{7b^2}{3a}$$

Ex. (*d*) Divide $2a + 3b + 4c$ by y.

$$\frac{2a + 3b + 4c}{y} = \frac{2a}{y} + \frac{3b}{y} + \frac{4c}{y}$$

The result in each case is the numerator divided by the denominator.

Ex. (*e*) Divide $x^2 - a^2b^2c^3$ by $a^2b^2c^3$.

$$\frac{x^2 + a^2b^2c^3}{a^2b^2c^3} = \frac{x^2}{a^2b^2c^3} - \frac{a^2b^2c^3}{a^2b^2c^3} = \frac{x^2}{a^2b^2c^3} - 1$$

Division of a symbol with an index by a like symbol and index is equal to 1.

887. What is the procedure for division of a polynomial by a polynomial?

Proceed as in long division in arithmetic.

EXAMPLE: Divide $a^2 + 4a - 45$ by $a - 5$.

$$\begin{array}{r|l} & a + 9 \\ a - 5) & a^2 + 4a - 45 \\ & a^2 - 5a \\ & \quad 9a - 45 \\ & \quad 9a - 45 \\ & \qquad 0 \end{array}$$

$(a + 9)$ Ans.

a of divisor goes into a^2 of dividend a times.
Now multiply a by $(a - 5)$, getting $a^2 - 5a$.
Subtract this from $a^2 + 4a$, getting $9a$.
Bring down -45, getting $9a - 45$ = remainder.
a of divisor goes into $9a$ 9 times.
Multiply 9 by $a - 5$, getting $9a - 45$.
Subtract this from $9a - 45$, getting zero.

888. What is the quotient of $a^2 + 2a^2b + 4ab + 2ab^2 + 3b^2$ divided by $a + 2ab + 3b$?

$$\begin{array}{r|l} & a + b \\ a + 2ab + 3b) & a^2 + 2a^2b + 4ab + 2ab^2 + 3b^2 \\ & a^2 + 2a^2b + 3ab \\ & \qquad ab + 2ab^2 + 3b^2 = \text{remainder} \\ & \qquad ab + 2ab^2 + 3b^2 \\ & \qquad\qquad 0 \end{array}$$

$(a + b)$ Ans.

889. What is the result of division of $a^3 - a^2b - 7ab^2 - 20b^3$ by $a - 4b$?

$$\begin{array}{r|l} & a^2 + 3ab + 5b^2 \\ a - 4b) & a^3 - a^2b - 7ab^2 - 20b^3 \\ & a^3 - 4a^2b \\ & \quad 3a^2b - 7ab^2 \\ & \quad 3a^2b - 12ab^2 \\ & \qquad + 5ab^2 - 20b^3 \\ & \qquad + 5ab^2 - 20b^3 \\ & \qquad\qquad 0 \end{array}$$

$a^2 + 3ab + 5b^2$ Ans.

890. What is the "common term" method of getting the factors of an expression?

Take any term which is a factor common to each term of the expression. Divide the expression by this common factor to get the other factor.

Ex. (*a*) Find the factors of $12x - 16$.

4 is a factor of $12x$ and -16. Divide by factor 4.

$$\frac{12x - 16}{4} = 3x - 4 = \text{other factor}$$

$$\therefore\ 12x - 16 = 4(3x - 4)$$

Check by multiplying factors together to get the original expression.

Ex. (*b*) Factor $36x^3y^2 - 12x^2y$.

$12x^2y$ is a factor of $36x^3y^2$ and $-12x^2y$. Divide by $12x^2y$.

$$\frac{36x^3y^2 - 12x^2y}{12x^2y} = 3xy - 1 = \text{other factor}$$

$$\therefore\ 36x^3y^2 - 12x^2y = 12x^2y(3xy - 1)$$

Ex. (*c*) Factor $ab + ac - ad$.

a is common to all three terms.

$$\therefore\ a(b + c - d) = \text{algebraic expression in terms of factors}$$

Ex. (*d*) Factor $4a^2b^4 - 8ab^2 + 10a^5b^3$.

$2ab^2$ is common to all terms.

$$\therefore\ 2ab^2(2ab^2 - 4 + 5a^4b) = \text{factors}$$

Ex. (*e*) Factor $5(a + b)b - 6(a + b)a$.

$(a + b)$ is common to both terms.

$$\therefore\ (a + b)(5b - 6a) = \text{factors}$$

Ex. (*f*) Factor $4b^2 - 12b^5$.

$4b^2$ is common to both terms.

$$\therefore\ 4b^2(1 - 3b^3) = \text{factors}$$

891. What is the "common parentheses" method of getting the factors of an expression?

Take out any parenthesis which is a factor common to the terms of the expression and use this as a factor. Divide by this to get the other factor or factors.

Ex. (*a*) Factor $2a(3x + y) + 3a(3x + y)$.

$(3x + y)$ is common to both parts. Divide by $(3x + y)$.

$$\therefore\ (3x + y)(2a + 3a) = \text{factors}$$

Ex. (*b*) $2a(3x + y) - 3a(3x + y) = (3x + y)(2a - 3a) =$ factors.

Ex. (*c*) $b(2y + x) + k(x + 2y) = (x + 2y)(b + k) =$ factors.

892. What is the procedure for factoring by the combination of the common term and the common parentheses methods?

First find a common term factor.
Then find a common parentheses factor.

Ex. (*a*) Factor $2x^2 + 3xy + 2xz + 3yz$.

Take out a common term $x(2x + 3y) + z(2x + 3y)$.
Take out the common parentheses $(2x + 3y)(x + z) =$ factors.

Ex. (*b*) Factor $2x^2 - 3xy - 2xz + 3yz$.

$x(2x - 3y) - z(2x - 3y)$

$$\therefore (2x - 3y)(x - z) = \text{factors}$$

Ex. (*c*) Factor $2x^2 + 4xy - x - 2y$.

$2x(x + 2y) - 1(x + 2y)$

$$\therefore (x + 2y)(2x - 1) = \text{factors}$$

893. What is the "product of two binomials" method of getting the factors of a three-term expression?

(*a*) Write, to the left within each parenthesis, two factors of the first term of the expression.

(*b*) Write, to the right within each parenthesis, two factors of the last term of the expression.

(*c*) Multiply together the extreme terms of the parentheses, multiply together the middle terms of the parentheses, and add the two products. Check to see that this equals the middle term of the expression, and try another set if these do not give the middle term.

Ex. (*a*) Factor $x^2 + 17x + 60$.

$$(x + 12)(x + 5) = \text{factors}$$

Factor x^2 into $(x + \quad)(x + \quad)$
Factor 60 into $(\quad + 12)(\quad + 5)$
Multiply extremes x and 5 $= 5x$
Multiply middle terms 12 and x $= 12x$
Sum $= 17x =$ middle term; correct

Ex. (*b*) Factor $x^2 + 6x + 8$.

x and x are the factors of the first term.
Now try 8 and 1 as the factors of the last term.

$$\therefore (x + 1)(x + 8)$$

Extremes $x \times 8 = 8x$
Middle terms $1 \times x = 1x$
Sum $= 9x$ does not equal middle term

Now try 4 and 2 as the factors of the last term.

$$\therefore (x + 4)(x + 2)$$

Extremes $x \times 2 = 2x$
Middle terms $4 \times x = 4x$
Sum $= 6x$ = middle term; correct

Ex. (*c*) Factor $x^2 - 8x - 20$.

Try $(x + 10)(x - 2)$ $-2x + 10x = 8x$ (No good; wrong sign)

Try $(x - 10)(x + 2)$ $-10x + 2x = -8x$ = middle term; correct

Ex. (*d*) Factor $x^2 - 6x + 9$.

Try $(x + 3)(x + 3)$ $3x + 3x = 6x$ (No good; wrong sign)

Try $(x - 3)(x - 3)$ $-3x - 3x = -6x$ = middle term; correct

Ex. (*e*) Factor $8x^2 - 18x - 18$.

Try $(4x + 9)(2x - 2)$ $-8x + 18x = 10x$ (No good)

Try $(4x - 6)(2x + 3)$ $12x - 12x = 0$ (No good)

Try $(4x + 3)(2x - 6)$ $-24x + 6x = -18x$ = middle term; correct

894. What are the factors when the expression is recognized as a perfect square?

When the middle term is twice the product of the square roots of the other two terms, then we have a perfect square.

Ex. (*a*) Factor $9a^2 - 30ab + 25b^2$.

Square root of $9a^2$ is $3a$.
Square root of $25b^2$ is $5b$.

Twice their product is $2 \times 3 \times 5ab = 30ab$ = middle term.

$$\therefore\ (3a - 5b)(3a - 5b) = (3a - 5b)^2 = \text{factors}$$

Ex. (*b*) Factor $x^2 + 6x + 9$.

$$(x + 3)(x + 3) = (x + 3)^2 = \text{factors}$$

895. What are the factors when the expression is in the form of the difference of two squares?

One factor is the square root of the first term minus the square root of the second term, and the other factor is the square root of the first term plus the square root of the second term.

Ex. (*a*) Factor $x^2 - 16$.

$$(x - 4)(x + 4) = \text{factors}$$

Ex. (*b*) Factor $x^8 - 625$.

$$(x^4 - 25)(x^4 + 25)$$

But $(x^4 - 25)$ is also the difference of two squares.

$$\therefore\ (x^2 - 5)(x^2 + 5) = (x^4 - 25) = \text{factors}$$

Ex. (*c*) Factor $256a^8b^8 + c^8$.

$$(16a^4b^4 - c^4)(16a^4b^4 + c^4)$$

$$(4a^2b^2 - c^2)(4a^2b^2 + c^2)(16a^4b^4 + c^4)$$

$$\therefore\ (2ab - c)(2ab + c)(4a^2b^2 + c^2)(16a^4b^4 + c^4) = \text{factors}$$

Ex. (*d*) Factor $(x + y)^2 - 1$.

$$(x + y - 1)(x + y + 1) = \text{factors}$$

Ex. (*e*) Factor $(x + y)^2 - (2a + 3b)^2$.

$$(x + y - 2a - 3b)(x + y + 2a + 3b) = \text{factors}$$

Since the [−] sign is in front of $(2a + 3b)$ the entire expression is minus for one of the factors.

Ex. (*f*) Factor $(x + y)^2 - (c - p)$.

$$(x + y - c + p)(x + y + c - p) = \text{factors}$$

Ex. (*g*) Factor $1 - 4x^2 + 8xy - 4y^2$, or $1 - (4x^2 - 8xy + 4y^2)$.

Try $(2x - 2)(2x - 2)$ $\quad -4x - 4x = -8x$ = middle term

$\therefore\ 1 - (2x - 2)^2$ and $(1 - 2x + 2)(1 + 2x - 2)$ = factors

Ex. (*h*) Factor $(9x^2 - 12xy + 4y^2 - 4c^2 + 4cd - d^2)$.

$$(9x^2 - 12xy + 4y^2) - (4c^2 - 4cd + d^2)$$

For the first part try $(3x - 2y)(3x - 2y)$

$$-6xy - 6xy = -12xy = \text{middle term}$$

For the second part try $(2c - d)(2c - d)$

$$-2cd - 2cd = -4cd = \text{middle term}$$

$$\therefore (3x - 2y)^2 - (2c - d)^2$$

or

$$(3x - 2y - 2c + d)(3x - 2y + 2c - d) = \text{factors}$$

896. What is the value of $\dfrac{16a^2 - 25b^2}{4a + 5b}$ when $a = 3$ and $b = 2$?

Factor the numerator into $(4a - 5b)(4a + 5b)$.

$$\frac{(4a - 5b)\cancel{(4a + 5b)}}{\cancel{(4a + 5b)}} = 4a - 5b$$

$$= 4 \times 3 - 5 \times 2 = 12 - 10 = 2 \text{ Ans.}$$

897. What is an equation?

An equation is a balancing of expressions or quantities on each side of an equals sign. Because the two sides must balance it resembles a set of balance scales with the equals sign as the pivot point.

Examples: $15 - x = 10$, $A = \pi r^2$, $H = \dfrac{bh}{2}$, $v = \sqrt{2gh}$, $A = \dfrac{B}{3}h$, $ay^2 + by + c = 0$ are all equations.

898. How can we show the balance-scale resemblance of an equation?

Ex. (*a*) If 10 lb. is on the right pan of the scales and $(x + 3)$ lb. on the left, and they balance, then $x + 3 = 10$ is the expression of the equation.

Now, if you take away 3 lb. from the left pan you must also take away 3 lb. from the right pan to keep the balance, or

$$(x + 3) - 3 = 10 - 3$$

and $x = 7$ lb., which is the solution of x. More simply

$$x + 3 = 10 \quad \therefore x = 10 - 3 = 7$$

Ex. (*b*)

(1) $15 - 5 = 10 = \text{balance} = \text{equation}$
(2) $15 = 10 - (-5) = 10 + 5 = \text{balance} = \text{equation}$
[-5 from equation (1) is moved to right side]

Ex. (*c*)

(1) $15 = 12 + 3$ = balance = equation
(2) $15 - 12 = 3$ = balance = equation
[12 from equation (1) is moved to left side]
(3) $15 - 3 = 12$ = balance = equation
[3 from equation (1) is moved to left side]

Ex. (*d*)

(1) $8 + 5 = 13$ = balance = equation
(2) $8 = 13 - 5$ = balance = equation
[5 from equation (1) is moved to right side]
(3) $5 = 13 - 8$ = balance = equation
[8 from equation (2) is moved to right side and 5 from equation (2) is moved to left side]

899. What is the chief use of an equation?

It is a means of finding an unknown number in a problem.

Ex. (*a*) Solve $3x = 21$.

$3x$ must balance 21. Then $\frac{1}{3}$ of $3x$ must balance $\frac{1}{3}$ of 21, or

$$x = 7 \text{ Ans.}$$

Ex. (*b*) Solve $\frac{3}{8}x = 36$.

$\frac{8}{3}$ of $3x$ must balance $\frac{8}{3}$ of 36, or

$$x = \frac{8}{\cancel{3}_{1}} \times \overset{12}{\cancel{36}} = 96 \text{ Ans.}$$

900. What is meant by the root of an equation?

The solution or the value of the unknown that makes the equation balance is the root. This may be expressed as an integer, a decimal, or a common fraction.

EXAMPLE: If the circumference of a circular tank is 260 ft., what is its diameter?

$C = \pi d = 260$ ft, where $\pi = 3.1416$, d = diameter, $\frac{1}{\pi}$ of πd must balance $\frac{1}{\pi}$ of 260, or

$$d = \frac{260}{\pi} = \frac{260}{3.1416} = 82.73 \text{ ft.} = \text{diameter. Ans.}$$

901. What is meant by an identity?

When the left part of the equation is *identical* with the part to the right of the equals sign, then we have an identity. The equilibrium is true for all values of the symbol (or of the variables).

Ex. (*a*) $a(b - c) = ab - ac$ (true for all values of a, b, and c)

Ex. (*b*) $2x + 3y = 3y + 2x$ (true for all values of x and y)

Ex. (*c*) $5a + 7a = 12a$ (true for any value of a)

902. What is meant by a conditional equation?

One that imposes a condition upon the number values of the letters in the equation. The equality is true for only one value of the variable or for a limited number of values.

Ex. (*a*) 12 oranges cost 60¢. The equation is $12x = 60$ if we let $x =$ cost of one orange.

Here, only one value of x makes the equation balance. The condition is that $x = 5$. Thus the equation is a conditional equation.

Since $12x = 60$, $\frac{1}{12}$ of $12x$ must balance $\frac{1}{12}$ of 60

$\therefore\ x = 5¢ =$ cost of one orange

Ex. (*b*) $5y = 60$ (true only for $y = 12$)

Ex. (*c*) $\frac{x}{5} = 25$ (true only for $x = 125$)

903. What is a linear or simple equation?

When the highest power of the variable is 1, the equation is called linear, simple, or first degree.

Only one value will make the equality true in a simple or first degree equation.

EXAMPLE: $3x + 4 = 22$ is a linear equation.

$3x = 22 - 4 = 18$; $\frac{1}{3}$ of $3x = \frac{1}{3}$ of 18, or $x = 6$ (the only value of x that will make the equation true).

There is thus one root or solution.

904. What is a quadratic equation?

When the highest power of the variable is [^{2}], then the equation is quadratic.

EXAMPLE: $4x^2 = 64$.

$\frac{1}{4}$ of $4x^2 = \frac{1}{4}$ of 64 for balance, or $x^2 = 16$ and $x = +4$ or -4 (two roots).

There are always 2 roots or solutions to a quadratic equation.

905. What may be done to both sides of an equation without affecting its balance?

(*a*) We may add the same quantity to both sides.
(*b*) We may subtract the same quantity from both sides.
(*c*) We may multiply both sides by the same quantity.
(*d*) We may divide both sides by the same quantity.
(*e*) We may raise both sides to the same power, or we may take the same root of both sides.

906. What is the rule of signs for moving terms from one side of the equals sign to the other?

On moving a term from one side of an equation to the other side, you must change its sign. If it is plus it becomes minus, and if it is minus it becomes plus.

Ex. (*a*) $x - 5 = 0$

$x = 0 + 5 = 5$ Move -5 to right and change it to $+5$

Ex. (*b*) $x + 5 = 12$

$x = 12 - 5$ Move $+5$ to right side and change it to -5
$x = 7$

Ex. (*c*) $x - 7 = 8$

$x = 8 + 7$ Move -7 to right side and change it to $+7$
$x = 15$

907. What is the result when both sides of an equation are multiplied or divided by the same quantity?

Another equivalent equation results.

Ex. (*a*) Solve $\frac{1}{7}x = 19$

$7 \times \frac{1}{7}x = 7 \times 19$ Multiply both sides by 7

$\therefore x = 133$ Ans.

Ex. (*b*) Solve $\frac{4}{9}x = 64$

$9 \times \frac{4}{9}x = 9 \times 64$ Multiply both sides by 9
$4x = 9 \times 64$

$$\frac{4x}{4} = \frac{9 \times \overset{16}{\cancel{64}}}{\underset{1}{\cancel{4}}}$$ Divide both sides by 4

$\therefore x = 144$ Ans.

Ex. (*c*) If $\frac{3}{8}$ of a number is 18, what is the number?

Let x = the number.

Then $\frac{3}{8}x = 18$ (1)

$$8 \times \frac{3}{8}x = 8 \times 18 \quad \text{Multiply both sides by 8}$$

$$3x = 8 \times 18$$

$$\frac{3x}{3} = \frac{8 \times \overset{6}{\cancel{18}}}{\underset{1}{\cancel{3}}} \quad \text{Divide both sides by 3}$$

$$\therefore\ x = 48 \text{ Ans.}$$

More directly, divide both sides of (1) by $\frac{3}{8}$.

$$\frac{8}{3} \times \frac{3}{8}x = \overset{6}{\cancel{18}} \times \frac{8}{\underset{1}{\cancel{3}}}$$

$$\therefore\ x = 48 \text{ Ans.}$$

Ex. (*d*) Solve $.06x = 18$.

$$\frac{100}{\underset{1}{\cancel{6}}} x \times \overset{.01}{.\cancel{06}} = \frac{100}{\underset{1}{\cancel{6}}} \times \overset{3}{\cancel{18}} \quad \text{Multiply each side by } \tfrac{100}{6}$$

$$\therefore\ x = 300 \text{ Ans.}$$

Ex. (*e*) Solve $3\frac{1}{3}x = 30$.

$$\tfrac{3}{10} \times \tfrac{10}{3}x = \tfrac{3}{10} \times 30 \quad \text{Multiply each side by } \tfrac{3}{10}$$

$$\therefore\ x = 9 \text{ Ans.}$$

Ex. (*f*) Solve $.08x = 1{,}000$.

$$\frac{.08x}{.08} = \frac{1{,}000}{.08} \quad \text{Divide both sides by .08}$$

$$\therefore\ x = \tfrac{10}{8} = \tfrac{5}{4} = 1\tfrac{1}{4} \text{ Ans.}$$

908. How can we solve simple equations by addition or subtraction?

Add or subtract an appropriate number or quantity if the equation cannot be solved by multiplication or division.

This process is similar to that of Question 906 for moving terms from one side of the equation to the other.

Ex. (*a*) Solve $x + 4 = 10$.

$$x + 4 - 4 = 10 - 4 \quad \text{Subtract 4 from each side}$$

$$\therefore\ x = 6 \text{ Ans.}$$

Ex. (*b*) Solve $16 = 7 + y$.

$$16 - 7 = 7 + y - 7 \quad \text{Subtract 7 from each side}$$

$$\therefore y = 9 \text{ Ans.}$$

Ex. (*c*) Solve $20 = y - 3$.

$$20 + 3 = y - 3 + 3 \quad \text{Add 3 to each side}$$

$$\therefore y = 23 \text{ Ans.}$$

Ex. (*d*) Solve $8 = 14 - x$.

$$8 + x = 14 - x + x \quad \text{Add } x \text{ to each side}$$
$$8 + x - 8 = 14 - 8 \quad \text{Subtract 8 from each side}$$

$$\therefore x = 6 \text{ Ans.}$$

Of course, this can be done more directly by the rule of signs for moving terms to the opposite side of the equals sign.

From $8 = 14 - x$ we get
$x = 14 - 8$ Move $-x$ to left and move 8 to right

909. What are the steps in the solution of an equation?

(*a*) Clear equation of fractions.
(*b*) Remove any parentheses.
(*c*) Collect all terms containing the unknown factor on the left (preferably) of the equals sign and all other terms on the right of the equals sign.
(*d*) Change the sign from [+] to [−] or from [−] to [+] when moving a term to the opposite side of the equals sign.
(*e*) Factor the expression containing the unknown to make all other values in the expression the coefficient of the unknown.
(*f*) Divide the entire equation by the coefficient of the unknown.

Ex. (*a*) Solve for x in $7x - 5 = 9 + 3x$.

$$7x - 3x = 9 + 5 \quad \text{All } x\text{'s on left, numbers on right}$$
$$4x = 14$$

$$\therefore x = \tfrac{14}{4} = \tfrac{7}{2} = 3\tfrac{1}{2} \text{ Ans.}$$

To check, substitute $\frac{7}{2}$ for x on each side.

Left side $7 \times \frac{7}{2} - 5 = \frac{49}{2} - 5 = 24\frac{1}{2} - 5 = 19\frac{1}{2}$

Right side $9 + 3 \times \frac{7}{2} = 9 + \frac{21}{2} = 9 + 10\frac{1}{2} = 19\frac{1}{2}$ Check

Ex. (*b*) Solve $\dfrac{2x}{5} - \dfrac{3x}{4} = 2 - \dfrac{4}{x}$.

Convert any whole number into a fraction. Find the lowest common denominator of all the denominators, and arrange each side of the equation on the new denominator.

You need not write down this common denominator in working equations because if a fraction of one quantity equals the same fraction of another quantity, then the quantities themselves must be equal. If $\frac{1}{8}$ of $a = \frac{1}{8}$ of b, then $a = b$.

$$\frac{2x}{5} - \frac{3x}{4} = \frac{2}{1} - \frac{x}{4}$$ Convert whole numbers into fractions L.C.M. $= 4 \times 5 = 20$

Then $$\frac{8x - 15x}{20} = \frac{40 - 5x}{20}$$

$$8x - 15x = 40 - 5x$$
$$8x - 15x + 5x = 40$$
$$-2x = 40$$
$$\therefore x = -20 \text{ Ans.}$$

Substitute $x = -20$:

Left side:

$$\frac{2x}{5} - \frac{3x}{4} = \frac{2 \times (-20)}{5} - \frac{3 \times (-20)}{4} = \frac{-40}{5} - \left(-\frac{60}{4}\right)$$

L.C.M. $= 5 \times 4 = 20$

$$\frac{-40 \times 4 + 5 \times 60}{20} = \frac{-160 + 300}{20} = \frac{140}{20} = 7$$

Right side:

$$\frac{2}{1} - \left(-\frac{20}{4}\right) = \frac{8 + 20}{4} = \frac{28}{4} = 7 \text{ (Check)}$$

Ex. (*c*) Solve $5(x - 6) = 9(x + 3)$.

$$5x - 30 = 9x + 27$$ Remove parentheses
$$5x - 9x = 27 + 30$$
$$-4x = 57$$

$$\therefore x = \frac{-57}{4} = -14\tfrac{1}{4} \text{ Ans.}$$

Give answer in form of $+x = -14\frac{1}{4}$.

910. What is the solution for y of $P = \frac{a(c - y)}{(t - y)}$?

Multiply both sides by $(t - y)$.

$$P(t - y) = a(c - y)$$

Then

$Pt - Py = ac - ay$	Remove parentheses
$-Py + ay = ac - Pt$	Transpose $-ay$ to left and Pt to right side
$y(a - P) = ac - Pt$	Factor left side so that y stands as a single factor times the binomial factor $(a - P)$
$\therefore y = \frac{ac - Pt}{a - P}$	Divide both sides by $(a - P)$, the coefficient of y

911. What is the solution for d in $A - pd = b - d$?

$A - pd = b - d$	
$d - pd = b - A$	Transpose $-d$ to left and A to right
$d(1 - p) = b - A$	Factor left side
$\therefore d = \frac{b - A}{1 - p}$	Divide both sides by $(1 - p)$ factor

912. What is the solution for W in $\frac{3}{4}W = T$?

$\frac{4}{3} \times \frac{3}{4}W = \frac{4}{3}T$ Multiply both sides by $\frac{4}{3}$

$W = \frac{4}{3}T = \frac{4T}{3}$ Ans.

913. What is the solution for x in $\frac{b + P}{x} = \frac{x}{16t}$?

First cross-multiply. This is the same as multiplying both sides by $16t$ and then dividing both sides by $b + P$.

$$16t(b + P) = x^2$$

$\therefore x = \pm\sqrt{16t(b + P} = \pm 4\sqrt{t(b + P)}$ Ans.

914. What is the solution of $x + 7 - 3x - 5 = 12 - 4x$?

$x - 3x + 4x = -7 + 5 + 12$ Bring all x terms on left side

$2x = 10$

$\therefore x = 5$ Ans.

915. What is the solution of $(x + 5)^2 - (x + 4)^2 = x + 12$?

$$(x^2 + 10x + 25) - (x^2 + 8x + 16) = x + 12$$
$$x^2 + 10x + 25 - x^2 - 8x - 16 = x + 12$$
$$2x + 9 = x + 12$$
$$2x - x = 12 - 9$$
$$\therefore\ x = 3 \text{ Ans.}$$

916. What is the solution of $7(x + 5) - 9(x - 2) = 8x + 3$?

$$7x + 35 - 9x + 18 = 8x + 3$$
$$7x - 9x - 8x = -35 - 18 + 3$$
$$-10x = -50$$
$$\therefore\ x = 5 \text{ Ans.}$$

917. What is the value of x in $8(x - 3)(x + 3) = x(8x - 8)$?

$$8(x^2 - 9) = 8x^2 - 8x$$
$$8x^2 - 72 = 8x^2 - 8x$$
$$8x^2 - 8x^2 + 8x = 72$$
$$8x = 72$$
$$\therefore\ x = 9 \text{ Ans.}$$

918. What is the value of x in $\dfrac{x^3 - 4x - 15}{x - 3} - (x - 2)^2 = 22$?

Divide $x^3 - 4x - 15$ by $x - 3$ to get $x^2 + 3x + 5$. Then

$$x^2 + 3x + 5 - (x^2 - 4x + 4) = 22$$
$$x^2 + 3x + 5 - x^2 + 4x - 4 = 22$$
$$7x + 1 = 22$$
$$7x = 22 - 1 = 21$$
$$\therefore\ x = 3 \text{ Ans.}$$

919. What is the procedure for solving equations involving decimals?

(*a*) Consider the term containing the largest number of decimal places.

(*b*) Make that a whole number by moving the decimal point to the right.

(*c*) Move the decimal point in each other term in the entire equation the same number of places to the right to balance the equation.

Note: $.6x$ contains 1 decimal place
$.65x$ contains 2 decimal places

EXAMPLE: Solve $.6x + .05 = 5 - .3x$.

Move 2 places to the right in each term. Then

$$60x + 5 = 500 - 30x$$
$$60x + 30x = 500 - 5$$
$$90x = 495$$
$$\therefore\ x = \tfrac{495}{90} = 5.5 \text{ Ans.}$$

To check:

$$\left.\begin{array}{ll}\text{Left side} & .6 \times 5.5 + .05 = 3.35 \\ \text{Right side} & 5 - .3 \times 5.5 \quad = 3.35\end{array}\right\} \text{Check}$$

920. What is a formula?

It is an algebraic expression giving the relation of mathematical facts about various quantities.

Ex. (*a*) $A = \pi r^2$ = formula expressing the area of any circle.

A = area of any circle, $\pi = 3.1416$ = constant, and r = radius of circle.

For every radius r, we have an area A to correspond.

Ex. (*b*) $v^2 = 2gh$ = formula for a falling body.

v = velocity in ft. per sec., h = height, in feet, from which body falls, and $g = 32.2$ = constant of gravity.

Ex. (*c*) $d = vt$ = distance covered by an object moving at a constant speed v for a time t.

What is the distance when the speed is 60 miles per hour, and the time is $2\frac{3}{4}$ hours?

$$d = 60 \times 2\tfrac{3}{4} = 60 \times \tfrac{11}{4} = 165 \text{ miles. Ans.}$$

What is the distance when $v = 44$ ft./sec. and $t = 10$ sec.?

$$d = 44 \times 10 = 440 \text{ ft. Ans.}$$

Ex. (*d*) If the distance s covered by a freely falling body in t seconds is given by formula $s = 16t^2$, what is s when $t = 5$, and when $t = 20$?

$$s = 16 \times \overline{5}^2 = 400 \text{ ft. Ans.} \quad (\text{when } t = 5)$$
$$s = 16 \times \overline{20}^2 = 6{,}400 \text{ ft. Ans.} \quad (\text{when } t = 20)$$

921. If the relation between the Fahrenheit temperature readings of a thermometer, and the Centigrade readings, is expressed as $F = \frac{9}{5}C + 32$, what is the Fahrenheit reading when (*a*) C = 50°, (*b*) C = 30°, (*c*) C = 10°?

(*a*) $F = \frac{9}{5} \times 50 + 32 = 122°F$

(*b*) $F = \frac{9}{5} \times 30 + 32 = 86°F$

(*c*) $F = \frac{9}{5} \times 10 + 32 = 50°F$

922. What is meant by solving for another variable in a formula?

Ex. (*a*) In the formula $A = l \times w$ where l (= length) and w (= width) are variables, and A = resulting area, we can easily find A when we know l and w, but to find l directly we have to solve for the variable l.

Divide both sides of $A = l \times w$ by w to get $l = \dfrac{A}{w}$.

Ex. (*b*) What is the expression for v and the expression for t in $d = vt$?

Divide each side by v to get $t = \dfrac{d}{v}$

Divide each side by t to get $v = \dfrac{d}{t}$

Ex. (*c*) In the formula $I = \dfrac{E}{R}$, where I = current in amperes, E = voltage in volts, and R = resistance in ohms, what is (1) E, (2) R?

(1) Multiply both sides by R to get $IR = E$

(2) Divide both sides of (1) by I to get $R = \dfrac{E}{I}$

Ex. (*d*) From $F = \frac{9}{5}C + 32$, find C.

$F - 32 = \frac{9}{5}C$ Move 32 to left side

$\frac{5}{9} \times (F - 32) = \frac{5}{9} \times \frac{9}{5}C$ Multiply both sides by $\frac{5}{9}$

$\therefore \frac{5}{9}(F - 32) = C$ Ans.

Ex. (*e*) From $s = gt^2/2$ find g and t.

$\dfrac{2s}{t^2} = \dfrac{gt^2}{2} \times \dfrac{2}{t^2}$ Multiply both sides by $\dfrac{2}{t^2}$, getting $\dfrac{2s}{t^2} = g$

$\dfrac{2s}{g} = \dfrac{2}{g} \times \dfrac{gt^2}{2}$ Multiply both sides by $\dfrac{2}{g}$, getting $\dfrac{2s}{g} = t^2$

Then $t = \sqrt{\dfrac{2s}{g}}$

923. What is the general procedure for putting words into equation form to express simple equations with one unknown?

(*a*) Express each phrase or sentence that states a condition, and then combine these to form one or more expressions.

(*b*) Represent the unknown by a letter from the end of the alphabet.

(*c*) Express each statement pertaining to the unknown and any other unknown in terms of this letter.

(*d*) Expressions representing statements of equal value are then placed equal to each other.

EXAMPLE: If you multiply a number by 5 and, after taking 9 from the result, 16 remains, what is the number?

Let x = the number.

$5x$ = the number times 5

$5x - 9$ take away 9

$5x - 9 = 16$ remainder is 16

$5x = 16 + 9 = 25$

$\therefore x = \frac{25}{5} = 5$ Ans.

924. If the sum of three consecutive even numbers is 90, what are the numbers?

$2x$ = first even number

$2x + 2$ = next even number

$2x + 4$ = next even number

Then $2x + (2x + 2) + (2x + 4) = 90$

$$6x + 6 = 90$$

$$6x = 90 - 6 = 84$$

$$x = \frac{84}{6} = 14$$

$$2x = 28 \quad 2x + 2 = 30 \quad 2x + 4 = 32$$

$\therefore$ the numbers are 28, 30, and 32. Ans.

925. If a tank is $\frac{5}{8}$ full of water, and after running off 300 gallons it is $\frac{1}{4}$ full, what is the capacity of the tank?

Let x = capacity of the tank.

$\frac{5}{8}x$ gallons of water $-$ 300 gallons $= \frac{x}{4}$ gallons

$\frac{5}{8}x - \frac{x}{4} = 300$ Transpose and change signs

$$\tfrac{3}{8}x = 300$$

$$\tfrac{8}{3} \times \tfrac{3}{8}x = \tfrac{8}{3} \times 300$$

$\therefore x = 800$ gallons. Ans.

926. If you are 45 years old, and your son is 12 years old, (*a*) when will your son be half your age, (*b*) how long ago were you 5 times as old as your son?

(*a*) Let x = number of years until your son will be half your age. At that time your son will be $x + 12$ years old; and you will be $x + 45$ years old. Then

$$x + 45 = 2(x + 12) = 2x + 24$$
$$x - 2x = 24 - 45$$
$$-x = -21$$
$$\therefore\ x = 21$$

In 21 years' time your son will be half your age; then you will be $45 + 21 = 66$ and your son $12 + 21 = 33$ years old. Ans.

(*b*) Let y = the number of years ago when you were 5 times your son's age. Then

$$45 - y = 5(12 - y) = 60 - 5y$$
$$-y + 5y = 60 - 45$$
$$4y = 15$$
$$\therefore\ y = \tfrac{15}{4} = 3\tfrac{3}{4} \text{ years ago}$$

Then you were $45 - 3\frac{3}{4} = 41\frac{1}{4}$ years old; and your son $12 - 3\frac{3}{4} = 8\frac{1}{4}$ years old. Ans.

$$5 \times 8\tfrac{1}{4} = 41\tfrac{1}{4} \text{ years} \quad \text{Check}$$

927. If two machine operators punch out 1,400 plastic parts per hour, and one produces $\frac{3}{4}$ as many parts as the other, what is the production of each?

Let x = parts produced by faster worker = base.
Then $\frac{3}{4}x$ = parts produced by slower worker.
And $x + \frac{3}{4}x$ = parts produced by both workers per hour.

$$x + \tfrac{3}{4}x = 1{,}400$$
$$\tfrac{7}{4}x = 1{,}400$$

$$\therefore\ x = \frac{4}{\cancel{7}} \times \overset{200}{\cancel{1{,}400}} = 800 \text{ parts per hour by faster worker}$$

$$\tfrac{3}{4} \times 800 = 600 \text{ parts per hour by slower worker}$$

928. If you and your wife together hold \$7,800 in United States government bonds, and your share is \$1,100 more than your wife's, how much do you each have?

Let x = your wife's share.
Then $x + 1{,}100$ = your share.

$$\begin{aligned} \text{And} \quad x + x + 1{,}100 &= \text{combined holdings} \\ 2x + 1{,}100 &= 7{,}800 \\ 2x &= 7{,}800 - 1{,}100 = 6{,}700 \\ \therefore x &= \tfrac{6700}{2} = \$3{,}350 = \text{your wife's share} \\ x + 1{,}100 = 3{,}350 + 1{,}100 &= \$4{,}450 = \text{your share} \end{aligned}$$

929. If you bought 3 suits for \$226, and the first cost twice as much as the second, while the third cost \$10 more than the second, what is the cost of each suit?

Let x = cost of second suit = base.
Then $2x$ = cost of first suit.
And $x + 10$ = cost of third suit.

$$\begin{aligned} x + 2x + x + 10 &= \$226 \\ 4x = 226 - 10 &= \$216 \\ \therefore x = \tfrac{216}{4} &= \$54 = \text{cost of second suit} \\ 2x = 2 \times 54 &= \$108 = \text{cost of first suit} \\ x + 10 = 54 + 10 &= \$64 = \text{cost of third suit} \end{aligned}$$

930. If you have \$2.45 in nickels and dimes, and you have 30 coins in all, how many of each do you have?

Let x = number of nickels.
Then $30 - x$ = number of dimes.
$5x$ = number of cents represented by the nickels.
$10(30 - x)$ = number of cents represented by the dimes.

$$\begin{aligned} 5x + 10(30 - x) &= 245 \\ 5x + 300 - 10x &= 245 \\ -5x &= 245 - 300 = -55 \\ \therefore x &= \frac{-55}{-5} = 11 \text{ nickels} \\ 30 - 11 &= 19 \text{ dimes} \end{aligned}$$

931. At what time between 4 and 5 o'clock are the hands of a watch opposite each other?

Let x = distance or number of minute spaces traveled by the minute-hand from 4 o'clock to the required time.

Now the hour-hand is 20 minute spaces ahead of the minute-hand at exactly 4 o'clock, and when the hands are opposite each other it will be 30 minute spaces away from the minute-hand.

Thus the hour-hand will have traveled $30 + 20 = 50$ minute spaces *less* than the minute-hand. Therefore $x - 50$ = number of minute spaces or distance traveled by the hour-hand from 4 o'clock up to the time when the hands are opposite each other.

But the minute-hand travels 12 times as much (or 12 times the distance) as the hour-hand.

$$x = 12(x - 50) = 12x - 600$$
$$x - 12x = -600$$
$$-11x = -600$$
$$\therefore x = \tfrac{600}{11} = 54\tfrac{6}{11}$$ minutes after 4 o'clock, when the hands are opposite each other

932. If you want to sale price 300 lb. of coffee at 78¢ a lb., and you have one kind that normally sells for 90¢ a lb., and another that sells for 70¢ a lb., how many lb. of each must you mix so that you will not lose money?

Let x = number of lb. of the 90¢ kind = base
Then $.90x$ = sales value of this kind.
And $300 - x$ = number of lb. of the 70¢-a-lb. kind.
And $.70(300 - x)$ = sales value of this kind.

$$.90x + .70(300 - x) = 300 \times .78$$
$$.9x + 210 - .7x = 234$$
$$.2x = 234 - 210 = 24$$
$$\therefore x = \frac{24}{.2} = \frac{240}{2} = 120 \text{ lb. of the 90¢ kind}$$
$$300 - 120 = 180 \text{ lb. of the 70¢ kind}$$

933. If you sell 3 taxicabs and buy 2 new ones for $7,800, and you then have $2,400 left, how much did you get for each taxicab you sold?

Let x = amount received per taxicab sold.
Then $3x$ = amount received for 3 taxicabs.
And $3x$ − $7,800 = amount left after buying 2 new taxicabs.

$$3x - 7{,}800 = 2{,}400$$
$$3x = 2{,}400 + 7{,}800 = 10{,}200$$
$$\therefore x = \tfrac{10{,}200}{3} = \$3{,}400 \text{ per taxicab sold. Ans.}$$

934. During the year you, your wife, and your daughter saved a total of $1,200. You saved $100 less than twice your daughter's savings, and your daughter saved $10 more than twice your wife's. How much did each save?

Let x = your wife's savings = base.
Then $2x + 10$ = your daughter's savings.

And $2(2x + 10) - 100$ = your savings.

$$x + 2x + 10 + 2(2x + 10) - 100 = \$1{,}200$$
$$x + 2x + 10 + 4x + 20 - 100 = \$1{,}200$$
$$7x = 1{,}200 - 10 - 20 + 100 = \$1{,}270$$
$$\therefore x = \tfrac{1270}{7} = \$181.43 = \text{wife's savings}$$
$$2x + 10 = 2 \times 181.43 + 10 = \$372.86 = \text{daughter's savings}$$
$$2(2x + 10) - 100$$
$$= 2 \times 372.86 - 100 = \$645.72 = \text{your savings}$$
$$181.43 + 372.86 + 645.72 = \$1{,}200.01 \text{ Check}$$

The one penny more is due to fractional manipulations of the figures.

935. What is the number which when multiplied by 4 equals the original number plus 36?

Let x = the number. Then

$$4x = x + 36$$
$$4x - x = 36$$
$$3x = 36$$
$$\therefore x = \tfrac{36}{3} = 12 \text{ Ans.}$$

936. If a train leaves Washington, D.C., for Chicago and travels at the rate of 50 miles per hour, and $\frac{1}{2}$ hour later an auto leaves for Chicago from Washington traveling at the rate of 55 miles per hour, how long will it take the auto to overtake the train?

Let x = travel time in hours of auto until it overtakes train.
Then $x + \frac{1}{2}$ = travel time of train.
Now $5x$ = distance auto travels (mph × hours = distance), and $50(x + \frac{1}{2})$ = distance train travels.
Both have traveled the same distance at meeting point. Then

$$55x = 50(x + \tfrac{1}{2}) = 50\left(\frac{2x + 1}{2}\right) = 25(2x + 1) = 50x + 25$$
$$55x - 50x = 25$$
$$5x = 25$$
$$\therefore x = \tfrac{25}{5} = 5 \text{ hours. Ans.}$$

Auto travels 5 hours before overtaking train.

937. You start out to walk to your friend's house at the rate of 4 mph. Your friend starts at the same time for your house at 3 mph. You live 14 miles from each other. How far does each of you walk before meeting?

Let x = time of walking for each before meeting.
Then $4x$ = number of miles you walk.
And $3x$ = number of miles your friend walks.

$$4x + 3x = 14 \text{ miles total distance}$$
$$7x = 14$$
$$x = \frac{14}{7} = 2 \text{ hours}$$

$\therefore\ 4 \times 2 = 8$ miles. You walk 8 miles
$3 \times 2 = 6$ miles. Your friend walks 6 miles

PROBLEMS

1. What are the factors of the product $6cdp$?

2. What is the numerical coefficient of $36k$?

3. What are the literal factors of $20xyz^2$?

4. Is $-8ab$ a specific number?

5. What are the coefficients of the product $15y(a - b)$?

6. What is the difference between $8 + 7$ and $7 + 8$, $c + d$ and $d + c$?

7. Is there a difference between $b \cdot k$ and $k \cdot b$?

8. Is $6(5 - 4) = 6 \times 5 - 6 \times 4$?

9. In what ways may $\frac{3}{8}y$ be written?

10. Translate the following verbal expressions into algebraic symbols:

(*a*) Six times a number
(*b*) One-third the base B
(*c*) Seven times a number increased by 8
(*d*) A number less one-eighth of itself
(*e*) Cost plus 10
(*f*) The difference of two numbers
(*g*) Weight w less 20
(*h*) Three times the sum of any two numbers
(*i*) One-sixth the difference of any two numbers
(*j*) Eight times a number less three times another number
(*k*) The product of any four numbers
(*l*) The next even numbers above and below $2x$
(*m*) The next higher number after x
(*n*) Five consecutive numbers of which x is the middle number
(*o*) Five times the product of any two numbers divided by a third number
(*p*) The square root of the product of two numbers

11. Convert the following symbols to verbal expressions:

(*a*) $b - 6$ (*b*) $a + 7$ (*c*) $9pg$

(*d*) $7x + 3y$ (*e*) $4w - 8$ (*f*) $\frac{ab}{7}$

(*g*) $3a + b - 6c$ (*h*) $\frac{5}{8}l$ (*i*) $9(c + 5)$

(*j*) $\frac{l}{6}(A + 4B + C)$ (*k*) $\frac{c + d}{4}$ (*l*) $\frac{1}{2}bh$

(*m*) $\frac{mv^2}{2}$ (*n*) $A = P(1 + r)^n$ (*o*) $d\sqrt{a^2 + b^2}$

(*p*) $\frac{1}{2}h(a + b)$

12. What is the algebraic expression for the total value of sales when a merchant sells *a* pairs of trousers at \$15 a pair and *b* pairs of trousers at \$18.95 a pair?

13. What is the difference between $4a$ and a^4?

14. What is:

(*a*) $5y$ squared? (*b*) $5y$ cubed? (*c*) $5y$ to the fourth power?

15. What is the result of:

(*a*) $(a + b)^2 + (a + b)^6$? (*b*) $x^4 \times x^7$?
(*c*) $5x \times 6x^2$? (*d*) $7x^4 \cdot c^3 \times 8x \cdot c^4$?
(*e*) $9a^{3a+2} \times 5a^{4a-1}$?

16. Evaluate:

(*a*) $\sqrt{y}$ (*b*) $\sqrt{y^2}$ (*c*) $\sqrt{y^3}$ (*d*) $\sqrt{y^4}$ (*e*) $\sqrt{9^{1/2}}$
(*f*) $\sqrt{y^9}$ (*g*) $\sqrt{5ac^3x^5}$

17. Evaluate: (*a*) $\sqrt[3]{x9}$ (*b*) $\sqrt[3]{27x^9}$ (*c*) $\sqrt[3]{a^6b^5}$

18. Divide:

(*a*) y^6 by x^2 (*b*) y^9 by y^3 (*c*) $8y^5$ by $2y^2$ (*d*) y^5 by y^5

19. What is the result of:

(*a*) $\frac{a^7b^4c^3}{a^3b^2c}$? (*b*) $\frac{x^4y^4z^2}{x^3z}$? (*c*) $\frac{x^4y^2z^3}{xy}$?

(*d*) $\frac{yb^2}{y^3}$? (*e*) $\frac{27x^6y^5z^3}{3x^5y^3z^5}$?

20. Are $\frac{1}{x^5}$ and x^{-5} the same? Why?

21. Are a^3c^2 and c^2a^3 alike? Is $c^3b^2a^5$ like $c^3b^5a^2$?

22. If you went down in a mine 2,400 ft. and came up 1,100 ft., what would be your position algebraically?

23. (*a*) What is your financial position algebraically, if you spend \$50, then \$25, and then get a check for \$200?

(*b*) If you have $50x$ dollars and you owe $75x$ dollars, what is the algebraic sum?

24. Subtract: (*a*) 9 from 16 (*b*) -9 from 16 (*c*) $-6ab$ from $13ab$

25. What is the result of:

(*a*) $+5 \times +9$? (*b*) -5×-9? (*c*) $+5 \times -9$?
(*d*) $-5 \times +9$? (*e*) $2a \times 3b$? (*f*) $-2a \times -3b$?
(*g*) $+2a \times -3b$? (*h*) $-2a \times +3b$?

26. What is the result of:

(*a*) $\frac{+40}{+5}$? (*b*) $\frac{+40}{-5}$? (*c*) $\frac{-40}{-5}$? (*d*) $\frac{-40}{+5}$?

(*e*) $\frac{15ab}{3ab}$? (*f*) $\frac{-15ab}{-3ab}$? (*g*) $\frac{+15ab}{-3ab}$? (*h*) $\frac{-15ab}{+3ab}$?

27. What is the result of $9(a - b) - 15(2a - 5b)$?

28. What is the result of $8[5a - 6(4b - 7a) - 3b]$?

29. What is the result of $4[5x - \{(3x + 2y) + 7(5x + 2y)\} - 3y]$?

30. If $a = 2$, $b = 3$, $c = -4$, and $x = 8$, what is the value of

(*a*) $2a^3 - 3a^2x^2 + x^3$? (*b*) $\frac{x^7}{b^2}$? (*c*) $\frac{5x^2b^3\sqrt{b}}{a^4}$?

31. Find the algebraic sum of: $10x$, $-12y$, $9z$, $-15x$, $14y$, $11z$, $19y$, $-23x$, $15z$, $9x$, $-8z$, and $-3y$.

32. Add $7a$, $4bc$, $5a^2d$, $-5a$, $-3bc$, $9a^2d$, $8a$, $-116bcd$.

33. From $12a - 6b + 8c$ subtract $4a + 6b + 2d$.

34. What is the value of:

(*a*) $7 + (12 - 5) - (8 + 4)$? (*b*) $7a + (9 - [5a - 10])$?
(*c*) $8a - (11 - [4a - 9])$?

35. Multiply:

(*a*) $5a + 8b$ by $3a$ (*b*) $5a + 9b$ by $3a - 2b$
(*c*) $-4a^3b^2$ by $6a^4c^2$ (*d*) $5a^2 + 7ab - 9c^2$ by $8a^4$
(*e*) $3a^2b^4 - 5b^3c^2 + 6c^3d^2 - 5a^3b^2cd^3$ by $-7a^3b^2c^4$
(*f*) $8a^2 + 4b^2$ by $5a + 3b^2$

36. What is the result of:

(*a*) $(a + 4)^2$? (*b*) $(4x + 6y)^2$? (*c*) $(a - 4)^2$?
(*d*) $(9x - 4y)^2$? (*e*) $(a - 7)(a + 7)$? (*f*) $(6x - 2)(6x + 2)$?
(*g*) $(8x - 3y)(8x + 3y)$?

37. Multiply $(5x + 3y)$ by $(4x - 6y)$ directly as shown in text.

38. Simplify:

(*a*) $3x(x + 8y) + 4y(x + 7y)$
(*b*) $(a + 3)(a + 5) + (a + 2)(a + 4) + (a + 2)(a + 7)$
(*c*) $(3a - 2b)(3a + 5b) - (3a + 4x)(3a - 6x)$

39. Divide:

(*a*) $48x^4 - 36x^3 + 12x^2 - 6x$ by $3x^2$
(*b*) $70x - 42y - 56z$ by -7
(*c*) $24a^5b^4 - 15a^2b^3 + 16ab^2$ by $4a^2b^3$
(*d*) $3a + 4b + 5c$ by x

(*e*) $y^3 - a^3b^3c^5$ by $a^3b^2c^2$
(*f*) $a^2 - 2a - 35$ by $a - 7$
(*g*) $6a^2 + 8a^2b + 17ab + 12ab^2 + 12b^2$ by $3a + 4ab + 4b$
(*h*) $6a^3 - 11a^2b - 2ab^2 - 20b^3$ by $2a - 5b$

40. Factor:
(*a*) $16x - 20$ (*b*) $24x^4y^3 - 6x^3y^2$
(*c*) $2a + 3ac - 4ad$ (*d*) $8a^3b^5 - 4a^2b^3 + 12a^6b^2$
(*e*) $6(a + 2b)a - 7(a + 2b)b$ (*f*) $6b^3 - 18b^7$
(*g*) $3a(4x + 2y) + 5a(4x + 2y)$ (*h*) $3a(4x + 2y) - 5a(4x + 2y)$
(*i*) $c(2x + 3y) + p(3y + 2x)$ (*j*) $6x^2 + 10xy + 12xz + 20yz$
(*k*) $6x^2 - 30xy - xz + 5yz$ (*l*) $8x^2 + 4xy - 2x - 4y$

41. Factor by the product of two binomials method:
(*a*) $10x^2 + 14x - 24$ (*b*) $x^2 + 12x + 35$ (*c*) $x^2 - 9x - 36$
(*d*) $x^2 - 11x + 28$ (*e*) $18x^2 - 18x - 20$

42. Factor by perfect square method:
(*a*) $4a^2 - 12ab + 9b^2$ (*b*) $x^2 + 16x + 64$

43. Factor by the difference of two squares method:
(*a*) $x^4 - 25$ (*b*) $y^2 - 49$ (*c*) $225a^4b^2 - c^6$
(*d*) $(2x + 3y)^2 - 1$ (*e*) $(a - b)^2 - (3a - 2b)^2$
(*f*) $(x + y)^4 - (k - l)^8$ (*g*) $1 - 9y^2 + 24y - 16$
(*h*) $25x^2 - 30xy + 9y^2 - 16c^2 - 16cd - 4d^2$

44. What is the value of $18a^2 - 20b^2$ when $a = 4$ and $b = 3$?

45. Solve:
(*a*) $5x = 35$ (*b*) $\frac{5}{8}x - 20$ (*c*) $C = \pi d$ when $d = 12$, $\pi = 3.1416$
(*d*) $6x + 7 = 25$ (*e*) $16x^2 = 96$ (*f*) $x - 7 = 0$
(*g*) $x + 8 = 15$ (*h*) $x - 8 = 15$ (*i*) $\frac{2}{5}x = 28$ (*j*) $\frac{3}{9}x = 24$

46. If $\frac{7}{8}$ of a number is 49, what is the number? Use the equation method.

47. Solve:
(*a*) $.08x = 24$ (*b*) $5\frac{1}{6}x = 62$ (*c*) $.07x = 22{,}400$
(*d*) $x + 3 = 12$ (*e*) $26 = 8 + y$ (*f*) $18 = y - 5$
(*g*) $7 = 12 - x$ (*h*) $9x - 6 = 11 + 4x$
(*i*) $\frac{3x}{7} - \frac{5x}{9} = 3 - \frac{x}{5}$ (*j*) $6(x - 7) = 8(x + 4)$
(*k*) $M = \frac{b(t - w)}{(p - w)}$ (for w) (*l*) $B - sm = c - m^5$ (for m)
(*m*) $\frac{5}{8}P = K$ (for P) (*n*) $\frac{a + C}{y} = \frac{y}{10k}$ (for y)
(*o*) $x + 9 - 4x - 7 = 12 - 5x$
(*p*) $(x + 9)^2 - (x + 7)^2 = x + 6$
(*q*) $6(x + 4) - 8(x - 1) = 9x + 2$
(*r*) $6(x - 4)(x + 4) = x(6x - 6)$

(s) $\dfrac{x^3 + x^2 - 14x - 24}{x - 4} - (x - 3)^2 = 14$

(t) $.8x + .09 = 9 - .4x$

48. If $d = vt$, what is the distance d when v is 30 mph, and $t = 4$ hr.?

49. If $F = \frac{9}{5}C + 32°$, what is F when $C = -4°$?

50. If $\dfrac{v_t^2 - v_0^2}{2a} = s$, what is a?

51. If you multiply a number by 7, and 55 remains after you have taken away 15 from the result, what is the number?

52. If the sum of three consecutive even numbers is 48, what are the numbers?

53. If a tank is $\frac{3}{4}$ full of water, and after running off 250 gallons, it is $\frac{1}{8}$ full, what is the capacity of the tank?

54. If you are 30 years old, and your son is 8 years old, (*a*) when will your son be half your age, (*b*) how long ago were you 7 times as old as your son?

55. If two machine operators punch 2,600 plastic parts per hour, and one produces $\frac{5}{8}$ as many parts as the other, what is the production of each?

56. If you and your son together have $12,000 in bonds, and your share is $2,500 more than your son's, how much do you each have?

57. If you bought 3 suits of clothes for $277, and the first cost $2\frac{1}{2}$ times as much as the second, while the third cost $25 more than the second, what is the cost of each suit?

58. If you have $4.45 in dimes and quarters, and you have 25 coins in all, how many of each do you have?

59. At what time between 2 and 3 o'clock are the hands of a watch opposite each other?

60. If you want to sale price 400 lb. of groats at 30¢ a lb., and you have one grade that sells for 35¢ a lb., and another that sells for 25¢ a lb., how many lb. of each must you mix so that you will not lose money?

61. If you sell 3 safes and buy two new ones for $26,000, and then have $7,000 left, how much did you get for each safe you sold?

62. If *A*, *B*, and *C* saved $6,001 total, and *A* saved $500 less than twice *C*'s savings, while *C* saved $200 more than twice *B*'s, how much did each save?

63. What is the number which when multiplied by 5 will be equal to the original number increased by 44?

64. If a train leaves Washington, D.C., for Chicago, traveling at the rate of 52 miles per hour, and $\frac{3}{4}$ of an hour later an auto leaves for Chicago from Washington, D.C., traveling at the rate of 58 miles per hour, how long will it take the auto to overtake the train?

65. You start out to walk to your friend's house at the rate of $4\frac{1}{2}$ mph. Your friend starts at the same time for your house at $3\frac{1}{2}$ mph. You live 16 miles apart. How far does each of you walk before meeting?

APPENDIX A
ANSWERS TO PROBLEMS

Introduction (pp. 10–12)

1. 3, 7, 9.

3. 1, 9; 3, 7; 4, 6; 7, 2; 9, 6.

5. 7 bundles of hundreds, 6 bundles of tens, 5 bundles of units; 2 bundles of hundreds, 3 bundles of tens, 4 bundles of units; etc.

7. 7 bundles of thousands, 4 bundles of hundreds, 8 bundles of tens, 6 bundles of units; 8 bundles of thousands, 0 bundles of hundreds, 9 bundles of tens, 0 bundles of units; etc.

9. 6 bundles of ten thousands, 0 bundles of thousands, 3 bundles of hundreds, 0 bundles of tens, 8 bundles of units; 4 bundles of ten thousands, 6 bundles of thousands, 9 bundles of hundreds, 5 bundles of tens, 1 bundle of units; etc.

11. 3 bundles of hundred thousands, 6 bundles of ten thousands, 9 bundles of thousands, 2 bundles of hundreds, 4 bundles of tens, 3 bundles of units; etc.

13. 1 bundle of millions, 7 bundles of hundred thousands, 5 bundles of ten thousands, 3 bundles of thousands, 0 bundles of hundreds, 0 bundles of tens, 2 bundles of units; 75 bundles of millions (may also be called 7 bundles of ten millions and 5 bundles of millions), 2 bundles of hundred thousands, 0 bundles of ten thousands, 6 bundles of thousands, 0 bundles of hundreds, 0 bundles of tens, 8 bundles of units; etc.

15. 27 bundles of billions (may also be called 2 bundles of ten billions and 7 bundles of billions), 3 bundles of hundred millions, 9 bundles of ten millions, 2 bundles of millions, 4 bundles of hundred thousands, 9 bundles of ten thousands, 6 bundles of thousands, 0 bundles of hundreds, tens, and units; etc.

17. (*a*) 0.73586 (*b*) 8,000.008
(*c*) 0.5; 0.3; 2.1 (*d*) 7.009; 0.000012
(*e*) 0.235; 0.491; 0.0006; 300.03
(*f*) 4.7; 9.2; 0.86; 500.005
(*g*) 0.4; 0.9; 0.3; 0.6; 0.01; 0.007; 0.0008
(*h*) 0.364575 (*i*) 0.908006034

19. (*a*) Sixteen and five thousandths
(*b*) Fifty and six hundred seven thousandths
(*c*) Two ten-thousandths
(*d*) Eighty-seven and nine thousand three hundred seventy-five ten-thousandths
(*e*) Thirty-five and two hundred one thousandths

(*f*) Eighty-six and five thousand three hundred ninety-two ten-thousandths
(*g*) Two and three thousand four hundred forty-one ten-thousandths
(*h*) Two hundred and three thousand four hundred eighty-seven ten-thousandths
(*i*) Twenty and two thousand seventy-four ten-thousandths
(*j*) Two hundred six and ten thousand fifty-seven hundred-thousandths
(*k*) Thirty and five hundred sixty-four thousandths
(*l*) Ninety-seven and four thousand three hundred fifty-six ten-thousandths

21. Three hundred fifty-six dollars thirty-five cents six mills.

23. (*a*) \$0.66 (*b*) \$0.80 (*c*) \$0.47
(*d*) \$0.10 (*e*) \$1.20 (*f*) \$7.12

25. (*a*) 475 (*b*) 5,621 (*c*) 22 (*d*) 10,540
(*e*) 1,076.5 (*f*) 255.5 (*g*) 100 (*h*) 444.4

27. (*a*) XII (*b*) XVIII (*c*) XIX
(*d*) XLIII (*e*) XXXIII (*f*) XXVIII
(*g*) LVI (*h*) LXXXII (*i*) LXXVI
(*j*) XCVII (*k*) CXVII (*l*) CCCLXXXV
(*m*) CCXL (*n*) DXII (*o*) CDLXX
(*p*) DCCXLII (*q*) CDXXII (*r*) CMXLII
(*s*) MCDXXVI (*t*) MDCCCLXXIV
(*u*) $\overline{\text{V}}\text{DCCCLXXII}$ (*v*) $\overline{\text{XXIV}}\text{DCCLXIV}$
(*w*) $\overline{\text{CCLVII}}\text{DCCCXLVI}$ (*x*) $\overline{\text{MCDL}}\text{DCCXXIX}$
(*y*) $\overline{\text{MMMMDCCCXL}}\text{V}$
(*z*) $\overline{\text{MMMMMMMMMDLXII}}\text{CMXLII}$

Chapter I (pp. 21–23)

1. 3, 6, 9, 12... **3.** 6, 12, 18, 24...

5. 3, 5, 7, 9...; 3, 7, 11, 15...; 3, 9, 15, 21...; 3, 11, 19, 27...

7. 9, 13, 17, 21...; 9, 16, 23, 30...; 9, 18, 27, 36...; 9, 11, 13, 15...

9. 1,997 **11.** \$203.265 **13.** 60

15. 95 **17.** 21 **19.** 90

21. (*a*) 255 (*b*) 244 (*c*) 209
(*d*) 263 (*e*) 270 (*f*) 250

23. (*a*) 169 (*b*) 155 (*c*) 140 (*d*) 141
(*e*) 1,879 (*f*) 1,457 (*g*) 1,667 (*h*) 2,039

25. 907 gallons **27.** 1,039 miles **29.** \$525

31. (*a*) 280,778 (*b*) 295,263
(*c*) 292,690 (*d*) 242,893

33. (*a*) 195,564 (*b*) 293,220
(*c*) 208,675 (*d*) 142,415

Chapter II (pp. 34–36)

1. 22, 42, 19, 22, 25, 61, 24, 25
3. 633, 432, 392, 134, 320, 372
5. 1,522, 942, 1,505, 981, 1,720
7. 4 steps to the left or -4
9. $-4°$ lat.
11. $24

12. (*a*) 124,959 (*b*) 151,833 (*c*) 74,296
(*d*) 161,574 (*e*) $3,059.07 (*f*) $8,738.83
(*g*) $388,492.54 (*h*) $605,791.79

14. (*a*) 25,697 (*b*) 49,779 (*c*) 92,922
(*d*) $220,152.50 (*e*) $1,000,350.90 (*f*) $913,578.18

17. (*a*) 4,228 (*b*) 4,214 (*c*) 4,319
(*d*) 5,659 (*e*) 3,357 (*f*) 2,165

19. (*a*) 2,443,393 (*b*) 888 (*c*) 1,669 (*d*) 178,556

21. (*a*) 1,421 (*b*) 41,135

23. $407.46
25. 1,548,576

27. (*a*) 53, 51, 49, 47 . . . ; 53, 49, 45, 41 . . . ; 53, 47, 41, 35 . . . ; 53, 45, 37, 29 . . .
(*b*) 89, 86, 83, 80 . . . ; 89, 84, 79, 74 . . . ; 89, 82, 75, 68 . . . ; 89, 80, 71, 62 . . .
(*c*) 74, 69, 64, 59 . . . ; 74, 67, 60, 53 . . . ; 74, 71, 68, 65 . . . ; 74, 65, 56, 47 . . .

Chapter III (pp. 54–56)

1. 540; 5,400; 54,000
3. 17,620; 176,200; 1,762,000
5. 1,800; 18,000; 180,000; 18,000,000
7. 1,917,000

9. (*a*) 28,428 (*b*) 7,136 (*c*) 63,851
(*d*) 54,008 (*e*) 43,362 (*f*) 55,859
(*g*) 43,776 (*h*) 2,700,578 (*i*) 443,772
(*j*) 7,589,594 (*k*) 3,050,260 (*l*) 3,794,186
(*m*) 3,157,596 (*n*) 2,615,057 (*o*) 2,893,230
(*p*) 28,201,925 (*q*) 3,047,385 (*r*) 75,874,332
(*s*) 18,083,583 (*t*) 75,490,868 (*u*) 3,571,632
(*v*) 9,602,484 (*w*) 428,505 (*x*) 4,346,136
(*y*) 3,455,412 (*z*) 7,346,628

11. $144.25; $634.70
13. $1,886
15. 1,742,400 lb.
17. $420

19. (*a*) 238 (*b*) 272 (*c*) 306 (*d*) 304

21. (*a*) 7,395 (*b*) 2,352 (*c*) 3,074
(*d*) 1,184 (*e*) 4,355 (*f*) 9,306
(*g*) 5,328 (*h*) 728 (*i*) 306

23. (*a*) 945 (*b*) 8,295 (*c*) 6,435
(*d*) 630 (*e*) 4,005

25. (*a*) 2,709 (*b*) 2,625 (*c*) 1,316 (*d*) 3,149
(*e*) 3,364 (*f*) 2,016 (*g*) 2,236

27. (*a*) 4,275 (*b*) 4,875 (*c*) 5,525 (*d*) 1,925
(*e*) 3,325 (*f*) 4,125 (*g*) 1,225 (*h*) 6,375
(*i*) 9,425 (*j*) $60.75 (*k*) $123.75 (*l*) $204.25

29. (*a*) $\frac{1}{2}$ (*b*) $\frac{1}{8}$ (*c*) $\frac{3}{8}$ (*d*) $\frac{5}{8}$ (*e*) $\frac{7}{12}$
(*f*) $\frac{1}{6}$ (*g*) $\frac{5}{12}$ (*h*) $\frac{1}{3}$ (*i*) $\frac{1}{12}$

31. (*a*) $\frac{1}{4}$ (*b*) $\frac{3}{8}$ (*c*) $\frac{5}{8}$ (*d*) $\frac{1}{8}$

33. $14.00

35. (*a*) $212.50 (*b*) $123.25 (*c*) $28.75
(*d*) $12.00 (*e*) $12.00 (*f*) $21.00
(*g*) $18.00 (*h*) $416.00 (*i*) $9.00

37. $9,000

39. (*a*) 768 (*b*) 1,632 (*c*) 30,008 (*d*) 1,368

41. (*a*) 516,456 (*b*) 528,849 (*c*) 38,952
(*d*) 890,901 (*e*) 7,628,688

43. (*a*) 5,496 (*b*) 4,809 (*c*) 3,456
(*d*) 3,024 (*e*) 7,856 (*f*) 6,874

45. (*a*) 8,232 (*b*) 9,024 (*c*) 7,998
(*d*) 7,505 (*e*) 7,216 (*f*) 960,376

47. (*a*) 6,384 (*b*) 63,672 (*c*) 3,196 (*d*) 49,088
(*e*) 7,128 (*f*) 2,964 (*g*) 7,392 (*h*) 64,528

Chapter IV (pp. 72–74)

1. 7 **3.** 4 **5.** 20

7. 1 acre per man and $\frac{1}{2}$ acre per boy

9. (*a*) 321 (*b*) 221 (*c*) 231 (*d*) 216
(*e*) 72 (*f*) 64 (*g*) 91 (*h*) 95
(*i*) 52 (*j*) 126 (*k*) 137 (*l*) $34\frac{1}{7}$
(*m*) 1,824 (*n*) 1,077 (*o*) 8,912 (*p*) 5,072
(*q*) 10,586 (*r*) 10,534 (*s*) 6,801 (*t*) 2,647
(*u*) $4{,}684\frac{8}{12}$ (*v*) $5{,}569\frac{7}{15}$ (*w*) $6{,}657\frac{9}{11}$ (*x*) $9{,}731\frac{6}{9}$
(*y*) $6{,}974\frac{3}{12}$ (*z*) $6{,}550\frac{4}{15}$

11. 7 hours

13. (*a*) 214 (*b*) 402 (*c*) 428

17. (*a*) 3 (*b*) Yes; 2

19. (*a*) 29.58 (*b*) 60 (*c*) 808.68 (*d*) $365\frac{5}{7}$
(*e*) 16.80 (*f*) 6,912 (*g*) 72 (*h*) 42
(*i*) 139 (*j*) 36 (*k*) 11.2

21. (*a*) Subtract 2 (*b*) Subtract 1

23. 84.2 **25.** $1,015.22

Chapter V (pp. 80–81)

1. 2, 3, 6, 9; 2, 3, 5, 6, 10, 15; 2, 3, 4, 6, 12, 18; 3, 9, 27; 2, 3, 4, 5, 6, 8, 10, 12, 15, 20, 24, 30, 40, 60

3. 2, 3, 4, 6, 12

5. 1, 2, 3, 5, 7, 11, 13, 17, 19, 23, 29, 31, 37, 41, 43, 47, 53, 59, 61, 67, 71, 73, 79, 83, 89, 97

7. 2, 2; 5; 7; 2, 2, 2; 2, 5; 2, 2, 3; 13; 2, 7; 2, 2, 2, 2; 2, 3, 3; 3, 7; 2, 2, 2, 3; 5, 5; 2, 3, 5; 2, 17. (Note: 1 is a prime factor of all integers.)

9. (*a*) 2, 5, 31 (*b*) 3, 3, 3, 11 (*c*) 2, 5, 67
(*d*) 3, 13, 19 (*e*) 2, 7, 7 (*f*) 5, 7, 11
(*g*) 2, 5, 5, 53 (*h*) 3, 3, 3, 3, 3, 5 (*i*) 3, 107
(*j*) 3, 3, 5, 5, 7 (*k*) 3-seven times; 5 (*l*) 2, 2, 3, 5, 7
(*m*) 7, 71 (*n*) 2, 3, 3, 3, 7 (*o*) 2, 3, 7, 11
(*p*) 2, 3, 3, 3, 3, 3, 5 (*q*) 2-eight times; 3, 3, 11
(*r*) 2, 2, 3, 3, 5, 11, 37 (*s*) 2, 2, 3, 7, 11, 11, 59
(*t*) 2, 2, 3, 3, 3, 17, 149 (*u*) 5-six times
(*v*) 5, 5, 7, 61 (*w*) 5, 5, 5, 101 (*x*) 2, 2, 2, 2, 61
(*y*) 2, 5, 5, 7, 23 (*z*) 2, 2, 2, 13, 37

11. (*a*) 9, 18 (*b*) 35, 70 (*c*) 18, 36 (*d*) 21, 42 (*e*) 40, 80
(*f*) 6, 12 (*g*) 8, 16 (*h*) 72, 144 (*i*) 9, 18 (*j*) 24, 48

13. (*a*) 21 (*b*) 15 (*c*) 28 (*d*) 24 (*e*) 16

15. 36 lb. **17.** 18 days

Chapter VI (pp. 102–106)

1. $\frac{1}{4}$

3. One-third, one-sixth, one-sixteenth, one-twelfth, one-twentieth; the denominator

5. (*a*) all proper
(*b*) $\frac{2}{3}$ proper; $\frac{5}{4}$, $\frac{18}{16}$ improper; $4\frac{1}{8}$ mixed number
(*c*) $\frac{4}{7}$, $\frac{17}{19}$ proper; $\frac{19}{15}$, $\frac{8}{3}$ improper
(*d*) $\frac{3}{5}$ proper; $\frac{9}{2}$ improper; $8\frac{1}{5}$, $16\frac{1}{2}$ mixed numbers
(*e*) $\frac{9}{10}$ proper; $\frac{27}{4}$ improper; $8\frac{2}{7}$, $17\frac{3}{4}$ mixed numbers
(*f*) $\frac{27}{39}$, $\frac{14}{24}$ proper; $\frac{18}{5}$ improper; $6\frac{1}{2}$ mixed number

7. (*a*) 8 (*b*) 5 (*c*) 6 (*d*) $8\frac{1}{6}$ (*e*) $5\frac{14}{23}$
(*f*) 9 (*g*) $7\frac{1}{3}$ (*h*) $10\frac{3}{8}$ (*i*) $26\frac{1}{2}$ (*j*) $6\frac{3}{11}$
(*k*) $24\frac{1}{6}$ (*l*) $1\frac{71}{160}$ (*m*) 1 (*n*) 72 (*o*) $13\frac{12}{19}$

9. (*a*) 14 (*b*) 28 (*c*) 7 (*d*) No

11. (*a*) $\frac{15}{20}$ (*b*) $\frac{40}{64}$ (*c*) $\frac{24}{84}$ (*d*) $\frac{78}{96}$ (*e*) $\frac{45}{100}$ (*f*) $\frac{20}{24}$

13. (*a*) $\frac{15}{20}, \frac{8}{20}, \frac{6}{20}$ (*b*) $\frac{12}{24}, \frac{15}{24}, \frac{14}{24}$ (*c*) $\frac{42}{105}, \frac{45}{105}, \frac{70}{105}$
(*d*) $\frac{25}{40}, \frac{16}{40}, \frac{1}{40}$ (*e*) $\frac{18}{45}, \frac{5}{45}, \frac{9}{45}$ (*f*) $\frac{9}{132}, \frac{11}{132}$
(*g*) $\frac{301}{9546}, \frac{333}{9546}$ (*h*) $\frac{84}{180}, \frac{52}{180}, \frac{69}{180}$
(*i*) $\frac{693}{1089}, \frac{143}{1089}, \frac{81}{1089}$

15. (*a*) $\frac{15}{8}$ or $1\frac{7}{8}$ (*b*) $\frac{11}{10}$ or $1\frac{1}{10}$ (*c*) $\frac{175}{48}$ or $3\frac{31}{48}$

17. (*a*) $\frac{35}{72}$ (*b*) $\frac{20}{33}$ (*c*) $\frac{55}{279}$ (*d*) $\frac{13}{64}$ (*e*) $\frac{55}{108}$ (*f*) $\frac{45}{368}$

19. (*a*) 2 (*b*) $15\frac{5}{7}$ (*c*) $2\frac{1}{4}$ (*d*) $67\frac{11}{20}$
(*e*) $1\frac{1}{18}$ or $\frac{19}{18}$ (*f*) $1\frac{1}{21}$ or $\frac{22}{21}$ (*g*) $1\frac{17}{20}$ (*h*) $1\frac{13}{126}$
(*i*) $28\frac{3}{4}$ (*j*) $42\frac{173}{180}$ (*k*) $139\frac{1}{6}$ (*l*) $129\frac{7}{8}$

21. (*a*) $\frac{15}{32}$ (*b*) $\frac{51}{280}$ (*c*) $\frac{7}{33}$ (*d*) $\frac{9}{4}$
(*e*) $\frac{60}{7}$ or $8\frac{4}{7}$ (*f*) $\frac{119}{9}$ (*g*) $\frac{130}{9}$ (*h*) $27\frac{27}{32}$
(*i*) $60\frac{9}{16}$ (*j*) $\frac{4}{3}$ (*k*) $9\frac{3}{8}$ (*l*) $14\frac{6}{55}$

23. (*a*) 5, $2\frac{1}{2}$, $1\frac{2}{3}$, $1\frac{1}{4}$ (*b*) 14, 7, $4\frac{2}{3}$, $3\frac{1}{2}$, $2\frac{4}{5}$, $2\frac{1}{3}$

25. (*a*) $\frac{1}{8}$ (*b*) $\frac{8}{43}$ (*c*) $\frac{25}{68}$ (*d*) 16 (*e*) 2
(*f*) 216 (*g*) $\frac{1}{2}$ (*h*) $\frac{1}{4}$ (*i*) $\frac{2}{9}$ (*j*) $\frac{5}{19}$

27. (*a*) 100 (*b*) $688

29. $\dfrac{1}{4 + \dfrac{1}{7 + \frac{1}{2}}}$ **31.** $\frac{49}{87}$ **33.** $\frac{22}{7}$

35. $44 **37.** $1\frac{7}{9}$ oz. per slice **39.** 286 miles

41. $\frac{3}{8}$, $\frac{3}{8}$, $\frac{1}{4}$; $1,350, $1,350, $900

43. $246; $61.50, $92.25, $30.75, $20.50

45. $\frac{41}{200}$, $\frac{209}{1000}$, $\frac{21}{100}$ **47.** $17\frac{4}{99}$ rods

Chapter VII (pp. 121–124)

1. (*a*) 0.6, 0.4, 2.1 (*b*) 7.009, 9.053, 0.0003, 0.000011
(*c*) 0.155, 0.492, 0.0006, 300.04 (*d*) 6.7, 8.2, 0.86, 500.006
(*e*) $0.04\frac{3}{8}$, $0.036\frac{5}{7}$, $8.000\frac{2}{3}$, $8.004\frac{2}{3}$

3. Twelve and five hundred eighty-four million sixty-two thousand eighteen billionths

5. 1,000; 100,000 **7.** Ten

9. (*a*) .8 = .80 = .800 (*b*) .046 = .0460 = .04600
(*c*) .738 = .7380 = .73800 = 0.738

11. 0.04, 0.004 **13.** 24.6, 2.46

15. 2,460, 24,600, 246,000 **17.** 2,465.76, 24,657.6

19. (*a*) 0.32 (*b*) 0.625 (*c*) 0.14 (*d*) 0.392
(*e*) 0.1875 (*f*) 0.65 (*g*) 0.4 (*h*) 0.175
(*i*) 0.3125 (*j*) 0.115 (*k*) 0.46875 (*l*) 0.232

21. (*a*) $\frac{3}{5}$ (*b*) $\frac{43}{50}$ (*c*) $\frac{5}{8}$ (*d*) $\frac{3}{16}$
(*e*) $\frac{1}{80}$ (*f*) $\frac{3}{4}$ (*g*) $\frac{953}{2000}$ (*h*) $\frac{7}{75}$

23. (*a*) 127.4735 (*b*) $181.25608\frac{1}{3}$
(*c*) $221.35538\frac{1}{3}$ (*d*) $720.2238\frac{8}{9}$

25. (*a*) $42.34408\frac{1}{3}$ (*b*) 4.494375 (*c*) $3.8316\frac{2}{3}$
(*d*) 3.5425 (*e*) $55.3308\frac{1}{3}$

27. (*a*) 52.655625 (*b*) 2.582398 (*c*) 39.130222
(*d*) 2.012315 (*e*) 0.638027

29. (*a*) 15.895794 (*b*) 38.884176 (*c*) 17.517890
(*d*) 112.489886 (*e*) 54.923664 (*f*) 21.073016

31. (*a*) 59.77 (*b*) 59.76

33. 0.12

35. 29.18

37. (*a*) 0.078125 (*b*) 0.15625 (*c*) 0.375 (*d*) 0.3125
(*e*) 0.28125 (*f*) 0.171875 (*g*) 0.28 (*h*) 0.184

39. $0.42, $0.07

41. $5,687.50

43. $34,000; $7,480, $10,880, $12,240

45. $2.82

47. 0.968 lb.

49. 11 cents 8.32 mills.

51. 32 lb.

53. A. 0.750; B. 0.714

Chapter VIII (pp. 136–139)

1. 27%

3. $38\frac{8}{9}$%

5. 0.25%, 0.2%, 0.2%; 0.0025, 0.002, 0.002

7. (*a*) 900% (*b*) 60% (*c*) 25% (*d*) $6\frac{1}{4}$%
(*e*) $814\frac{2}{7}$% (*f*) 28% (*g*) 85% (*h*) $58\frac{1}{3}$%
(*i*) $3\frac{1}{8}$% (*j*) 16% (*k*) 0.4% (*l*) $87\frac{1}{2}$%
(*m*) 84% (*n*) $33\frac{1}{3}$% (*o*) 65% (*p*) 60%
(*q*) 80% (*r*) $87\frac{1}{2}$% (*s*) 7% (*t*) $4\frac{1}{4}$%
(*u*) $16\frac{2}{3}$%

9. 16%

11. (*a*) 25 (*b*) 64 (*c*) 100 (*d*) 325 (*e*) 30 (*f*) 420

13. $3,000; $11,040; $9,960

15. 1,904 votes

17. 20%

19. $1,323.89

21. (*a*) 40.6% (*b*) 13.1% (*c*) 12.78% (*d*) 4.0%
(*e*) $28\frac{1}{3}$% (*f*) 23.23% (*g*) 0.135% (*h*) 21.88%
(*i*) 166.2% (*j*) 36.4% (*k*) 7,150% (*l*) 44.42%
(*m*) 51.38%

23. $\frac{1}{200}$ acre = 217.8 sq. ft.

25. 2%

27. 400, 32

29. 50.667, 5.911

31. 190,000; 54

33. $291.43

35. $13,636

37. 21%

39. $14\frac{2}{7}$%

41. $18.38

43. $150

45. $352.35

47. 4,200 students, 26% smaller

49. $20,588

51. $33\frac{1}{3}$%, 25%

53. (*a*) 72 (*b*) 60 (*c*) 0.06696

55. (*a*) 304 (*b*) 720 (*c*) 23.00

Chapter IX (pp. 155–158)

1. (*a*) 6% (*b*) 6% (*c*) 6%

3. $928, $128.

5. (*a*) March 4 (*b*) March 3

7. (*a*) 249 (*b*) 84 (*c*) 118 (*d*) 248 (*e*) 142

9. $7.89 **11.** $240 **13.** $49.19

15. (*a*) $47.17 (*b*) $381.11 (*c*) $2.91
(*d*) $11.86 (*e*) $286.03 (*f*) $3.70
(*g*) $34.31 (*h*) $3.63 (*i*) $49.12

17. $4310 **19.** $14.38 **21.** $2.47

23. (*a*) 15, 6, 6 (*b*) 60, 15 (*c*) 30, 6, 3
(*d*) 60, 30, 6 (*e*) 30, 10 (*f*) 60, 15, 6, 6
(*g*) 60, 60, 6, 3 (*h*) 60, 30, 15 (*i*) 60, 60, 15, 10
(*j*) 15, 6 (*k*) 60, 60, 6 (*l*) 60, 30, 6, 3

25. $4.69 **27.** $1.50 **29.** $96.53

31. $4\frac{1}{2}$% **33.** 85 days

35. (*a*) 120 days (*b*) 140 days (*c*) 47 days (*d*) 229 days

37. (*a*) $1.20 (*b*) $0.68 (*c*) $8.29
(*d*) $2.40 (*e*) $0.28 (*f*) $4.25

39. $4,445.00

41. Accumulation factor for 8 years at 2% = 1.171659
Accumulation factor for 4 years at 4% = 1.169859

43. $2,600.00 **45.** $1,040.40 **47.** $16,436.15

49. $62.89 **51.** $26,937.06, $6,937.06

Chapter X (pp. 181–185)

1. (*a*) 1:3 (*b*) 3:1 (*c*) 1:7 (*d*) 4:3 (*e*) 5:6
(*f*) 6:5 (*g*) 1:2 (*h*) 3:4 (*i*) 5:6

3. 5:9 **5.** 11:6 **7.** 11:8; 8:11 **9.** 0.6

11. (*a*) 3:5 (*b*) 8:5 (*c*) 1:3 (*d*) 9:8

13. (*a*) 1:3 (*b*) 1:2 (*c*) 1:3 (*d*) 1:10
(*e*) 1:3.79 (*f*) 1:9 (*g*) 1:$2\frac{1}{11}$ (*h*) 1:6.25
(*i*) 1:40 (*j*) 1:5.71 (*k*) 1:11.5 (*l*) 1:22.2
(*m*) 1:4 (*n*) 1:65 (*o*) 1:60 (*p*) 1:13.6
(*q*) 1:1.14

15. $1\frac{5}{8}$:1 **17.** 4:1 **19.** 14 and 21

21. 50, 70, 80 **23.** 51, 34 **25.** 1:5

27. $14\frac{1}{4}$ inches **29.** $438.75 **31.** $3\frac{1}{4}$%

33. 22 ft. × $18\frac{1}{2}$ ft. **35.** 3:1

37. (*a*) 4 (*b*) 7 (*c*) $2\frac{1}{4}$

39. 8

41. (*a*) 6 (*b*) 2 (*c*) 18 (*d*) 2.4
(*e*) 18 (*f*) 3 (*g*) 12 (*h*) 3.2

43. $37.50 **45.** 72 feet. **47.** 1,057 lb.

49. (*a*) 10 (*b*) 15 (*c*) $\frac{1}{4}$

51. 217 lb. **53.** 42 men **55.** 8:21; \$3,528; \$1,536
57. $6\frac{3}{4}$ days **59.** 2:3 **61.** 0.82 ohm.
63. 400 feet **65.** $x = 6$ **67.** 90 p.s.i.
69. 66 men

Chapter XI (pp. 198–199)

1. \$92.67 **3.** 47 mph **5.** 85.62
7. 43.5 minutes. **9.** \$1,784 **11.** \$2,090.67
13. 59 **15.** (*a*) 13 (*b*) 19
17. \$340 **19.** \$300 to \$399 **21.** No
23. There are as many grades above 81 as there are below

Chapter XII (pp. 214–216)

1. (*a*) 39 inches (*b*) 12 feet (*c*) 33 yards
(*d*) $9\frac{1}{3}$ feet (*e*) 1,600 rods (*f*) 396 inches
(*g*) $29\frac{1}{3}$ yards (*h*) $2\frac{2}{3}$ yards (*i*) 5.576 rods
(*j*) $49\frac{1}{2}$ feet (*k*) 6.602 miles (*l*) 31,680 feet
3. 8 rods, 2 feet **5.** $20\frac{1}{4}$ cubic inches
7. 83,688 lb. of water **9.** 302.5 bbl. **11.** \$18.16
13. 49,280 lb. **15.** 366 **17.** 184
19. 42 doz. **21.** 30 years
23. (*a*) 288 sheets (*b*) 1,440 sheets (*c*) 1,920 sheets
(*d*) 14,400 sheets
25. (*a*) 735 dm. (*b*) 7.4126 meters
27. (*a*) 0.048261 sq. meters (*b*) 7,480 sq. dm.
29. 39,122 dg. **31.** 6,944 grains **33.** 102,058 cg.
35. 0.664 grains **37.** 240″, 360″, 7,200″
39. 392 pt. **41.** $\frac{13}{32}$ bu. **43.** 0.883 bu.
45. 0.0181 gal.
47. 1 yr. 9 mo. 18 days 4 hr. 44 min. 52 sec.
49. 3 A, 76 sq. rd. 13 sq. yd. 6 sq. ft. 108 sq. in.
51. 7,504.610 meters **53.** 79.76 meters **55.** 11,664 kg.

Chapter XIII (pp. 238–240)

1. (*a*) 25 (*b*) 64 (*c*) 400 (*d*) 1
(*e*) 121 (*f*) 1 (*g*) 1,000 (*h*) 81
(*i*) 625 (*j*) 4,913 (*k*) 571,787 (*l*) 1.953125
(*m*) 0.5625 (*n*) $\frac{9}{16}$ (*o*) $1\frac{61}{64}$ (*p*) $\frac{343}{512}$
(*q*) x^4 (*r*) $16x^2$ (*s*) $8b^3$ (*t*) 1,953.125

3. 4,000 sq. ft. **5.** 48 sq. yd.

7. (*a*) 256 (*b*) 19,683 (*c*) 16 (*d*) 3 (*e*) a^{x-y}
(*f*) a^{x+y} (*g*) 4,096 (*h*) 15,625 (*i*) 1 (*j*) 1
(*k*) 1 (*l*) 24 (*m*) $\frac{1}{3125}$ (*n*) $\frac{1}{8}$ (*o*) $\frac{1}{256}$

9. 2,176,782,336

11. (*a*) 784 (*b*) 4,489 (*c*) 5,776 (*d*) 7,921

13. 950,625

15. (*a*) 256 (*b*) 2,025 (*c*) 650.25

17. (*a*) 99,980,001 (*b*) 9,801 (*c*) 999,998,000,001

19. (*a*) 12 (*b*) $4b^4$ (*c*) $a^3b^{3/2}$ (*d*) x^2y^4
(*e*) $\frac{125}{4096}$ (*f*) $\frac{1}{5}$ (*g*) 8 (*h*) $2.646 = \sqrt{7}$
(*i*) $2.080 = \sqrt[3]{9}$

21. (*a*) $2\sqrt[3]{75}$ (*b*) 12 (*c*) $4\sqrt[4]{3}$

23. (*a*) $\dfrac{\sqrt{7}}{7}$ (*b*) $\dfrac{\sqrt{7} + \sqrt{3}}{4}$ (*c*) $\dfrac{\sqrt{a}}{a}$

25. (*a*) $\sqrt[10]{4}$ (*b*) $\sqrt[8]{9}$ (*c*) $\sqrt[12]{a^2}$

27. (*a*) $4a^2y^{5/4}$ (*b*) $x\sqrt{2x}$ (*c*) $\dfrac{\sqrt{14}}{4}$

29. 1,287 feet **31.** 6.314

33. (*a*) $\frac{11}{12}$ (*b*) $\frac{7}{8}$ (*c*) $\frac{4}{5}$ (*d*) $\dfrac{\sqrt{2}}{2}$
(*e*) $\dfrac{\sqrt{3}}{2}$ (*f*) 0.1334 (*g*) $\frac{55}{57}$ (*h*) 0.949
(*i*) $2\frac{3}{5}$ (*j*) 9.709 (*k*) 0.0255

35. $\frac{3}{4}$

Chapter XIV (pp. 258–261)

1. (*a*) 3 log 5 (*b*) 6 (*c*) −5 log 3
(*d*) $\sqrt{2}$ log 9 (*e*) 4 log 3 (*f*) −2 log 2

3. (*a*) $4^4 = 256$ (*b*) $x^b = a$ (*c*) $b^0 = 1$
(*d*) $10^{-6} = 0.000001$ (*e*) $10^4 = 10{,}000$ (*f*) $6^4 = 1{,}296$

5. 0, 1, 2, 3, 4, −1, −2, −3, −4

7. (*a*) 1,000 (*b*) 64 (*c*) −5
(*d*) 512 (*e*) $\frac{1}{3125}$ (*f*) 10
(*g*) $\frac{1}{125}$ (*h*) $\frac{3}{2}$ (*i*) 7

9. (*a*) 0 (*b*) 3 (*c*) −1 (*d*) 6
(*e*) 1 (*f*) −5 (*g*) 2 (*h*) 8
(*i*) −3 (*j*) 1 (*k*) 0 (*l*) 0
(*m*) −14 (*n*) −8 (*o*) 6 (*p*) −1

11. (*a*) 0.7740 (*b*) 2.9910 (*c*) 8.8075 − 10
(*d*) 7.9441 − 10 (*e*) 1.5790 (*f*) 0.1396
(*g*) 8.4857 − 10 (*h*) 5.8321 − 10 (*i*) 5.7539 − 10
(*j*) 1.8048

13. 2 log 7 + log 4 **15.** log 194.4

17. (*a*) log 4.32 + log 7.48 − log 5.66
(*b*) $\frac{1}{2}$[log 9.48 − log 72.4 − log 0.069]

19. (*a*) $\log \dfrac{6^4 \times 5^{1/6}}{8^7}$ (*b*) $\log \dfrac{25^{9/5} \times 9}{10^{7/9} \times 5^{1/7}}$

(*c*) $\log \left[\dfrac{10^6}{6^{1/3} \times 7^{3/8}}\right]^{3/5}$ (*d*) $\log \dfrac{24}{4^{1/9}}$

21. (*a*) 3.170 (*b*) 2.633 (*c*) 1.490 (*d*) 1.057
(*e*) 2.681 (*f*) −2.861 (*g*) −1.661 (*h*) 1.661
(*i*) 0.792 (*j*) 0.921 (*k*) −2.861 (*l*) 2.861

23. (*a*) 0.340 (*b*) 367.9 (*c*) 0.0036
(*d*) 4016 (*e*) 0.00027 (*f*) 1.64

Chapter XV (pp. 264–265)

1. (*a*) 5 miles west on the scale (*b*) At sea-level or Elev. 0
(*c*) At zero, or 0° on the scale
(*d*) Zero change, no gain and no loss

3. (*a*) 12 (*b*) 6 (*c*) $\frac{5}{8}$ (*d*) 1.6 (*e*) 350

5. (*a*) 146.4 (*b*) −378 (*c*) $\frac{343}{450}$ (*d*) 12 (*e*) 84 (*f*) −84

Chapter XVI (pp. 276–277)

1. (*a*) 24 (*b*) 4 (*c*) 16 (*d*) 5 (*e*) $\frac{1}{9}$
(*f*) 98 (*g*) 64 (*h*) 16

3. $a = 15$; S = 645 **5.** 3, $3\frac{5}{7}$, $4\frac{3}{7}$, $5\frac{1}{7}$, $5\frac{6}{7}$, $6\frac{4}{7}$, $7\frac{2}{7}$, 8

7. $\frac{19}{112}$ **9.** 250,500

11. $l = 39{,}366$; S = 29,524 **13.** $2\frac{2}{5}$

15. $\frac{843}{990}$ **17.** 2, $\frac{4}{3}$, $\frac{8}{9}$, $\frac{16}{27}$, $\frac{32}{81}$, $\frac{64}{243}$, $\frac{128}{729}$

19. 15 **21.** 2, $\frac{36}{13}$, $\frac{9}{2}$, 12, −18, $-\frac{36}{7}$, −3

23. $703.88

Chapter XVII (pp. 294–296)

13. 21.74¢ (average cost per quart) **15.** $54.25

Chapter XVIII (pp. 331–335)

1. $42 **3.** 38.46%

5. $3 **7.** $108, $1,692

9. \$273.54
11. 38.25%
13. \$1,412.21
15. \$54
17. \$26.93
19. 42.6%
21. \$44.21, \$2.69
23. \$48.15
25. \$37.08
27. \$400
29. \$2,296.65, \$103.35
31. \$2,133.59
33. \$30
35. \$26.50
37. $33\frac{1}{3}$%
39. 37.3%
41. \$233.00, 36%
43. \$3,000, \$1,000
45. \$7,990.56, \$1,990.56
47. 24.2%
49. 20.4%
51. 10.82%
53. 43.6%
55. 45.7%
57. \$959.40
59. \$225
61. \$287.12
63. 2.3075%, \$2.30$\frac{75}{100}$ per \$100, \$23.07$\frac{5}{10}$ per \$1,000
65. \$3,073,196

Chapter XIX (pp. 353–355)

1. 96 days
3. 60 days
5. $\frac{1}{4}$
7. $1\frac{29}{31}$ or 1.94 days
9. 40 gal.
11. 28.6%
13. $3\frac{3}{4}$ qt.
14. 11:19
17. 39.94 av. oz.
19. 125 cc.
21. $\frac{1}{12}$
23. 3 min.
25. 4,320 gal.
27. $17\frac{1}{2}''$
29. 1:253,440
31. 276 miles
33. 972.8 acres
35. 720°
37. 4″

Chapter XX (pp. 398–402)

1. 6, c, d, p
3. x, y, z, z
5. 15
7. No
9. $\frac{1}{8}\cdot 3\cdot y$, $\frac{1}{8}\cdot 3y$, $3y/8$, $3(y/8)$, $\frac{1}{8}(3y)$, etc.
11. (*a*) A number, *b*, less six
(*b*) A number, *a*, plus seven
(*c*) The product of nine, the quantity *p*, and the quantity *q*
(*d*) Seven times the quantity *x*, plus three times the quantity *y*
(*e*) Four times the quantity *w*, the product less eight
(*f*) One-seventh the product of two numbers, *a* and *b*
(*g*) Three times a number, *a*, plus a second number *b*, less six times a third number *c*

(*h*) Five-eighths of a certain length, l
(*i*) Nine times the sum of the number c and the number 5
(*j*) One-sixth of l times the sum of the three terms, A, four times B and C. (Prismoidal formula for volume)
(*k*) One-fourth the sum of c and d
(*l*) One-half the product of b and h. (Area of a triangle)
(*m*) One-half the product of m and the square of v. (Formula for kinetic energy)
(*n*) The compound amount (A) is equal to the principal (P) multiplied by a binomial, one plus the rate (r), said binomial having been multiplied by itself n times. (Compound interest formula)
(*o*) d times the square root of the binomial, a squared plus b squared
(*p*) One-half of h times the sum of a and b. (Area of a trapezoid)

13. $4a =$ the number $4 \times a$; $a^4 = a \times a \times a \times a$

15. (*a*) $(a + b)^8$ (*b*) x'' (*c*) $30x^3$ (*d*) $56x^5 \cdot c^7$ (*e*) $45a^{7a+1}$

17. (*a*) x^3 (*b*) $3x^3$ (*c*) $a^2b^{\frac{5}{3}}$

19. (*a*) $a^4b^2c^2$ (*b*) x^2y^4z (*c*) x^3yz^3 (*d*) $y^{-2}b^2$
(*e*) $9xy^2z^{-2}$

21. Yes; no

23. (*a*) $+\$125$ (*b*) $-\$25x$

25. (*a*) 45 (*b*) 45 (*c*) -45 (*d*) -45
(*e*) $6ab$ (*f*) $6ab$ (*g*) $-6ab$ (*h*) $-6ab$

27. $-21a + 66b$

29. $-132x - 76y$

31. $-19x + 18y + 27z$

33. $8a - 12b + 8c - 2d$

35. (*a*) $15a^2 + 24ab$
(*b*) $15a^2 + 17ab - 18b^2$
(*c*) $-24a^7b^2c^2$
(*d*) $40a^6 + 56a^5b - 72a^4c^2$
(*e*) $-21a^5b^6c^4 + 35a^3b^5c^6 - 42a^3b^2c^7d^2 + 35a^6b^4c^5d^3$
(*f*) $40a^3 + 24a^2b^2 + 20ab^2 + 12b^4$

37. $20x^2 - 18xy - 18y^2$

39. (*a*) $16x^2 - 12x + 4 - 2x^{-1}$
(*b*) $-10x + 6y + 8z$
(*c*) $6a^3b - \frac{15}{4} + 4a^{-1}b^{-1}$
(*d*) $\dfrac{3a}{x} + \dfrac{4b}{x} + \dfrac{5c}{x}$ or $\dfrac{3a + 4b + 5c}{x}$
(*e*) $\dfrac{y^3}{a^3b^2c^2} - bc^3$
(*f*) $a + 5$
(*g*) $2a + 3b$
(*h*) $3a^2 + 2ab + 4b^2$

41. (*a*) $2(5x + 12)(x - 1)$ (*b*) $(x + 5)(x + 7)$
(*c*) $(x + 3)(x - 12)$ (*d*) $(x - 7)(x - 4)$
(*e*) $2(3x - 5)(3x + 2)$

43. (*a*) $(x^2 - 5)(x^2 + 5)$ (*b*) $(y - 7)(y + 7)$
(*c*) $(15a^2b - c^3)(15a^2b + c^3)$ (*d*) $(2x + 3y - 1)(2x + 3y + 1)$
(*e*) $(-2a + b)(4a - 3b)$
(*f*) $[(x + y) - (k - l)^2][(x + y) + (k - l)^2][(x + y)^2 + (k - l)^4]$
(*g*) $-3(y - 1)(3y - 5)$
(*h*) $(5x - 3y - 4c - 2d)(5x - 3y + 4c + 2d)$

45. (*a*) $x = 7$ (*b*) $y = 32$ (*c*) $c = 37.699$
(*d*) $x = 3$ (*e*) $x = \pm\sqrt{6}$ (*f*) $x = 7$
(*g*) $x = 7$ (*h*) $x = 23$ (*i*) $x = 70$
(*j*) $x = 72$

47. (*a*) $x = 300$ (*b*) $x = 12$ (*c*) $x = 320{,}000$
(*d*) $x = 9$ (*e*) $y = 18$ (*f*) $y = 23$
(*g*) $x = 5$ (*h*) $x = \frac{17}{5}$ (*i*) $x = 41\frac{2}{23}$
(*j*) $x = -37$ (*k*) $w = \dfrac{bt - Mp}{b - M}$ (*l*) $m = \dfrac{c - B}{1 - s}$
(*m*) $P = \frac{8}{5}K$ (*n*) $y = \pm\sqrt{10k(a + C)}$
(*o*) $x = 5$ (*p*) $x = -8\frac{2}{3}$ (*q*) $x = 2\frac{8}{11}$
(*r*) $x = 16$ (*s*) $x = \frac{17}{11}$ (*t*) $x = 7.425$

49. 24.8°F.
51. 10
53. 400 gal.
55. 1,000, 1,600
57. 140, 56, 81
59. $2{:}43\frac{7}{11}$ o'clock
61. \$9,000
63. 11
65. 9 miles, 7 miles

APPENDIX B: TABLES

TABLE I

NUMBER OF EACH DAY OF THE YEAR**

Day of Mo.	Jan.	Feb.	Mar.	Apr.	May	Jun.	Jul.	Aug.	Sep.	Oct.	Nov.	Dec.	Day of Mo.
1	1	32	60	91	121	152	182	213	244	274	305	335	1
2	2	33	61	92	122	153	183	214	245	275	306	336	2
3	3	34	62	93	123	154	184	215	246	276	307	337	3
4	4	35	63	94	124	155	185	216	247	277	308	338	4
5	5	36	64	95	125	156	186	217	248	278	309	339	5
6	6	37	65	96	126	157	187	218	249	279	310	340	6
7	7	38	66	97	127	158	188	219	250	280	311	341	7
8	8	39	67	98	128	159	189	220	251	281	312	342	8
9	9	40	68	99	129	160	190	221	252	282	313	343	9
10	10	41	69	100	130	161	191	222	253	283	314	344	10
11	11	42	70	101	131	162	192	223	254	284	315	345	11
12	12	43	71	102	132	163	193	224	255	285	316	346	12
13	13	44	72	103	133	164	194	225	256	286	317	347	13
14	14	45	73	104	134	165	195	226	257	287	318	348	14
15	15	46	74	105	135	166	196	227	258	288	319	349	15
16	16	47	75	106	136	167	197	228	259	289	320	350	16
17	17	48	76	107	137	168	198	229	260	290	321	351	17
18	18	49	77	108	138	169	199	230	261	291	322	352	18
19	19	50	78	109	139	170	200	231	262	292	323	353	19
20	20	51	79	110	140	171	201	232	263	293	324	354	20
21	21	52	80	111	141	172	202	233	264	294	325	355	21
22	22	53	81	112	142	173	203	234	265	295	326	356	22
23	23	54	82	113	143	174	204	235	266	296	327	357	23
24	24	55	83	114	144	175	205	236	267	297	328	358	24
25	25	56	84	115	145	176	206	237	268	298	329	359	25
26	26	57	85	116	146	177	207	238	269	299	330	360	26
27	27	58	86	117	147	178	208	239	270	300	331	361	27
28	28	59	87	118	148	179	209	240	271	301	332	362	28
29	29	*	88	119	149	180	210	241	272	302	333	363	29
30	30		89	120	150	181	211	242	273	303	334	364	30
31	31		90		151		212	243		304		365	31

*** In leap years, after February 28, add 1 to the tabulated number.**

* * Reprinted with permission from *C.R.C. Standard Mathematical Tables*, 12th ed., Chemical Rubber Publishing Company, Cleveland, Ohio, 1959, p. 435.

TABLE 2

AMOUNT AT COMPOUND INTEREST $(1 + i)^n$*

Periods	Rate i				
n	.01 (1%)	.01125 (1⅛%)	.0125 (1¼%)	.015 (1½%)	.0175 (1¾%)
1	1.0100 0000	1.0112 5000	1.0125 0000	1.0150 0000	1.0175 0000
2	1.0201 0000	1.0226 2656	1.0251 5625	1.0302 2500	1.0353 0625
3	1.0303 0100	1.0341 3111	1.0379 7070	1.0456 7838	1.0534 2411
4	1.0406 0401	1.0457 6509	1.0509 4534	1.0613 6355	1.0718 5903
5	1.0510 1005	1.0575 2994	1.0640 8215	1.0772 8400	1.0906 1656
6	1.0615 2015	1.0694 2716	1.0773 8318	1.0934 4326	1.1097 0235
7	1.0721 3535	1.0814 5821	1.0908 5047	1.1098 4491	1.1291 2215
8	1.0828 5671	1.0936 2462	1.1044 8610	1.1264 9259	1.1488 8178
9	1.0936 8527	1.1059 2789	1.1182 9218	1.1433 8998	1.1689 8721
10	1.1046 2213	1.1183 6958	1.1322 7083	1.1605 4083	1.1894 4449
11	1.1156 6835	1.1309 5124	1.1464 2422	1.1779 4894	1.2102 5977
12	1.1268 2503	1.1436 7444	1.1607 5452	1.1956 1817	1.2314 3931
13	1.1380 9328	1.1565 4078	1.1752 6395	1.2135 5244	1.2529 8950
14	1.1494 7421	1.1695 5186	1.1899 5475	1.2317 5573	1.2749 1682
15	1.1609 6896	1.1827 0932	1.2048 2918	1.2502 3207	1.2972 2786
16	1.1725 7864	1.1960 1480	1.2198 8955	1.2689 8555	1.3199 2935
17	1.1843 0443	1.2094 6997	1.2351 3817	1.2880 2033	1.3430 2811
18	1.1961 4748	1.2230 7650	1.2505 7739	1.3073 4064	1.3665 3111
19	1.2081 0895	1.2368 3611	1.2662 0961	1.3269 5075	1.3904 4540
20	1.2201 9004	1.2507 5052	1.2820 3723	1.3468 5501	1.4147 7820
21	1.2323 9194	1.2648 2146	1.2980 6270	1.3670 5783	1.4395 3681
22	1.2447 1586	1.2790 5071	1.3142 8848	1.3875 6370	1.4647 2871
23	1.2571 6302	1.2934 4003	1.3307 1709	1.4083 7715	1.4903 6146
24	1.2697 3465	1.3079 9123	1.3473 5105	1.4295 0281	1.5164 4279
25	1.2824 3200	1.3227 0613	1.3641 9294	1.4509 4535	1.5429 8054
26	1.2952 5631	1.3375 8657	1.3812 4535	1.4727 0953	1.5699 8269
27	1.3082 0888	1.3526 3442	1.3985 1092	1.4948 0018	1.5974 5739
28	1.3212 9097	1.3678 5156	1.4159 9230	1.5172 2218	1.6254 1290
29	1.3345 0388	1.3832 3989	1.4336 9221	1.5399 8051	1.6538 5762
30	1.3478 4892	1.3988 0134	1.4516 1336	1.5630 8022	1.6828 0013
31	1.3613 2740	1.4145 3785	1.4697 5853	1.5865 2642	1.7122 4913
32	1.3749 4068	1.4304 5140	1.4881 3051	1.6103 2432	1.7422 1349
33	1.3886 9009	1.4465 4398	1.5067 3214	1.6344 7918	1.7727 0223
34	1.4025 7699	1.4628 1760	1.5255 6629	1.6589 9637	1.8037 2452
35	1.4166 0276	1.4792 7430	1.5446 3587	1.6838 8132	1.8352 8970
36	1.4307 6878	1.4959 1613	1.5639 4382	1.7091 3954	1.8674 0727
37	1.4450 7647	1.5127 4519	1.5834 9312	1.7347 7663	1.9000 8689
38	1.4595 2724	1.5297 6357	1.6032 8678	1.7607 9828	1.9333 3841
39	1.4741 2251	1.5469 7341	1.6233 2787	1.7872 1025	1.9671 7184
40	1.4888 6373	1.5643 7687	1.6436 1946	1.8140 1841	2.0015 9734
41	1.5037 5237	1.5819 7611	1.6641 6471	1.8412 2868	2.0366 2530
42	1.5187 8989	1.5997 7334	1.6849 6677	1.8688 4712	2.0722 6624
43	1.5339 7779	1.6177 7079	1.7060 2885	1.8968 7982	2.1085 3090
44	1.5493 1757	1.6359 7071	1.7273 5421	1.9253 3302	2.1454 3019
45	1.5648 1075	1.6543 7538	1.7489 4614	1.9542 1301	2.1829 7522
46	1.5804 5885	1.6729 8710	1.7708 0797	1.9835 2621	2.2211 7728
47	1.5962 6344	1.6918 0821	1.7929 4306	2.0132 7910	2.2600 4789
48	1.6122 2608	1.7108 4105	1.8153 5485	2.0434 7829	2.2995 9872
49	1.6283 4834	1.7300 8801	1.8380 4679	2.0741 3046	2.3398 4170
50	1.6446 3182	1.7495 5150	1.8610 2237	2.1052 4242	2.3807 8893

* Reprinted with permission from *C.R.C. Standard Mathematical Tables*, 12th ed., Chemical Rubber Publishing Company, Cleveland, Ohio, 1959; pp. 438–443.

AMOUNT AT COMPOUND INTEREST $(1+i)^n$

Periods	Rate i				
n	.01 (1%)	.01125 ($1\frac{1}{8}$%)	.0125 ($1\frac{1}{4}$%)	.015 ($1\frac{1}{2}$%)	.0175 ($1\frac{3}{4}$%)
50	1.6446 3182	1.7495 5150	1.8610 2237	2.1052 4242	2.3807 8893
51	1.6610 7814	1.7692 3395	1.8842 8515	2.1368 2106	2.4224 5274
52	1.6776 8892	1.7891 3784	1.9078 3872	2.1688 7337	2.4648 4566
53	1.6944 6581	1.8092 6564	1.9316 8670	2.2014 0647	2.5079 8046
54	1.7114 1047	1.8296 1988	1.9558 3279	2.2344 2757	2.5518 7012
55	1.7285 2457	1.8502 0310	1.9802 8070	2.2679 4398	2.5965 2785
56	1.7458 0982	1.8710 1788	2.0050 3420	2.3019 6314	2.6419 6708
57	1.7632 6792	1.8920 6684	2.0300 9713	2.3364 9259	2.6882 0151
58	1.7809 0060	1.9133 5259	2.0554 7335	2.3715 3998	2.7352 4503
59	1.7987 0960	1.9348 7780	2.0811 6676	2.4071 1308	2.7831 1182
60	1.8166 9670	1.9566 4518	2.1071 8135	2.4432 1978	2.8318 1628
61	1.8348 6367	1.9786 5744	2.1335 2111	2.4798 6807	2.8813 7306
62	1.8532 1230	2.0009 1733	2.1601 9013	2.5170 6609	2.9317 9709
63	1.8717 4443	2.0234 2765	2.1871 9250	2.5548 2208	2.9831 0354
64	1.8904 6187	2.0461 9121	2.2145 3241	2.5931 4442	3.0353 0785
65	1.9093 6649	2.0692 1087	2.2422 1407	2.6320 4158	3.0884 2574
66	1.9284 6015	2.0924 8949	2.2702 4174	2.6715 2221	3.1424 7319
67	1.9477 4475	2.1160 2999	2.2986 1976	2.7115 9504	3.1974 6647
68	1.9672 2220	2.1398 3533	2.3273 5251	2.7522 6896	3.2534 2213
69	1.9868 9442	2.1639 0848	2.3564 4442	2.7935 5300	3.3103 5702
70	2.0067 6337	2.1882 5245	2.3858 9997	2.8354 5629	3.3682 8827
71	2.0268 3100	2.2128 7029	2.4157 2372	2.8779 8814	3.4272 3331
72	2.0470 9931	2.2377 6508	2.4459 2027	2.9211 5796	3.4872 0990
73	2.0675 7031	2.2629 3994	2.4764 9427	2.9649 7533	3.5482 3607
74	2.0882 4601	2.2883 9801	2.5074 5045	3.0094 4996	3.6103 3020
75	2.1091 2847	2.3141 4249	2.5387 9358	3.0545 9171	3.6735 1098
76	2.1302 1975	2.3401 7659	2.5705 2850	3.1004 1059	3.7377 9742
77	2.1515 2195	2.3665 0358	2.6026 6011	3.1469 1674	3.8032 0888
78	2.1730 3717	2.3931 2675	2.6351 9336	3.1941 2050	3.8697 6503
79	2.1947 6754	2.4200 4942	2.6681 3327	3.2420 3230	3.9374 8592
80	2.2167 1522	2.4472 7498	2.7014 8494	3.2906 6279	4.0063 9192
81	2.2388 8237	2.4748 0682	2.7352 5350	3.3400 2273	4.0765 0378
82	2.2612 7119	2.5026 4840	2.7694 4417	3.3901 2307	4.1478 4260
83	2.2838 8390	2.5308 0319	2.8040 6222	3.4409 7492	4.2204 2984
84	2.3067 2274	2.5592 7473	2.8391 1300	3.4925 8954	4.2942 8737
85	2.3297 8997	2.5880 6657	2.8746 0191	3.5449 7838	4.3694 3740
86	2.3530 8787	2.6171 8232	2.9105 3444	3.5981 5306	4.4459 0255
87	2.3766 1875	2.6466 2562	2.9469 1612	3.6521 2535	4.5237 0584
88	2.4003 8494	2.6764 0016	2.9837 5257	3.7069 0723	4.6028 7070
89	2.4243 8879	2.7065 0966	3.0210 4948	3.7625 1084	4.6834 2093
90	2.4486 3267	2.7369 5789	3.0588 1260	3.8189 4851	4.7653 8080
91	2.4731 1900	2.7677 4867	3.0970 4775	3.8762 3273	4.8487 7496
92	2.4978 5019	2.7988 8584	3.1357 6085	3.9343 7622	4.9336 2853
93	2.5228 2869	2.8303 7331	3.1749 5786	3.9933 9187	5.0199 6703
94	2.5480 5698	2.8622 1501	3.2146 4483	4.0532 9275	5.1078 1645
95	2.5735 3755	2.8944 1492	3.2548 2789	4.1140 9214	5.1972 0324
96	2.5992 7293	2.9269 7709	3.2955 1324	4.1758 0352	5.2881 5429
97	2.6252 6565	2.9599 0559	3.3367 0716	4.2384 4057	5.3806 9699
98	2.6515 1831	2.9932 0452	3.3784 1600	4.3020 1718	5.4748 5919
99	2.6780 3349	3.0268 7807	3.4206 4620	4.3665 4744	5.5706 6923
100	2.7048 1383	3.0609 3045	3.4634 0427	4.4320 4565	5.6681 5594

The table reproduced here includes rates of 1 to 8 per cent. For lesser rates see *C.R.C. Standard Mathematical Tables*, pp. 436–437.

AMOUNT AT COMPOUND INTEREST $(1+i)^n$

Periods	Rate i				
n	.02 (2%)	.0225 (2¼%)	.025 (2½%)	.0275 (2¾%)	.03 (3%)
1	1.0200 0000	1.0225 0000	1.0250 0000	1.0275 0000	1.0300 0000
2	1.0404 0000	1.0455 0625	1.0506 2500	1.0557 5625	1.0609 0000
3	1.0612 0800	1.0690 3014	1.0768 9063	1.0847 8955	1.0927 2700
4	1.0824 3216	1.0930 8332	1.1038 1289	1.1146 2126	1.1255 0881
5	1.1040 8080	1.1176 7769	1.1314 0821	1.1452 7334	1.1592 7407
6	1.1261 6242	1.1428 2544	1.1596 9342	1.1767 6836	1.1940 5230
7	1.1486 8567	1.1685 3901	1.1886 8575	1.2091 2949	1.2298 7387
8	1.1716 5938	1.1948 3114	1.2184 0290	1.2423 8055	1.2667 7008
9	1.1950 9257	1.2217 1484	1.2488 6297	1.2765 4602	1.3047 7318
10	1.2189 9442	1.2492 0343	1.2800 8454	1.3116 5103	1.3439 1638
11	1.2433 7431	1.2773 1050	1.3120 8666	1.3477 2144	1.3842 3387
12	1.2682 4179	1.3060 4999	1.3448 8882	1.3847 8378	1.4257 6089
13	1.2936 0663	1.3354 3611	1.3785 1104	1.4228 6533	1.4685 3371
14	1.3194 7876	1.3654 8343	1.4129 7382	1.4619 9413	1.5125 8972
15	1.3458 6834	1.3962 0680	1.4482 9817	1.5021 9896	1.5579 6742
16	1.3727 8571	1.4276 2146	1.4845 0562	1.5435 0944	1.6047 0644
17	1.4002 4142	1.4597 4294	1.5216 1826	1.5859 5595	1.6528 4763
18	1.4282 4625	1.4925 8716	1.5596 5872	1.6295 6973	1.7024 3306
19	1.4568 1117	1.5261 7037	1.5986 5019	1.6743 8290	1.7535 0605
20	1.4859 4740	1.5605 0920	1.6386 1644	1.7204 2843	1.8061 1123
21	1.5156 6634	1.5956 2066	1.6795 8185	1.7677 4021	1.8602 9457
22	1.5459 7967	1.6315 2212	1.7215 7140	1.8163 5307	1.9161 0341
23	1.5768 9926	1.6682 3137	1.7646 1068	1.8663 0278	1.9735 8651
24	1.6084 3725	1.7057 6658	1.8087 2595	1.9176 2610	2.0327 9411
25	1.6406 0599	1.7441 4632	1.8539 4410	1.9703 6082	2.0937 7793
26	1.6734 1811	1.7833 8962	1.9002 9270	2.0245 4575	2.1565 9127
27	1.7068 8648	1.8235 1588	1.9478 0002	2.0802 2075	2.2212 8901
28	1.7410 2421	1.8645 4499	1.9964 9502	2.1374 2682	2.2879 2768
29	1.7758 4469	1.9064 9725	2.0464 0739	2.1962 0606	2.3565 6551
30	1.8113 6158	1.9493 9344	2.0975 6758	2.2566 0173	2.4272 6247
31	1.8475 8882	1.9932 5479	2.1500 0677	2.3186 5828	2.5000 8035
32	1.8845 4059	2.0381 0303	2.2037 5694	2.3824 2138	2.5750 8276
33	1.9222 3140	2.0839 6034	2.2588 5086	2.4479 3797	2.6523 3524
34	1.9606 7603	2.1308 4945	2.3153 2213	2.5152 5626	2.7319 0530
35	1.9998 8955	2.1787 9356	2.3732 0519	2.5844 2581	2.8138 6245
36	2.0398 8734	2.2278 1642	2.4325 3532	2.6554 9752	2.8982 7833
37	2.0806 8509	2.2779 4229	2.4933 4870	2.7285 2370	2.9852 2668
38	2.1222 9879	2.3291 9599	2.5556 8242	2.8035 5810	3.0747 8348
39	2.1647 4477	2.3816 0290	2.6195 7448	2.8806 5595	3.1670 2698
40	2.2080 3966	2.4351 8897	2.6850 6384	2.9598 7399	3.2620 3779
41	2.2522 0046	2.4899 8072	2.7521 9043	3.0412 7052	3.3598 9893
42	2.2972 4447	2.5460 0528	2.8209 9520	3.1249 0546	3.4606 9589
43	2.3431 8936	2.6032 9040	2.8915 2008	3.2108 4036	3.5645 1677
44	2.3900 5314	2.6618 6444	2.9638 0808	3.2991 3847	3.6714 5227
45	2.4378 5421	2.7217 5639	3.0379 0328	3.3898 6478	3.7815 9584
46	2.4866 1129	2.7829 9590	3.1138 5086	3.4830 8606	3.8950 4372
47	2.5363 4352	2.8456 1331	3.1916 9713	3.5788 7093	4.0118 9503
48	2.5870 7039	2.9096 3961	3.2714 8956	3.6772 8988	4.1322 5188
49	2.6388 1179	2.9751 0650	3.3532 7680	3.7784 1535	4.2562 1944
50	2.6915 8803	3.0420 4640	3.4371 0872	3.8823 2177	4.3839 0602

AMOUNT AT COMPOUND INTEREST $(1+i)^n$

Periods	Rate i				
n	.02 (2%)	.0225 (2¼%)	.025 (2½%)	.0275 (2¾%)	.03 (3%)
50	2.6915 8803	3.0420 4640	3.4371 0872	3.8823 2177	4.3839 0602
51	2.7454 1979	3.1104 9244	3.5230 3644	3.9890 8562	4.5154 2320
52	2.8003 2819	3.1804 7852	3.6111 1235	4.0987 8547	4.6508 8590
53	2.8563 3475	3.2520 3929	3.7013 9016	4.2115 0208	4.7904 1247
54	2.9134 6144	3.3252 1017	3.7939 2491	4.3273 1838	4.9341 2485
55	2.9717 3067	3.4000 2740	3.8887 7303	4.4463 1964	5.0821 4859
56	3.0311 6529	3.4765 2802	3.9859 9236	4.5685 9343	5.2346 1305
57	3.0917 8859	3.5547 4990	4.0856 4217	4.6942 2975	5.3916 5144
58	3.1536 2436	3.6347 3177	4.1877 8322	4.8233 2107	5.5534 0098
59	3.2166 9685	3.7165 1324	4.2924 7780	4.9559 6239	5.7200 0301
60	3.2810 3079	3.8001 3479	4.3997 8975	5.0922 5136	5.8916 0310
61	3.3466 5140	3.8856 3782	4.5097 8449	5.2322 8827	6.0683 5120
62	3.4135 8443	3.9730 6467	4.6225 2910	5.3761 7620	6.2504 0173
63	3.4818 5612	4.0624 5862	4.7380 9233	5.5240 2105	6.4379 1379
64	3.5514 9324	4.1538 6394	4.8565 4464	5.6759 3162	6.6310 5120
65	3.6225 2311	4.2473 2588	4.9779 5826	5.8320 1974	6.8299 8273
66	3.6949 7357	4.3428 9071	5.1024 0721	5.9924 0029	7.0348 8222
67	3.7688 7304	4.4406 0576	5.2299 6739	6.1571 9130	7.2459 2868
68	3.8442 5050	4.5405 1939	5.3607 1658	6.3265 1406	7.4633 0654
69	3.9211 3551	4.6426 8107	5.4947 3449	6.5004 9319	7.6872 0574
70	3.9995 5822	4.7471 4140	5.6321 0286	6.6792 5676	7.9178 2191
71	4.0795 4939	4.8539 5208	5.7729 0543	6.8629 3632	8.1553 5657
72	4.1611 4038	4.9631 6600	5.9172 2806	7.0516 6706	8.4000 1727
73	4.2443 6318	5.0748 3723	6.0651 5876	7.2455 8791	8.6520 1778
74	4.3292 5045	5.1890 2107	6.2167 8773	7.4448 4158	8.9115 7832
75	4.4158 3546	5.3057 7405	6.3722 0743	7.6495 7472	9.1789 2567
76	4.5041 5216	5.4251 5396	6.5315 1261	7.8599 3802	9.4542 9344
77	4.5942 3521	5.5472 1993	6.6948 0043	8.0760 8632	9.7379 2224
78	4.6861 1991	5.6720 3237	6.8621 7044	8.2981 7869	10.0300 5991
79	4.7798 4231	5.7996 5310	7.0337 2470	8.5263 7861	10.3309 6171
80	4.8754 3916	5.9301 4530	7.2095 6782	8.7608 5402	10.6408 9056
81	4.9729 4794	6.0635 7357	7.3898 0701	9.0017 7751	10.9601 1727
82	5.0724 0690	6.2000 0397	7.5745 5219	9.2493 2639	11.2889 2079
83	5.1738 5504	6.3395 0406	7.7639 1599	9.5036 8286	11.6275 8842
84	5.2773 3214	6.4821 4290	7.9580 1389	9.7650 3414	11.9764 1607
85	5.3828 7878	6.6279 9112	8.1569 6424	10.0335 7258	12.3357 0855
86	5.4905 3636	6.7771 2092	8.3608 8834	10.3094 9583	12.7057 7981
87	5.6003 4708	6.9296 0614	8.5699 1055	10.5930 0696	13.0869 5320
88	5.7123 5402	7.0855 2228	8.7841 5832	10.8843 1465	13.4795 6180
89	5.8266 0110	7.2449 4653	9.0037 6228	11.1836 3351	13.8839 4865
90	5.9431 3313	7.4079 5782	9.2288 5633	11.4911 8322	14.3004 6711
91	6.0619 9579	7.5746 3688	9.4595 7774	11.8071 9076	14.7294 8112
92	6.1832 3570	7.7450 6621	9.6960 6718	12.1318 8851	15.1713 6556
93	6.3069 0042	7.9193 3020	9.9384 6886	12.4655 1544	15.6265 0652
94	6.4330 3843	8.0975 1512	10.1869 3058	12.8083 1711	16.0953 0172
95	6.5616 9920	8.2797 0921	10.4416 0385	13.1605 4584	16.5781 6077
96	6.6929 3318	8.4660 0267	10.7026 4395	13.5224 6085	17.0755 0559
97	6.8267 9184	8.6564 8773	10.9702 1004	13.8943 2852	17.5877 7076
98	6.9633 2768	8.8512 5871	11.2444 6530	14.2764 2255	18.1154 0388
99	7.1025 9423	9.0504 1203	11.5255 7693	14.6690 2417	18.6588 6600
100	7.2446 4612	9.2540 4630	11.8137 1635	15.0724 2234	19.2186 3198

AMOUNT AT COMPOUND INTEREST $(1+i)^n$

Periods	Rate i				
n	.035 (3½%)	.04 (4%)	.045 (4½%)	.05 (5%)	.055 (5½%)
1	1.0350 0000	1.0400 0000	1.0450 0000	1.0500 0000	1.0550 0000
2	1.0712 2500	1.0816 0000	1.0920 2500	1.1025 0000	1.1130 2500
3	1.1087 1788	1.1248 6400	1.1411 6613	1.1576 2500	1.1742 4138
4	1.1475 2300	1.1698 5856	1.1925 1860	1.2155 0625	1 2388 2465
5	1.1876 8631	1.2166 5290	1.2461 8194	1.2762 8156	1.3069 6001
6	1.2292 5533	1.2653 1902	1.3022 6012	1.3400 9564	1.3788 4281
7	1.2722 7926	1.3159 3178	1.3608 6183	1.4071 0042	1.4546 7916
8	1.3168 0904	1.3685 6905	1.4221 0061	1.4774 5544	1.5346 8651
9	1.3628 9735	1.4233 1181	1.4860 9514	1.5513 2822	1.6190 9427
10	1.4105 9876	1.4802 4428	1.5529 6942	1.6288 9463	1.7081 4446
11	1.4599 6972	1.5394 5406	1.6228 5305	1.7103 3936	1.8020 9240
12	1.5110 6866	1.6010 3222	1.6958 8143	1.7958 5633	1.9012 0749
13	1.5639 5606	1 6650 7351	1.7721 9610	1.8856 4914	2.0057 7390
14	1.6186 9452	1.7316 7645	1.8519 4492	1.9799 3160	2.1160 9146
15	1.6753 4883	1.8009 4351	1.9352 8244	2.0789 2818	2.2324 7649
16	1.7339 8604	1.8729 8125	2.0223 7015	2.1828 7459	2.3552 6270
17	1.7946 7555	1.9479 0050	2.1133 7681	2.2920 1832	2.4848 0215
18	1.8574 8920	2.0258 1652	2.2084 7877	2.4066 1923	2.6214 6627
19	1.9225 0132	2.1068 4918	2.3078 6031	2.5269 5020	2.7656 4691
20	1.9897 8886	2.1911 2314	2.4117 1402	2.6532 9771	2.9177 5749
21	2.0594 3147	2.2787 6807	2.5202 4116	2.7859 6259	3.0782 3415
22	2.1315 1158	2.3699 1879	2.6336 5201	2.9252 6072	3.2475 3703
23	2.2061 1448	2.4647 1554	2.7521 6635	3.0715 2376	3.4261 5157
24	2.2833 2849	2.5633 0416	2.8760 1383	3.2250 9994	3.6145 8990
25	2.3632 4498	2.6658 3633	3.0054 3446	3.3863 5494	3.8133 9235
26	2.4459 5856	2.7724 6978	3.1406 7901	3.5556 7269	4.0231 2893
27	2.5315 6711	2.8833 6858	3.2820 0956	3.7334 5632	4.2444 0102
28	2.6201 7196	2.9987 0332	3.4296 9999	3.9201 2914	4.4778 4307
29	2.7118 7798	3.1186 5145	3.5840 3649	4.1161 3560	4.7241 2444
30	2.8067 9370	3.2433 9751	3.7453 1813	4.3219 4238	4.9839 5129
31	2.9050 3148	3.3731 3341	3.9138 5745	4.5380 3949	5.2580 6861
32	3.0067 0759	3.5080 5875	4.0899 8104	4.7649 4147	5.5472 6238
33	3.1119 4235	3.6483 8110	4.2740 3018	5.0031 8854	5.8523 6181
34	3.2208 6033	3.7943 1634	4.4663 6154	5.2533 4797	6.1742 4171
35	3.3335 9045	3.9460 8899	4.6673 4781	5.5160 1537	6.5138 2501
36	3.4502 6611	4.1039 3255	4.8773 7846	5.7918 1614	6.8720 8538
37	3.5710 2543	4.2680 8986	5.0968 6049	6.0814 0694	7.2500 5008
38	3.6960 1132	4.4388 1345	5.3262 1921	6.3854 7729	7.6488 0283
39	3.8253 7171	4.6163 6599	5.5658 9908	6.7047 5115	8.0694 8699
40	3.9592 5972	4.8010 2063	5.8163 6454	7.0399 8871	8.5133 0877
41	4.0978 3381	4.9930 6145	6.0781 0094	7.3919 8815	8.9815 4076
42	4.2412 5799	5.1927 8391	6.3516 1548	7.7615 8756	9.4755 2550
43	4.3897 0202	5.4004 9527	6.6374 3818	8.1496 6693	9.9966 7940
44	4.5433 4160	5.6165 1508	6.9361 2290	8.5571 5028	10.5464 9677
45	4.7023 5855	5.8411 7568	7.2482 4843	8.9850 0779	11.1265 5409
46	4.8669 4110	6.0748 2271	7.5744 1961	9.4342 5818	11.7385 1456
47	5.0372 8404	6.3178 1562	7.9152 6849	9.9059 7109	12.3841 3287
48	5.2135 8898	6.5705 2824	8.2714 5557	10.4012 6965	13.0652 6017
49	5.3960 6459	6.8333 4937	8.6436 7107	10.9213 3313	13.7838 4948
50	5.5849 2686	7.1066 8335	9.0326 3627	11.4673 9979	14.5419 6120

AMOUNT AT COMPOUND INTEREST $(1+i)^n$

Periods	Rate i				
n	.06 (6%)	.065 (6½%)	.07 (7%)	.075 (7½%)	.08 (8%)
1	1.0600 0000	1.0650 0000	1.0700 0000	1.0750 0000	1.0800 0000
2	1.1236 0000	1.1342 2500	1.1449 0000	1.1556 2500	1.1664 0000
3	1.1910 1600	1.2079 4963	1.2250 4300	1.2422 9688	1.2597 1200
4	1.2624 7696	1.2864 6635	1.3107 9601	1.3354 6914	1 3604 8896
5	1.3382 2558	1.3700 8666	1.4025 5173	1.4356 2933	1.4693 2808
6	1.4185 1911	1.4591 4230	1.5007 3035	1.5433 0153	1.5868 7432
7	1.5036 3026	1.5539 8655	1.6057 8148	1.6590 4914	1.7138 2427
8	1.5938 4807	1.6549 9567	1.7181 8618	1.7834 7783	1.8509 3021
9	1.6894 7896	1.7625 7039	1.8384 5921	1.9172 3866	1.9990 0463
10	1.7908 4770	1.8771 3747	1.9671 5136	2.0610 3156	2.1589 2500
11	1.8982 9856	1.9991 5140	2.1048 5195	2.2156 0893	2.3316 3900
12	2.0121 9647	2.1290 9624	2.2521 9159	2.3817 7960	2.5181 7012
13	2.1329 2826	2.2674 8750	2.4098 4500	2.5604 1307	2.7196 2373
14	2.2609 0396	2.4148 7418	2.5785 3415	2.7524 4405	2.9371 9362
15	2.3965 5819	2.5718 4101	2.7590 3154	2.9588 7735	3.1721 6911
16	2.5403 5168	2.7390 1067	2.9521 6375	3.1807 9315	3.4259 4264
17	2.6927 7279	2.9170 4637	3.1588 1521	3.4193 5264	3.7000 1805
18	2.8543 3915	3.1066 5438	3.3799 3228	3.6758 0409	3.9960 1950
19	3.0255 9950	3.3085 8691	3.6165 2754	3.9514 8940	4.3157 0106
20	3.2071 3547	3.5236 4506	3.8696 8446	4.2478 5110	4.6609 5714
21	3.3995 6360	3.7526 8199	4.1405 6237	4.5664 3993	5.0338 3372
22	3.6035 3742	3.9966 0632	4.4304 0174	4.9089 2293	5.4365 4041
23	3.8197 4966	4.2563 8573	4.7405 2986	5.2770 9215	5.8714 6365
24	4.0489 3464	4.5330 5081	5.0723 6695	5.6728 7406	6.3411 8074
25	4.2918 7072	4.8276 9911	5.4274 3264	6.0983 3961	6.8484 7520
26	4.5493 8296	5.1414 9955	5.8073 5292	6.5557 1508	7.3963 5321
27	4.8223 4594	5.4756 9702	6.2138 6763	7.0473 9371	7.9880 6147
28	5.1116 8670	5.8316 1733	6.6488 3836	7.5759 4824	8.6271 0639
29	5.4183 8790	6.2106 7245	7.1142 5705	8.1441 4436	9.3172 7490
30	5.7434 9117	6.6143 6616	7.6122 5504	8.7549 5519	10.0626 5689
31	6.0881 0064	7.0442 9996	8.1451 1290	9.4115 7683	10.8676 6944
32	6.4533 8668	7.5021 7946	8.7152 7080	10.1174 4509	11.7370 8300
33	6.8405 8988	7.9898 2113	9.3253 3975	10.8762 5347	12.6760 4964
34	7.2510 2528	8.5091 5950	9.9781 1354	11.6919 7248	13.6901 3361
35	7.6860 8679	9.0622 5487	10.6765 8148	12.5688 7042	14.7853 4429
36	8.1472 5200	9.6513 0143	11.4239 4219	13.5115 3570	15.9681 7184
37	8.6360 8712	10.2786 3603	12.2236 1814	14.5249 0088	17.2456 2558
38	9.1542 5235	10.9467 4737	13.0792 7141	15.6142 6844	18.6252 7563
39	9.7035 0749	11.6582 8595	13.9948 2041	16.7853 3858	20.1152 9768
40	10.2857 1794	12.4160 7453	14.9744 5784	18.0442 3897	21.7245 2150
41	10.9028 6101	13.2231 1938	16.0226 6989	19.3975 5689	23.4624 8322
42	11.5570 3267	14.0826 2214	17.1442 5678	20.8523 7366	25.3394 8187
43	12.2504 5463	14.9979 9258	18.3443 5475	22.4163 0168	27.3666 4042
44	12.9854 8191	15.9728 6209	19.6284 5959	24.0975 2431	29.5559 7166
45	13.7646 1083	17.0110 9813	21.0024 5176	25.9048 3863	31.9204 4939
46	14.5904 8748	18.1168 1951	22.4726 2338	27.8477 0153	34.4740 8534
47	15.4659 1673	19.2944 1278	24.0457 0702	29.9362 7915	37.2320 1217
48	16.3938 7173	20.5485 4961	25.7289 0651	32.1815 0008	40.2105 7314
49	17.3775 0403	21.8842 0533	27.5299 2997	34.5951 1259	43.4274 1899
50	18.4201 5427	23.3066 7868	29.4570 2506	37.1897 4603	46.9016 1251

TABLE 3

FOUR-PLACE COMMON LOGARITHMS

N.	0	1	2	3	4	5	6	7	8	9
10	0000	0043	0086	0128	0170	0212	0253	0294	0334	0374
11	0414	0453	0492	0531	0569	0607	0645	0682	0719	0755
12	0792	0828	0864	0899	0934	0969	1004	1038	1072	1106
13	1139	1173	1206	1239	1271	1303	1335	1367	1399	1430
14	1461	1492	1523	1553	1584	1614	1644	1673	1703	1732
15	1761	1790	1818	1847	1875	1903	1931	1959	1987	2014
16	2041	2068	2095	2122	2148	2175	2201	2227	2253	2279
17	2304	2330	2355	2380	2405	2430	2455	2480	2504	2529
18	2553	2577	2601	2625	2648	2672	2695	2718	2742	2765
19	2788	2810	2833	2856	2878	2900	2923	2945	2967	2989
20	3010	3032	3054	3075	3096	3118	3139	3160	3181	3201
21	3222	3243	3263	3284	3304	3324	3345	3365	3385	3404
22	3424	3444	3464	3483	3502	3522	3541	3560	3579	3598
23	3617	3636	3655	3674	3692	3711	3729	3747	3766	3784
24	3802	3820	3838	3856	3874	3892	3909	3927	3945	3962
25	3979	3997	4014	4031	4048	4065	4082	4099	4116	4133
26	4150	4166	4183	4200	4216	4232	4249	4265	4281	4298
27	4314	4330	4346	4362	4378	4393	4409	4425	4440	4456
28	4472	4487	4502	4518	4533	4548	4564	4579	4594	4609
29	4624	4639	4654	4669	4683	4698	4713	4728	4742	4757
30	4771	4786	4800	4814	4829	4843	4857	4871	4886	4900
31	4914	4928	4942	4955	4969	4983	4997	5011	5024	5038
32	5051	5065	5079	5092	5105	5119	5132	5145	5159	5172
33	5185	5198	5211	5224	5237	5250	5263	5276	5289	5302
34	5315	5328	5340	5353	5366	5378	5391	5403	5416	5428
35	5441	5453	5465	5478	5490	5502	5514	5527	5539	5551
36	5563	5575	5587	5599	5611	5623	5635	5647	5658	5670
37	5682	5694	5705	5717	5729	5740	5752	5763	5775	5786
38	5798	5809	5821	5832	5843	5855	5866	5877	5888	5899
39	5911	5922	5933	5944	5955	5966	5977	5988	5999	6010
40	6021	6031	6042	6053	6064	6075	6085	6096	6107	6117
41	6128	6138	6149	6160	6170	6180	6191	6201	6212	6222
42	6232	6243	6253	6263	6274	6284	6294	6304	6314	6325
43	6335	6345	6355	6365	6375	6385	6395	6405	6415	6425
44	6435	6444	6454	6464	6474	6484	6493	6503	6513	6522
45	6532	6542	6551	6561	6571	6580	6590	6599	6609	6618
46	6628	6637	6646	6656	6665	6675	6684	6693	6702	6712
47	6721	6730	6739	6749	6758	6767	6776	6785	6794	6803
48	6812	6821	6830	6839	6848	6857	6866	6875	6884	6893
49	6902	6911	6920	6928	6937	6946	6955	6964	6972	6981
50	6990	6998	7007	7016	7024	7033	7042	7050	7059	7067
51	7076	7084	7093	7101	7110	7118	7126	7135	7143	7152
52	7160	7168	7177	7185	7193	7202	7210	7218	7226	7235
53	7243	7251	7259	7267	7275	7284	7292	7300	7308	7316
54	7324	7332	7340	7348	7356	7364	7372	7380	7388	7396
N.	0	1	2	3	4	5	6	7	8	9

Common Logarithms.—(*Continued*)

N.	0	1	2	3	4	5	6	7	8	9
55	7404	7412	7419	7427	7435	7443	7451	7459	7466	7474
56	7482	7490	7497	7505	7513	7520	7528	7536	7543	7551
57	7559	7566	7574	7582	7589	7597	7604	7612	7619	7627
58	7634	7642	7649	7657	7664	7672	7679	7686	7694	7701
59	7709	7716	7723	7731	7738	7745	7752	7760	7767	7774
60	7782	7789	7796	7803	7810	7818	7825	7832	7839	7846
61	7853	7860	7868	7875	7882	7889	7896	7903	7910	7917
62	7924	7931	7938	7945	7952	7959	7966	7973	7980	7987
63	7993	8000	8007	8014	8021	8028	8035	8041	8048	8055
64	8062	8069	8075	8082	8089	8096	8102	8109	8116	8122
65	8129	8136	8142	8149	8156	8162	8169	8176	8182	8189
66	8195	8202	8209	8215	8222	8228	8235	8241	8248	8254
67	8261	8267	8274	8280	8287	8293	8299	8306	8312	8319
68	8325	8331	8338	8344	8351	8357	8363	8370	8376	8382
69	8388	8395	8401	8407	8414	8420	8426	8432	8439	8445
70	8451	8457	8463	8470	8476	8482	8488	8494	8500	8506
71	8513	8519	8525	8531	8537	8543	8549	8555	8561	8567
72	8573	8579	8585	8591	8597	8603	8609	8615	8621	8627
73	8633	8639	8645	8651	8657	8663	8669	8675	8681	8686
74	8692	8698	8704	8710	8716	8722	8727	8733	8739	8745
75	8751	8756	8762	8768	8774	8779	8785	8791	8797	8802
76	8808	8814	8820	8825	8831	8837	8842	8848	8854	8859
77	8865	8871	8876	8882	8887	8893	8899	8904	8910	8915
78	8921	8927	8932	8938	8943	8949	8954	8960	8965	8971
79	8976	8982	8987	8993	8998	9004	9009	9015	9020	9025
80	9031	9036	9042	9047	9053	9058	9063	9069	9074	9079
81	9085	9090	9096	9101	9106	9112	9117	9122	9128	9133
82	9138	9143	9149	9154	9159	9165	9170	9175	9180	9186
83	9191	9196	9201	9206	9212	9217	9222	9227	9232	9238
84	9243	9248	9253	9258	9263	9269	9274	9279	9284	9289
85	9294	9299	9304	9309	9315	9320	9325	9330	9335	9340
86	9345	9350	9355	9360	9365	9370	9375	9380	9385	9390
87	9395	9400	9405	9410	9415	9420	9425	9430	9435	9440
88	9445	9450	9455	9460	9465	9469	9474	9479	9484	9489
89	9494	9499	9504	9509	9513	9518	9523	9528	9533	9538
90	9542	9547	9552	9557	9562	9566	9571	9576	9581	9586
91	9590	9595	9600	9605	9609	9614	9619	9624	9628	9633
92	9638	9643	9647	9652	9657	9661	9666	9671	9675	9680
93	9685	9689	9694	9699	9703	9708	9713	9717	9722	9727
94	9731	9736	9741	9745	9750	9754	9759	9763	9768	9773
95	9777	9782	9786	9791	9795	9800	9805	9809	9814	9818
96	9823	9827	9832	9836	9841	9845	9850	9854	9859	9863
97	9868	9872	9877	9881	9886	9890	9894	9899	9903	9908
98	9912	9917	9921	9926	9930	9934	9939	9943	9948	9952
99	9956	9961	9965	9969	9974	9978	9983	9987	9991	9996
N.	0	1	2	3	4	5	6	7	8	9

TABLE 4

PRESENT VALUE $1/(1 + i)^n$*

Periods	Rate i				
n	.035 (3½%)	.04 (4%)	.045 (4½%)	.05 (5%)	.055 (5½%)
1	.9661 8357	.9615 3846	.9569 3780	.9523 8095	.9478 6730
2	.9335 1070	.9245 5621	.9157 2995	.9070 2948	.8984 5242
3	.9019 4271	.8889 9636	.8762 9660	.8638 3760	.8516 1366
4	.8714 4223	.8548 0419	.8385 6134	.8227 0247	.8072 1674
5	.8419 7317	.8219 2711	.8024 5105	.7835 2617	.7651 3435
6	.8135 0064	.7903 1453	.7678 9574	.7462 1540	.7252 4583
7	.7859 9096	.7599 1781	.7348 2846	.7106 8133	.6874 3681
8	.7594 1156	.7306 9021	.7031 8513	.6768 3936	.6515 9887
9	.7337 3097	.7025 8674	.6729 0443	.6446 0892	.6176 2926
10	.7089 1881	.6755 6417	.6439 2768	.6139 1325	.5854 3058
11	.6849 4571	.6495 8093	.6161 9874	.5846 7929	.5549 1050
12	.6617 8330	.6245 9705	.5896 6386	.5568 3742	.5259 8152
13	.6394 0415	.6005 7409	.5642 7164	.5303 2135	.4985 6068
14	.6177 8179	.5774 7508	.5399 7286	.5050 6795	.4725 6937
15	.5968 9062	.5552 6450	.5167 2044	.4810 1710	.4479 3305
16	.5767 0591	.5339 0818	.4944 6932	.4581 1152	.4245 8109
17	.5572 0378	.5133 7325	.4731 7639	.4362 9669	.4024 4653
18	.5383 6114	.4936 2812	.4528 0037	.4155 2065	.3814 6590
19	.5201 5569	.4746 4242	.4333 0179	.3957 3396	.3615 7906
20	.5025 6588	.4563 8695	.4146 4286	.3768 8948	.3427 2896
21	.4855 7090	.4388 3360	.3967 8743	.3589 4236	.3248 6158
22	.4691 5063	.4219 5539	.3797 0089	.3418 4987	.3079 2567
23	.4532 8563	.4057 2633	.3633 5013	.3255 7131	.2918 7267
24	.4379 5713	.3901 2147	.3477 0347	.3100 6791	.2766 5656
25	.4231 4699	.3751 1680	.3327 3060	.2953 0277	.2622 3370
26	.4088 3767	.3606 8923	.3184 0248	.2812 4073	.2485 6275
27	.3950 1224	.3468 1657	.3046 9137	.2678 4832	.2356 0450
28	.3816 5434	.3334 7747	.2915 7069	.2550 9364	.2233 2181
29	.3687 4815	.3206 5141	.2790 1502	.2429 4632	.2116 7944
30	.3562 7841	.3083 1867	.2670 0002	.2313 7745	.2006 4402
31	.3442 3035	.2964 6026	.2555 0241	.2203 5947	.1901 8390
32	.3325 8971	.2850 5794	.2444 9991	.2098 6617	.1802 6910
33	.3213 4271	.2740 9417	.2339 7121	.1998 7254	.1708 7119
34	.3104 7605	.2635 5209	.2238 9589	.1903 5480	.1619 6321
35	.2999 7686	.2534 1547	.2142 5444	.1812 9029	.1535 1963
36	.2898 3272	.2436 6872	.2050 2817	.1726 5741	.1455 1624
37	.2800 3161	.2342 9685	.1961 9921	.1644 3563	.1379 3008
38	.2705 6194	.2252 8543	.1877 5044	.1566 0536	.1307 3941
39	.2614 1250	.2166 2061	.1796 6549	.1491 4797	.1239 2362
40	.2525 7247	.2082 8904	.1719 2870	.1420 4568	.1174 6314
41	.2440 3137	.2002 7793	.1645 2507	.1352 8160	.1113 3947
42	.2357 7910	.1925 7493	.1574 4026	.1288 3962	.1055 3504
43	.2278 0590	.1851 6820	.1506 6054	.1227 0440	.1000 3322
44	.2201 0231	.1780 4635	.1441 7276	.1168 6133	.0948 1822
45	.2126 5924	.1711 9841	.1379 6437	.1112 9651	.0898 7509
46	.2054 6787	.1646 1386	.1320 2332	.1059 9668	.0851 8965
47	.1985 1968	.1582 8256	.1263 3810	.1009 4921	.0807 4849
48	.1918 0645	.1521 9476	.1208 9771	.0961 4211	.0765 3885
49	.1853 2024	.1463 4112	.1156 9158	.0915 6391	.0725 4867
50	.1790 5337	.1407 1262	.1107 0965	.0872 0373	.0687 6652

* Reprinted with permission from *C.R.C. Standard Mathematical Tables*, 12th ed., Chemical Rubber Publishing Company, Cleveland, Ohio, 1959, pp. 450–451.

PRESENT VALUE $1/(1+i)^n$

Periods	Rate i				
n	.06 (6%)	.065 (6½%)	.07 (7%)	.075 (7½%)	.08 (8%)
1	.9433 9623	.9389 6714	.9345 7944	.9302 3256	.9259 2593
2	.8899 9644	.8816 5928	.8734 3873	.8653 3261	.8573 3882
3	.8396 1928	.8278 4909	.8162 9788	.8049 6057	.7938 3224
4	.7920 9366	.7773 2309	.7628 9521	.7488 0053	.7350 2985
5	.7472 5817	.7298 8084	.7129 8618	.6965 5863	.6805 8320
6	.7049 6054	.6853 3412	.6663 4222	.6479 6152	.6301 6963
7	.6650 5711	.6435 0621	.6227 4974	.6027 5490	.5834 9040
8	.6274 1237	.6042 3119	.5820 0910	.5607 0223	.5402 6888
9	.5918 9846	.5673 5323	.5439 3374	.5215 8347	.5002 4897
10	.5583 9478	.5327 2604	.5083 4929	.4851 9393	.4631 9349
11	.5267 8753	.5002 1224	.4750 9280	.4513 4319	.4288 8286
12	.4969 6936	.4696 8285	.4440 1196	.4198 5413	.3971 1376
13	.4688 3902	.4410 1676	.4149 6445	.3905 6198	.3676 9792
14	.4423 0096	.4141 0025	.3878 1724	.3633 1347	.3404 6104
15	.4172 6506	.3888 2652	.3624 4602	.3379 6602	.3152 4170
16	.3936 4628	.3650 9533	.3387 3460	.3143 8699	.2918 9047
17	.3713 6442	.3428 1251	.3165 7439	.2924 5302	.2702 6895
18	.3503 4379	.3218 8969	.2958 6392	.2720 4932	.2502 4903
19	.3305 1301	.3022 4384	.2765 0833	.2530 6913	.2317 1206
20	.3118 0473	.2837 9703	.2584 1900	.2354 1315	.2145 4821
21	.2941 5540	.2664 7608	.2415 1309	.2189 8897	.1986 5575
22	.2775 0510	.2502 1228	.2257 1317	.2037 1067	.1839 4051
23	.2617 9726	.2349 4111	.2109 4688	.1894 9830	.1703 1528
24	.2469 7855	.2206 0198	.1971 4662	.1762 7749	.1576 9934
25	.2329 9863	.2071 3801	.1842 4918	.1639 7906	.1460 1790
26	.2198 1003	.1944 9579	.1721 9549	.1525 3866	.1352 0176
27	.2073 6795	.1826 2515	.1609 3037	.1418 9643	.1251 8682
28	.1956 3014	.1714 7902	.1504 0221	.1319 9668	.1159 1372
29	.1845 5674	.1610 1316	.1405 6282	.1227 8761	.1073 2752
30	.1741 1013	.1511 8607	.1313 6712	.1142 2103	.0993 7733
31	.1642 5484	.1419 5875	.1227 7301	.1062 5212	.0920 1605
32	.1549 5740	.1332 9460	.1147 4113	.0988 3918	.0852 0005
33	.1461 8622	.1251 5925	.1072 3470	.0919 4343	.0788 8893
34	.1379 1153	.1175 2042	.1002 1934	.0855 2877	.0730 4531
35	.1301 0522	.1103 4781	.0936 6294	.0795 6164	.0676 3454
36	.1227 4077	.1036 1297	.0875 3546	.0740 1083	.0626 2458
37	.1157 9318	.0972 8917	.0818 0884	.0688 4729	.0579 8572
38	.1092 3885	.0913 5134	.0764 5686	.0640 4399	.0536 9048
39	.1030 5552	.0857 7590	.0714 5501	.0595 7580	.0497 1341
40	.0972 2219	.0805 4075	.0667 8038	.0554 1935	.0460 3093
41	.0917 1905	.0756 2512	.0624 1157	.0515 5288	.0426 2123
42	.0865 2740	.0710 0950	.0583 2857	.0479 5617	.0394 6411
43	.0816 2962	.0666 7559	.0545 1268	.0446 1039	.0365 4084
44	.0770 0908	.0626 0619	.0509 4643	.0414 9804	.0338 3411
45	.0726 5007	.0587 8515	.0476 1349	.0386 0283	.0313 2788
46	.0685 3781	.0551 9733	.0444 9859	.0359 0961	.0290 0730
47	.0646 5831	.0518 2848	.0415 8747	.0334 0428	.0268 5861
48	.0609 9840	.0486 6524	.0388 6679	.0310 7375	.0248 6908
49	.0575 4566	.0456 9506	.0363 2410	.0289 0582	.0230 2693
50	.0542 8836	.0429 0616	.0339 4776	.0268 8913	.0213 2123

The table reproduced here includes rates of 3½ to 8 per cent. For lesser rates see *C.R.C. Standard Mathematical Tables*, pp. 444–449.

INDEX

a.c.: *see* arithmetical complement
abscissa, 292
  axis of, 291
absolute value, 262
abstract number, 37
accumulation factor, 151
accurate method (of simple interest), 142
addition, 2, 13–20
  algebraic rules of, 371, 372
  associative law of, 360
  by multiplication of an average, 15, 16
  checking correctness of, 17–20
  cumulative law of, 360
  decimalized, 17
  of decimals, 112
  of denominate numbers, 210
  of fractions, 89–91
  of per cents, 135
  of positive and negative numbers, 263
  rule for, 14
  symbol of, 2
    in algebra, 359
algebra, 356–398
  symbols for operations in, 359
algebraic expression, 358
algebraic quantity, 358
algebraic symbol raised to a power, 218
aliquot parts, 45–48, 146, 224
  fractional equivalent of, 46
  in division, 69
  in multiplication, 47, 48
alternation, proportion by, 176
amount
  compound, 149
  in interest, 141, 142
  in percentage, 131
  tax, 329
angle measurement, 351–353
antecedent (in ratios), 159
antilogarithm (antilog), 248
apothecaries' weights, 204
approximation of decimals, 115–117
Arabic numeral system, 1
arc, 352
arithmetic, 1 and *passim*
  fundamental operations of, 1, 2
arithmetical complement (a.c.), 31
  use in subtraction, 31, 32
arithmetic mean, 270
  see also average
arithmetic progression, 266, 268–271
ascending progression (series), 266
assessed valuation, 329
assessment, 329
associative law
  for addition, 360
  for multiplication, 360
Austrian method of subtraction, 29, 32
average, 70, 71, 186–197
  advantages of, 195
  deviation from, 187
  disadvantages of, 196
  how to simplify, 192
  two general classes of, 187
  weighted, 190, 191
avoirdupois weights, 204, 345
axis
  of abscissas, 291
  of coordinates, 291
  x, 291
  y, 291

bank discount, 300
bankers' method (of simple interest), 142, 143
bar chart
  divided, 279, 282
  100 per cent, 282
  long, 285
bar graph, 279–282
  horizontal, 279–281
  vertical, 279, 281, 282
base
  defined as factor raised to power, 217, 218
  in percentage, 129
  in profit and loss, 298
  in taxation, 329
  of logarithm, 241
basic numbers, 244
binomial, 357

block graph, 279, 285
Boyle's law, 179
Briggs system of logs, 242
British money, 206
broken-line graph, 279, 285–287
bundles of units, 5, 6
business, 297–331
  uses of percentage in, 134
buying commission, 298

calculation, 1
cancellation, 79
carrying charge, 310
Cartesian coordinates, 291
cash discount, 297
casting out elevens, 20
  in subtraction, 33
casting out nines
  in checking addition, 18–20
  in subtraction, 32, 33
  to check multiplication, 53, 54
chain fractions, 98–102
characteristic (of log), 245
  negative, 246
charge
  carrying, 310
  financing, 311
chart
  divided bar, 279, 282
  100 per cent bar, 282
  long bar, 285
  see also graph
checking correctness
  in addition, 17–20
  in algebra, 370
  in division, 63, 65
  in multiplication, 41, 53, 54
  in subtraction, 32, 33
check number (figure), 18
cipher, 1
circle graph, 279, 282–284
circular measure, 209
circulating decimal, 118
circumference, 352
coefficient, 268, 357–359
cologarithm (colog), 252–255
commission, 299
  buying, 298
  sales, 298
common divisor, 75
  greatest, 76, 77
common factor, 75
  greatest, 76
common fractions, 82–102
  powers of, 219
common log, 252
common multiple, 78
  least, 78
common parentheses method of factoring, 378
common system of logs, 242
common term method of factoring, 378
complement, 51
  arithmetical, 31, 32
complement multiplication, 51, 52
complex decimal, 109
complex fraction, 85
composite number, 75
composition, proportion by, 177
compound amount, 149
compound-amount-of-1 tables, 155
compound fraction, 85
compound interest, 141, 149–155, 418–423
  accumulation factor, 151
compound proportion, 173
compound ratio, 167
computation, 1
concrete number, 37
conditional equation, 384
consequent (in ratios), 159
constant, 179, 356
constant-ratio method for installment interest rate, 319–325
continued fraction, 98
conversion
  of common fractions and decimals, 110, 111, 118, 119
  of decimals into powers of ten, 226
  of interest, 149
    frequency of, 149
  of per cents into fractions and decimals, 126, 127
conversion period (of interest), 149
coordinates
  axes of, 291
  Cartesian, 291
cost
  gross, 297
  net, 297
  prime, 297
counting measures, 205
cross multiplication, 44, 52, 53
  in addition and subtraction of fractions, 91

cube, 217, 363
cube root, 227, 228, 236, 237
  extraction of, 236
cubic measure, 203
  in metric system, 208
cumulative law
  for addition, 360
  for multiplication, 360
curve(d) graph, 279, 287

decimal, 7, 107–121
  addition of, 112
  and U.S. money, 120
  approximation of, 115–117
  circulating, 118
  complex, 109
  conversion of
    to common fractions, 110, 111, 118, 119
    to per cent, 126, 127
  division of, 110, 114, 115
  equations with, 390
  multiplication of, 109, 113
  powers of, 219
  recurring, 118
  repeating, 118, 273
  simple, 109
  subtraction of, 112
decimal division, 107
decimal fraction: *see* decimal
decimalization in subtraction, 30
decimalized addition, 17
decimal place, 7, 107
decimal point, 1, 6, 107
degree, 352
denominate numbers, 200–214
  addition of, 210
  reduction of, 200
    ascending, 200
    descending, 200
  subtraction of, 211
denominator, 82
  lowest common, 89
depreciation, 297
descending progression (series), 266
deviation from average, 187
diagram
  line, 285–287
  staircase, 279, 288
difference
  in percentage, 131
  in subtraction, 25
difference method of comparing like quantities, 159
digit, 1
directed number, 358
direction
  concept of, 25
  negative, 292
  positive, 292
direct proportion, 170
direct tax, 329
direct variation, 179
discount, 299
  bank, 300
  cash, 297
  trade, 297
  true, 308
distribution laws for multiplication, 360
divided bar chart, 279, 282
dividend, 58
divisibility by various numbers, 65–68
division, 2, 57–72
  algebraic rules for, 369
  by logs, 249
  checking correctness of, 63, 65
  decimal, 107
  factoring-of-the-divisor method of, 63, 64
  how to simplify, 68–70
  long, 61, 62
  of decimals, 109, 114, 115
  of fractions, 95, 96
  of per cents, 135
  of polynomials, 376, 377
  of positive and negative numbers, 264
  of powers, 220, 237
  of powers of ten, 227
  of same kind of symbols, 365
  of U.S. money, 63
  proportion by, 177
  pure proof of, 63
  short, 60, 61
  symbol of, 2
    in algebra, 359
division sign, 2
divisor, 58
  common, 75
  greatest common, 76, 77
  trial, 62
dry measure, 203
  in metric system, 208

effective rate of interest, 153, 154
eleven as a check number, 19, 20

emptying problems, 346–348
"equal additions" method of subtraction, 28
equals (to), 2
equals sign, 2
equation, 382–391
  conditional, 384
  linear, 384
  quadratic, 384
  root of, 383
  simple, 384
  solution of, 386–391
  with decimals, 390
even number, 65
evolution, 2, 3, 227
  symbol of, 2, 227
exact method (of simple interest), 142, 143
excess-of-nines method of checking division, 65
exponent, 217
  fractional, 229, 230, 237
  laws of, 248
  logarithm defined as, 241
  negative, 222, 237
  raised to a power, 221, 237
  sign of, 221
  zero, 221, 237
  see also powers
expression, algebraic, 358
extrapolation, 290
"extremes" (of proportion), 168

factor, 37, 75, 357
  common, 75
  greatest common, 76
  literal, 357, 358
  prime, 75, 76
factoring, 75
  common parentheses method, 378
  common term method, 378
  in algebra, 378–382
  product of two binomials method, 379
factoring-of-the-divisor method in division, 63, 64
filling problems, 346–348
finance, 297–331
financing charge, 311
fluid ounces, 345
formula, 391
fourth root, 227, 228
fraction, 82–102
  addition of, 89–91
  chain, 98–102
  common: *see* fraction
  complex, 85
  compound, 85
  continued, 98
  conversion of
    to decimals, 110, 111, 118, 119
    to per cents, 126, 127
  decimal: *see* decimal
  division of, 95, 96
  improper, 85
  multiplication of, 91–94
  powers of, 219
  proper, 85
  reduction to lowest terms, 86, 87
  root of, 236
  simple, 84
  subtraction of, 89–91
  unit, 83
  vulgar, 83
fractional equivalent of aliquot parts, 46
fractional exponent, 229, 230, 237
fractional places, 7
French money, 206
frequency distribution graph, 279, 288, 289
frequency polygon, 288
future value (worth), 307

G.C.D. (g.c.d.): *see* greatest common divisor
geometric mean, 273
geometric progression, 266, 271–274
German money, 206
gram, 206
graph, 278–294
  advantages and disadvantages of, 278
  bar, 279–282
    horizontal, 279–281
    vertical, 279, 281, 282
  block, 279, 285
  broken-line, 279, 285–287
  circle, 279, 282–284
  curve(d), 279, 287
  frequency distribution, 279, 288, 289
  of quadratic formula, 294
  pie, 279, 282–284
  rectangle, 279, 282
  smooth-line, 279, 287

greatest common divisor (G.C.D.), 76, 77
rule for finding, 77
greatest common factor, 76
gross cost, 297
gross profit, 297
gross purchases, 297
gross sales, 297

harmonic mean, 275
harmonic progression, 266, 274, 275
Hooke's law, 179
horizontal bar graph, 279–281
100 per cent bar chart, 282
"hundreds" position, 4

identity, 384
imaginary number, 255
imperfect power, 228
improper fraction, 85
income tax, 329
index, 228, 359
index number, 289
indirect tax, 329
initial line (of angle), 352
installment purchase problems, 275, 276, 310–325
constant-ratio method of, 319–325
integer, 3
integral number, 3
interest, 140–155
compound, 149–155, 418–423
accumulation factor in, 151
conversion of, 149
rate of, 140
effective, 153, 154
nominal, 153, 154
simple, 140, 141
formula for, 141
methods of figuring, 142
six-day 6 per cent method of, 148, 149
sixty-day 6 per cent method of, 145–147
interest cost, 140
interest earned, 140
interpolation, 290
inverse proportion, 171
inverse ratio, 166
inverse variation, 180
inversion, proportion by, 176
inverted multiplication, 45
inverted subtraction, 30
involution, 2, 3, 218
symbol of, 2
irrational number, 242

joint variation, 180

key number (figure), 18, 19

L.C.D.: *see* lowest common denominator
language of variation, 418
laws
of addition: *see* addition
of multiplication: *see* multiplication
least common multiple (L.C.M.), 78
left-hand multiplication, 45
left-hand subtraction, 30
lever, principle of, 178
licence, 329
like terms, 367
line
initial (of angle), 352
terminal (of angle), 352
linear equation, 384
linear measure, 200–202
in metric system, 207
line diagram, 285–287
liquid measure, 203
in metric system, 208
liter, 206
literal factor, 357, 358
literal number, 356
loans, 316–318
logarithm (log), 241–258
accuracy of computation by, 258
Briggs system of, 242
characteristic of, 245
negative, 246
common, 252
common system of, 242
division by, 249
extraction of roots by, 250
mantissa of, 245
multiplication by, 249
Napierian system of, 242
natural, 252
natural system of, 242
proportional part of, 247
raising to powers by, 250
table of, 246–248, 424–425
long bar chart, 285
long division, 61, 62
rule for, 62
loss, 298

lowest common denominator (L.C.D.), 89

making change, method of: *see* Austrian method
mantissa, 245
maps, 349–351
  statistical, 279, 291
margin of profit, 297
mean
  arithmetic, 270
    see also average
  geometric, 273
  harmonic, 275
  proportional, 170
  square of, 225
"means" (of proportion), 168
mean value, 187, 188
measure
  circular, 209
  counting, 205
  cubic, 203
  dry, 203
  linear, 200–202
  liquid, 203
  metric system of, 206–208
  paper, 206
  square, 202
  time, 205
median, 193
  advantages of, 196
  disadvantages of, 196, 197
merchants' rule in partial payment problems, 326, 327
meter, 206
metric system of weights and measures, 206–209
mill in tax matters, 135
minuend, 25
minus sign, 2, 263
  in algebra, 367
minute (part of degree), 352
miscellaneous series, 267
mixed number, 85
mixtures, 339–346
mode, 193, 194
  advantages of, 197
  disadvantages of, 197
models, scale, 348, 349
money
  British, 120, 206
  French, 206
  German, 206
  United States, 206
    and decimals, 120
    division of, 63
    how written, 8
monomial, 357
multiple, 77
  common, 78
  least common, 78
multiplicand, 37
multiplication, 2, 37–54
  algebraic rules for, 369
  associative law for, 360
  by logs, 249
  checking correctness of, 41, 53, 54
  complement, 51, 52
  cross, 44, 52, 53, 91
  cumulative law of, 360
  distributive laws for, 360
  how to simplify, 41–45, 49–53
  inverted, 45
  left-hand, 45
  of decimals, 110, 113
  of fractions, 91–94
  of per cents, 135
  of polynomials, 373
  of positive and negative numbers, 264
  of powers, 220, 237
  of powers of ten, 227
  rule for like and unlike signs in, 38
  symbol of, 2
    in algebra, 359
multiplication sign, 2
multiplication table, 38
multiplier, 37

Napierian system of logs, 242
natural logs, 252
natural system of logs, 242
negative direction, 292
negative exponent, 222, 237
negative numbers, 25, 262–264
  addition of, 263
  division of, 264
  multiplication of, 264
  subtraction of, 263
net cost, 297
net profit, 297
net purchases, 297
net sales, 297
nine as check number
  in addition, 18, 19
  see also casting out nines
nominal rate of interest, 153, 154

nought, 1
number, 3
  abstract, 37
  Arabic, 1
  basic, 244
  composite, 75
  concrete, 37
  denominate, 200–214
    addition of, 210
    reduction of, 200
    subtraction of, 211
  directed, 358
  even, 65
  imaginary, 255
  index, 289
  integral, 3
  irrational, 242
  literal, 356
  mixed, 85
  negative, 25, 262–264
    addition of, 263
    division of, 264
    multiplication of, 264
    subtraction of, 263
  odd, 75
  positive, 262–264
    addition of, 263
    division of, 264
    multiplication of, 264
    subtraction of, 263
  prime, 75, 76
  real, 255
  Roman, 9
  signed, 262–264, 358
  specific, 357, 358
  whole, 3
number scale, 262
numerator, 82

odd number, 75
"on time": *see* installment purchase problems
operations of arithmetic, fundamental, 1, 2
  direct, 2
  inverse, 2
  symbols of, 2, 3
orders, 4
ordinary method (of simple interest), 142, 143
ordinate, 292
  axis of, 291
origin, 291

paper measure, 206
parabola, 294
parentheses, 3, 360, 361, 372
partial payments, 325–329
  merchants' rule for, 326, 327
  U.S. rule for, 325–329
per cent (percentage), 125–136
  addition of, 135
  business uses of, 134
  conversion to decimals and fractions, 126, 127
  division of, 135
  in profit and loss, 298
  less than 1 per cent, 134
  multiplication of, 135
  relation to ratio, 161
  subtraction of, 135
perfect power, 228
period, 4
pictograph, 279, 287
pie graph (chart), 279, 282–284
places
  decimal, 7, 107
  fractional, 7
plotting
  graph of quadratic formula, 294
  straight line relationship, 293
plus sign, 2, 263
  in algebra, 367
point, decimal, 1, 6, 107
poll tax, 329
polygon, frequency, 288
polynomial, 358
  multiplication of, 373
positive direction, 292
positive numbers, 262–264
  addition of, 263
  division of, 264
  multiplication of, 264
  subtraction of, 263
powers, 217–229, 364
  division of, 220, 237
  imperfect, 228
  multiplication of, 220, 237
  of common fractions, 219
  of decimals, 219
  of ten
    converting decimals into, 226
    division of, 227
    multiplication of, 227
  perfect, 228
  raising to, by logs, 250

powers, zero, 366
 see also exponent
present value (worth), 308, 426–427
price, selling, 297
prime cost, 297
prime factor, 75, 76
prime number, 75, 76
principal, 308
 in interest, 140
product, 37
"product of two binomials" method of factoring, 379
profit, 298
 gross, 297
 margin of, 297
 net, 297
progression, 266–276
 arithmetic, 266, 268–271
 ascending, 266
 descending, 266
 geometric, 266, 271–274
 harmonic, 266, 274, 275
 see also series
proof, pure, of division, 63
proper fraction, 85
property tax, 329
proportion, 168–181
 by alternation, 176
 by composition, 177
 by division, 177
 by inversion, 176
 compound, 173
 direct, 170
 inverse, 171
proportional mean, 170
proportional part of log, 247
protractor, 352
purchases
 gross, 297
 net, 297
 return, 297

quadrants, 292
quadratic equation, 384
quadratic formula, graph of, 294
quantity
 algebraic, 358
 constant, 356
 variable, 356
quotient, 58

radical
 reduced to simplest form, 230, 231
 similar, 229
radical sign, 3, 227
radicand, 228
rate
 in percentage, 129
 in profit and loss, 298
 of interest, 140
  effective, 153, 154
  nominal, 153, 154
 tax, 329
 working, of speed, 336–339
ratio, 159–181
 compound, 167
 how to simplify, 161
 in a series, 266
 inverse, 166
 relation to per cent, 161
 rules for calculation of, 166, 167
 symbol of, 160
ratio method of comparing like quantities, 159
real number, 255
receptacles, 346–348
reciprocal, 93, 94, 266
rectangle graph, 279, 282
recurring decimal, 118
reduction
 of denominate numbers, 200
  ascending, 200
  descending, 200
 of fractions, 86, 87
 of radicals to simplest form, 230, 231
remainder
 in division, 59
 in subtraction, 25
repeating decimal, 118, 273
return purchases, 297
Roman numeral system, 9
root, 227, 365
 cube, 227, 236, 237
  extraction of, 236
 extraction of, 228, 237
  by logs, 250
 fourth, 227
 of equation, 383
 of fraction, 236
 square, 227, 231–236, 365
  extraction of, 232

sales
 gross, 297
 net, 297
sales commission, 298

scale
  for models and maps, 348–351
  number, 262
second (part of degree), 352
selling price, 297
series, 266–276
  ascending, 266
  descending, 266
  miscellaneous, 267
  sum to infinity, 272
  see also progression
short division, 60, 61
sign, 359
  of exponents, 221
  rule for, 369
signed number, 262–264, 358
similar radicals, 229
simple decimal, 109
simple equation, 384
simple fraction, 84
simple interest, 140, 141
  bankers' method for, 142, 143
  exact method for, 142, 143
  formula for, 141
  ordinary method for, 142, 143
simplifying
  algebraic expressions, 375, 376
  averages, 192
  division, 68–70
  multiplication, 41–45, 49–53
  ratios, 161
  squaring of numbers, 223–226
  subtraction, 29, 30
smooth-line graph, 279, 287
solution of equations, 386–391
solutions (mixtures), 339–346
solving for variable in formula, 392
"so much per hundred," 134
specific number, 357, 358
speed, working rates of, 336–339
square, 363
  of a number, 217
  of the mean, 225
square measure, 202
  in metric system, 207
square root, 227, 231–236, 365
  extraction of, 232
squaring of numbers, how to simplify, 223–226
staircase diagram, 279, 288
statistical map, 279, 291
statistics, 186
straight line relationship, 293
subtraction, 2, 24–33
  algebraic rules for, 371, 372
  Austrian method of, 29, 32
  checking correctness of, 32, 33
  how to simplify, 29, 30
  inverted, 30
  left-hand, 30
  method of "equal additions" in, 28
  of decimals, 112
  of denominate numbers, 211
  of fractions, 89–91
  of per cents, 135
  of plus quantities, 369
  of positive and negative numbers, 263
  rule for, 27, 28
  symbol of, 2
    in algebra, 359
subtraction table, 26
subtrahend, 25
sum, 13
  of series to infinity, 272
surtax, 329
symbol, 359
  of fundamental operations, 2, 3

tables
  compound-amount-of-1, 155, 418–423
  multiplication, 38
  of logs, 246, 248, 424–425
  subtraction, 26
tanks, 346–348
tax
  direct, 329
  income, 329
  indirect, 329
  poll, 329
  property, 329
  total, 331
tax amount, 329
tax matters, 135
tax rate, 329
ten, powers of: *see* powers of ten
"tens" position, 4
term (algebraic), 357
  like, 367
terminal line (of angle), 352
"therefore," symbol, 2, 3
time
  in interest, 140
  measurement of, 205
"times" sign, 2
total tax, 331

trade discount, 297
trial divisor, 62
troy weights, 204
true discount, 308

unit, 3
United States money: *see* money, United States
United States rule in partial payment problems, 325–329
United States weights: *see* weights, United States
unit fraction, 83
"units" position, 4

valuation, assessed, 329
value
  absolute, 262
  future, 307
  mean, 187, 188
  present, 308, 426–427
variable, 356
  solving for in formula, 392
variation
  direct, 179
  inverse, 180
  joint, 180
  language of, 418
vertex (of angle), 352
vertical bar graph, 279, 281, 282
vulgar fraction, 83

weighted average, 190, 191
weights
  metric system, 206, 208, 209
  United States, 203
    apothecaries', 204
    avoirdupois, 204
    troy, 204
whole number, 3
working rates of speed, 336–339
worth
  future, 307
  present, 308

x axis, 291

y axis, 291

zero, 1
  effect on decimals, 8
  effect on numbers, 5
zero power (exponent), 221, 237, 366

A CATALOG OF SELECTED

DOVER BOOKS

IN ALL FIELDS OF INTEREST

SUNDIALS, Albert Waugh. Far and away the best, most thorough coverage of ideas, mathematics concerned, types, construction, adjusting anywhere. Over 100 illustrations. 230pp. 5⅜ × 8½. 22947-5 Pa. $5.95

PICTURE HISTORY OF THE NORMANDIE: With 190 Illustrations, Frank O. Braynard. Full story of legendary French ocean liner: Art Deco interiors, design innovations, furnishings, celebrities, maiden voyage, tragic fire, much more. Extensive text. 144pp. 8⅞ × 11¾. 25257-4 Pa. $10.95

THE FIRST AMERICAN COOKBOOK: A Facsimile of "American Cookery," 1796, Amelia Simmons. Facsimile of the first American-written cookbook published in the United States contains authentic recipes for colonial favorites—pumpkin pudding, winter squash pudding, spruce beer, Indian slapjacks, and more. Introductory Essay and Glossary of colonial cooking terms. 80pp. 5⅜ × 8½. 24710-4 Pa. $3.50

101 PUZZLES IN THOUGHT AND LOGIC, C. R. Wylie, Jr. Solve murders and robberies, find out which fishermen are liars, how a blind man could possibly identify a color—purely by your own reasoning! 107pp. 5⅜ × 8½. 20367-0 Pa. $2.95

ANCIENT EGYPTIAN MYTHS AND LEGENDS, Lewis Spence. Examines animism, totemism, fetishism, creation myths, deities, alchemy, art and magic, other topics. Over 50 illustrations. 432pp. 5⅜ × 8½. 26525-0 Pa. $8.95

ANTHROPOLOGY AND MODERN LIFE, Franz Boas. Great anthropologist's classic treatise on race and culture. Introduction by Ruth Bunzel. Only inexpensive paperback edition. 255pp. 5⅜ × 8½. 25245-0 Pa. $7.95

THE TALE OF PETER RABBIT, Beatrix Potter. The inimitable Peter's terrifying adventure in Mr. McGregor's garden, with all 27 wonderful, full-color Potter illustrations. 55pp. 4¼ × 5½. (Available in U.S. only) 22827-4 Pa. $1.75

THREE PROPHETIC SCIENCE FICTION NOVELS, H. G. Wells. *When the Sleeper Wakes, A Story of the Days to Come* and *The Time Machine* (full version). 335pp. 5⅜ × 8½. (Available in U.S. only) 20605-X Pa. $8.95

APICIUS COOKERY AND DINING IN IMPERIAL ROME, edited and translated by Joseph Dommers Vehling. Oldest known cookbook in existence offers readers a clear picture of what foods Romans ate, how they prepared them, etc. 49 illustrations. 301pp. 6⅛ × 9¼. 23563-7 Pa. $7.95

SHAKESPEARE LEXICON AND QUOTATION DICTIONARY, Alexander Schmidt. Full definitions, locations, shades of meaning of every word in plays and poems. More than 50,000 exact quotations. 1,485pp. 6½ × 9¼. 22726-X, 22727-8 Pa., Two-vol. set $31.90

THE WORLD'S GREAT SPEECHES, edited by Lewis Copeland and Lawrence W. Lamm. Vast collection of 278 speeches from Greeks to 1970. Powerful and effective models; unique look at history. 842pp. 5⅜ × 8½. 20468-5 Pa. $12.95

PLANTS OF THE BIBLE, Harold N. Moldenke and Alma L. Moldenke. Standard reference to all 230 plants mentioned in Scriptures. Latin name, biblical reference, uses, modern identity, much more. Unsurpassed encyclopedic resource for scholars, botanists, nature lovers, students of Bible. Bibliography. Indexes. 123 black-and-white illustrations. 384pp. 6 × 9. 25069-5 Pa. $8.95

FAMOUS AMERICAN WOMEN: A Biographical Dictionary from Colonial Times to the Present, Robert McHenry, ed. From Pocahontas to Rosa Parks, 1,035 distinguished American women documented in separate biographical entries. Accurate, up-to-date data, numerous categories, spans 400 years. Indices. 493pp. 6½ × 9¼. 24523-3 Pa. $10.95

THE FABULOUS INTERIORS OF THE GREAT OCEAN LINERS IN HISTORIC PHOTOGRAPHS, William H. Miller, Jr. Some 200 superb photographs capture exquisite interiors of world's great "floating palaces"—1890s to 1980s: *Titanic, Ile de France, Queen Elizabeth, United States, Europa,* more. Approx. 200 black-and-white photographs. Captions. Text. Introduction. 160pp. 8⅞ × 11¾. 24756-2 Pa. $9.95

THE GREAT LUXURY LINERS, 1927–1954: A Photographic Record, William H. Miller, Jr. Nostalgic tribute to heyday of ocean liners. 186 photos of *Ile de France, Normandie, Leviathan, Queen Elizabeth, United States,* many others. Interior and exterior views. Introduction. Captions. 160pp. 9 × 12. 24056-8 Pa. $12.95

A NATURAL HISTORY OF THE DUCKS, John Charles Phillips. Great landmark of ornithology offers complete detailed coverage of nearly 200 species and subspecies of ducks: gadwall, sheldrake, merganser, pintail, many more. 74 full-color plates, 102 black-and-white. Bibliography. Total of 1,920pp. 8⅝ × 11¼. 25141-1, 25142-X Cloth., Two-vol. set $100.00

THE SEAWEED HANDBOOK: An Illustrated Guide to Seaweeds from North Carolina to Canada, Thomas F. Lee. Concise reference covers 78 species. Scientific and common names, habitat, distribution, more. Finding keys for easy identification. 224pp. 5⅜ × 8½. 25215-9 Pa. $6.95

THE TEN BOOKS OF ARCHITECTURE: The 1755 Leoni Edition, Leon Battista Alberti. Rare classic helped introduce the glories of ancient architecture to the Renaissance. 68 black-and-white plates. 336pp. 8⅜ × 11¼. 25239-6 Pa. $14.95

MISS MACKENZIE, Anthony Trollope. Minor masterpieces by Victorian master unmasks many truths about life in 19th-century England. First inexpensive edition in years. 392pp. 5⅜ × 8½. 25201-9 Pa. $8.95

THE RIME OF THE ANCIENT MARINER, Gustave Doré, Samuel Taylor Coleridge. Dramatic engravings considered by many to be his greatest work. The terrifying space of the open sea, the storms and whirlpools of an unknown ocean, the ice of Antarctica, more—all rendered in a powerful, chilling manner. Full text. 38 plates. 77pp. 9¼ × 12. 22305-1 Pa. $4.95

THE EXPEDITIONS OF ZEBULON MONTGOMERY PIKE, Zebulon Montgomery Pike. Fascinating firsthand accounts (1805–6) of exploration of Mississippi River, Indian wars, capture by Spanish dragoons, much more. 1,088pp. 5⅜ × 8½. 25254-X, 25255-8 Pa., Two-vol. set $25.90

CHRISTMAS CUSTOMS AND TRADITIONS, Clement A. Miles. Origin, evolution, significance of religious, secular practices. Caroling, gifts, yule logs, much more. Full, scholarly yet fascinating; non-sectarian. 400pp. 5⅜ × 8½.
23354-5 Pa. $7.95

THE HUMAN FIGURE IN MOTION, Eadweard Muybridge. More than 4,500 stopped-action photos, in action series, showing undraped men, women, children jumping, lying down, throwing, sitting, wrestling, carrying, etc. 390pp. 7⅞ × 10⅝.
20204-6 Cloth. $24.95

THE MAN WHO WAS THURSDAY, Gilbert Keith Chesterton. Witty, fast-paced novel about a club of anarchists in turn-of-the-century London. Brilliant social, religious, philosophical speculations. 128pp. 5⅜ × 8½. 25121-7 Pa. $3.95

A CÉZANNE SKETCHBOOK: Figures, Portraits, Landscapes and Still Lifes, Paul Cézanne. Great artist experiments with tonal effects, light, mass, other qualities in over 100 drawings. A revealing view of developing master painter, precursor of Cubism. 102 black-and-white illustrations. 144pp. 8¾ × 6⅜. 24790-2 Pa. $6.95

AN ENCYCLOPEDIA OF BATTLES: Accounts of Over 1,560 Battles from 1479 B.C. to the Present, David Eggenberger. Presents essential details of every major battle in recorded history, from the first battle of Megiddo in 1479 B.C. to Grenada in 1984. List of Battle Maps. New Appendix covering the years 1967–1984. Index. 99 illustrations. 544pp. 6½ × 9¼. 24913-1 Pa. $14.95

AN ETYMOLOGICAL DICTIONARY OF MODERN ENGLISH, Ernest Weekley. Richest, fullest work, by foremost British lexicographer. Detailed word histories. Inexhaustible. Total of 856pp. 6½ × 9¼.
21873-2, 21874-0 Pa., Two-vol. set $19.90

WEBSTER'S AMERICAN MILITARY BIOGRAPHIES, edited by Robert McHenry. Over 1,000 figures who shaped 3 centuries of American military history. Detailed biographies of Nathan Hale, Douglas MacArthur, Mary Hallaren, others. Chronologies of engagements, more. Introduction. Addenda. 1,033 entries in alphabetical order. xi + 548pp. 6½ × 9¼. (Available in U.S. only)
24758-9 Pa. $13.95

LIFE IN ANCIENT EGYPT, Adolf Erman. Detailed older account, with much not in more recent books: domestic life, religion, magic, medicine, commerce, and whatever else needed for complete picture. Many illustrations. 597pp. 5⅜ × 8½.
22632-8 Pa. $9.95

HISTORIC COSTUME IN PICTURES, Braun & Schneider. Over 1,450 costumed figures shown, covering a wide variety of peoples: kings, emperors, nobles, priests, servants, soldiers, scholars, townsfolk, peasants, merchants, courtiers, cavaliers, and more. 256pp. 8⅜ × 11¼. 23150-X Pa. $9.95

THE NOTEBOOKS OF LEONARDO DA VINCI, edited by J. P. Richter. Extracts from manuscripts reveal great genius; on painting, sculpture, anatomy, sciences, geography, etc. Both Italian and English. 186 ms. pages reproduced, plus 500 additional drawings, including studies for *Last Supper, Sforza* monument, etc. 860pp. 7⅞ × 10¾. (Available in U.S. only) 22572-0, 22573-9 Pa., Two-vol. set $35.90